03-CTS-9T

IDENTIFYING AND CONTROLLING ODOR IN THE MUNICIPAL WASTEWATER ENVIRONMENT PHASE 3:

BIOSOLIDS PROCESSING MODIFICATIONS FOR CAKE ODOR REDUCTION

by:

Gregory A. Adams, P.E.
LACSD (Co-Principal Investigator)

Jay Witherspoon, P.E.,
CH2M HILL (Co-Principal Investigator)

Zeynep Erdal, Ph.D.
Bob Forbes, P.E.
David McEwen, P.E.
CH2M HILL

Ron Hargreaves, P.E.
LACSD

Matthew Higgins, P.E., Ph.D.
Bucknell University

John Novak, P.E., Ph.D.
Virginia Tech

2008

WERF

The Water Environment Research Foundation, a not-for-profit organization, funds and manages water quality research for its subscribers through a diverse public-private partnership between municipal utilities, corporations, academia, industry, and the federal government. WERF subscribers include municipal and regional water and wastewater utilities, industrial corporations, environmental engineering firms, and others that share a commitment to cost-effective water quality solutions. WERF is dedicated to advancing science and technology addressing water quality issues as they impact water resources, the atmosphere, the lands, and quality of life.

For more information, contact:
Water Environment Research Foundation
635 Slaters Lane, Suite 300
Alexandria, VA 22314-1177
Tel: (703) 684-2470
Fax: (703) 299-0742
www.werf.org
werf@werf.org

This report was co-published by the following organization. For non-subscriber sales information, contact:

IWA Publishing
Alliance House, 12 Caxton Street
London SW1H 0QS, United Kingdom
Tel: +44 (0) 20 7654 5500
Fax: +44 (0) 20 7654 5555
www.iwapublishing.com
publications@iwap.co.uk

Library of Congress Catalog Card Number: 2007926980
IWAP ISBN: 978-1-84339-790-8

LACSD, CH2M HILL

ACKNOWLEDGMENTS

Biosolids Processing Modifications for Cake Odor Reduction, representing Phase 3 of the WERF project, *Identifying and Controlling the Municipal Wastewater Odor Environment* (WERF Research Project 00-HHE-5), was a significant effort undertaken by a large number of project participants. The names and affiliations of the principal investigators and other project team members are listed below. All organizations affiliated with the project team members are acknowledged here for their support to the project team members.

In addition to the project team, the following agencies are recognized as having provided their wastewater treatment plants as test sites for the project:

- Hampton Roads Sanitation District, Virginia
- King County Department of Natural Resources, Washington
- Sanitation Districts of Los Angeles County, California
- Philadelphia Water Department, Pennsylvania
- Public Utilities Commission, City and County of San Francisco, California
- Charlotte Mecklenburg Utilities District, North Carolina
- Western Lake Superior Sanitation District, Minnesota
- Inland Empire Utilities Agency, California

The researchers would also like to acknowledge and thank the directors and staff of the Water Environment Research Foundation for their support.

Report Preparation

Co-Principal Investigators:
Gregory M. Adams, M.S., *LACSD*
Jay Witherspoon, M.S., *CH2M HILL*

Project Team:
Jim Barry, M.S.
Ron Hargreaves, M.S.
Rob Morton
Los Angeles County Sanitation Districts

Zeynep Erdal, Ph.D.
Bob Forbes, M.S.
David McEwen, M.S.
CH2M HILL

Matt Higgins, Ph.D.
Bucknell University

John Novak, Ph.D.
Virginia Tech

Technical Advisory Committee (TAC):
Mohammed Abu-Orf, Ph.D., *Metcalf and Eddy*
Tim Shea, Ph.D., *CH2M HILL*
Phil Wolstenholme, B.S., *Brown and Caldwell*

Project Subcommittee

Steve Rogowski
Metro Wastewater Reclamation District of Denver

Tim Ellis, Ph.D., P.E.
Iowa State University

Donald Gray (Gabb), Ph.D., P.E., DEE
East Bay Municipal Utility District

Chris Kaakaty, P.E.
Dallas Water Utilities

Mark Lang, P.E.
Stantec, New York

John M. Walker, Ph.D.
U.S. EPA (Retired)

Water Environmental Research Foundation Staff

Director of Research: Daniel M. Woltering, Ph.D.
Program Manager: Alan Hais, M.S.

ABSTRACT AND BENEFITS

Abstract:

This project was undertaken in response to needs by the wastewater treatment industry to better understand the generation of odors from biosolids produced by wastewater treatment plants (WWTPs). Its primary objective is to test relationships between WWTP process parameters and biosolids odors in the laboratory and at full-scale, developed in Phase 2 of the overall project. The goal of this research was to develop effective techniques for minimizing biosolids odors produced following anaerobic digestion, thereby reducing negatively perceived impacts on the environment or to the public when beneficially used on land.

Laboratory and full-scale investigations conducted in Phase 3 included: 1) the role of cations in odor production from anaerobically digested biosolids; 2) the effect of anaerobic digestion parameters such as mixing, detention time, temperature, and phasing on biosolids cake odor production; 3) effects of dewatering processes and their control parameters on cake odors; and 4) effects of chemical addition on cake odors.

This research produced several conclusions regarding biosolids odor generation and minimization. Odors from anaerobically digested and dewatered biosolids correlate strongly with total volatile organic sulfur (TVOSC) concentrations in bottle-headspace samples of dewatered cake. TVOSC emissions also increased with increasing iron content of the raw sludge feeding digestion. Shear during centrifugation subsequently releases the iron-bound biopolymer, which leads to increased odors in the biosolids cake. The reverse trend was found with aluminum content, in that TVOSC emissions decreased with increasing aluminum content in the raw sludge feeding digestion. Side-by-side comparisons of different centrifuges dewatering the same feed biosolids showed that different centrifuge characteristics and operating parameters (bowl speed, torque, etc) affected TVOSC emissions from the dewatered biosolids cakes.

Tests were also conducted investigating lengthening solids retention times (SRT) during mesophilic anaerobic digestion, which resulted in lower TVOSC emissions. Advanced digestion processes were tested in the laboratory and in the field with regard to lowering biosolids odors. Some promising results were observed with the MicroSludge™ Process (a cell lysing process incorporated into digestion) and other advanced digestion processes. Some TVOSC reduction was also noted in addition of aluminum sulfate (alum) at laboratory scale to samples prior to dewatering. Full-scale results on alum addition showed mixed results regarding TVOSC reduction, and further research is planned.

Recommendations were made for Phase 4, including a search for stronger evidence on the causes of odors in biosolids at WWTPs, refinement of full-scale trials on alum addition, and efforts to solve a real-world, biosolids odor problem at a WWTP. Additionally, WERF would like to apply some of the encouraging laboratory findings from Phase 3 to the field in a biosolids facility where odors are currently a problem. Ultimately, the fourth phase of the biosolids odor study should provide an enhanced “roadmap” that WWTPs can use to reduce bisosolids odors from their facility.

Benefits:

- ♦ Helps the wastewater treatment industry understand and manage biosolids odor and its impacts on surrounding communities, by understanding more completely the chain of events involved in the generation of biosolids odors.
- ♦ Identifies gaps in scientific knowledge regarding mechanisms of odor generation in WWTP biosolids.
- ♦ Shows that some biosolids stability parameters may be misleading with respect to their impact on odors produced from biosolids.
- ♦ Provides a reference guide for the wastewater treatment industry and a starting point in identifying the causes of biosolids odors.
- ♦ Provides laboratory-based and field-based data that demonstrate means of odor reduction from biosolids that are stabilized by anaerobic digestion.
- ♦ Emphasizes the importance of whole plant management for reduction and control of biosolids odors.

Keywords: Anaerobic digestion, advanced digestion, alum, biosolids, cations, dewatering, iron, odors, olfactometry, solids retention time, total volatile organic sulfur compounds, wastewater, WWTP.

TABLE OF CONTENTS

Note: The appendices are included in the online version of this report. See the Publications section of the WERF website at www.werf.org.

LIST OF TABLES

LIST OF FIGURES

LIST OF ACRONYMS

% Al dw	Percentage aluminum on a dry weight basis
°C	Degrees Celsius
AGMM	Acid-gas mesophilic-mesophilic
AGMT	Acid-gas mesophilic-thermophilic
AGMTM	Acid-gas mesophilic-thermophilic-mesophilic
Al	Aluminum
alum	Aluminum sulfate
AWMA	Air and Waste Management Association
BESA	Bromoethanesulfonic acid
BRC	Biosolids Recycling Center
Ca	Calcium
CEBA	Chemical, Enzymatic, and/or Biological Agent
CER	Cation exchange resin
CFR	Code of Federal Regulations
CMU	Charlotte Mecklenburg Utilities
CST	Capillary suction time
D/T	Dilutions to threshold
DMDS	Dimethyl disulfide
DMS	Dimethyl sulfide
DS	Dry solids
DT	Detection Threshold
EPS	Exocellular polymeric substances
Fe	Iron
$FeCl_3$	Ferric chloride
FeS	Ferric sulfide
FID	Flame ionization detector
G	Gradient
g/L	Grams per liter
GC/MS	Gas chromatograph/mass spectrometer
GCTP	General CEBA Test Protocol
gpm	Gallons per minute

H_2S	Hydrogen sulfide
HP	Hewlett Packard
HSCSP	High solids centrifuge simulation process
ICP	Inductively coupled plasma
IEUA	Inland Empire Utilities Agency
JWPCP	Joint Water Pollution Control Plant
K	Potassium
LACSD	Los Angeles County Sanitation Districts
M/D	Monovalent to divalent ratio
meq/meq	Milliequivalent/milliequivalent
Mg	Magnesium
mg/g DS	Milligrams per gram dry solids
mg/gTS	Milligrams per gram total solids
mg/L	Milligrams per liter
mg/m^3	Milligrams per cubic meter
MSDS	Material Safety Data Sheet
MT	Methane thiol
Na	Sodium
NACWA	National Association of Clean Water Agencies
NE	Northeast
OPD	Optimum polymer dose
PBS	Phosphate buffer saline
ppbv	Parts per billion by volume
psi	Pounds per square inch
PWD	Philadelphia Water Department
QA/QC	Quality Assurance/Quality Control
RBA	Residual biological activity
RFI	Request for Information
RFP	Requests for Proposal
RP	Regional Plant
rpm	Revolutions per minute
SCTP	Specific CEBA Test Protocol
SPME	Solid-phase microextraction

SRT	Sludge retention time
SW	Southwest
TCR	Targeted Collaborative Research
TMA	Trimethylamines
TPAD	Temperature-phased anaerobic digestion
TS	Total solids
TVOSC	Total volatile organic sulfur compounds
VAR	Vector Attraction Reduction
VFA	Volatile fatty acid
VOSC	Volatile organic sulfur compound
VS	Volatile solids
VSCs	Volatile sulfur compounds
VT	Virginia Polytechnic Institute & State University
WAS	Waste activated sludge
WEF	Water Environment Federation
WEFTEC	Water Environment Federation Technical Exposition and Conference
WLSSD	Western Lake Superior Sanitary District
WWMF	Wastewater Management Facility
WWTP	Wastewater Treatment Plant

EXECUTIVE SUMMARY

ES.1 Introduction and Background

The project summarized in this report, *Biosolids Processing Modifications for Cake Odor Reduction,* represents Phase 3 of a larger series of projects funded by the Water Environment Research Foundation (WERF) called *Identifying and Controlling the Municipal Wastewater Odor Environment*), which began in April 2000. The research to date has been led jointly by the Los Angeles County Sanitation Districts (LACSD) and CH2M HILL as Co-Principal Investigators in three major study phases.

The Phase 1 Final Report, a review of literature related to odors in the wastewater industry, was entitled *Phase 1: Literature Search and Review*, and published by WERF in 2003 under WERF Stock No. 00HHE5A. At a WERF-sponsored specialty workshop at the 2000 Water Environment Federation Technical Exposition and Conference (WEFTEC) in Anaheim, California, the project team presented a preliminary summary of the Phase 1 literature search and findings. In this meeting, participants voted biosolids odor issues the most important odor problem facing the wastewater industry and the highest priority for future research. Anaerobicly-digested and dewatered biosolids represent the majority of biosolids generated and processed in the U.S. and Canada, about 70% of the total biosolids mass produced at municipal wastewater treatment plants (WWTPs) in those countries.

Based on the direction from the WEFTEC 2000 meeting, Phase 2 of the research focused on impacts of in-plant parameters on biosolids cake odors. This phase comprised a field and laboratory study of plant process parameters related to odors from anaerobically digested and dewatered biosolids. Eleven WWTP sites were studied using a comprehensive analytical protocol that resulted in over 12,000 process sampling points at the eleven WWTPs that were tested, their main purpose being to characterize the odors from biosolids throughout the process. The Phase 2 Final Report, entitled *Impacts of In-Plant Operational Parameters on Biosolids Odor Quality,* was published in 2003 under WERF Project 00HHE5T.

An addendum to the Phase 2 research project was a literature study and analysis of the health effects from biosolids odors, which was published in late 2004 under WERF Project 00HHE5C and entitled *Health Effects of Biosolids Odors: A Literature Review.*

Phase 3 (WERF Research Project 03CTS9T and the subject of this Final Report) was developed to study the most promising Phase 2 hypotheses in more depth, beginning with laboratory (bench-scale) studies and continuing on to targeted investigations at selected WWTPs, where the WERF team investigated various equipment and processes at pilot or full scale to identify and quantify means of reducing biosolids cake odors.

Table ES-1 lists each of the Phase 2 hypotheses regarding factors which potentially impact biosolids cake odor production and that were investigated in Phase 3, and the portion of Phase 3 research that focused on each hypothesis. Those Phase 2 hypotheses that were clearly rejected by the Phase 2 data were not studied further in Phase 3.

Table ES-1. Summary of Phase 2 Hypotheses Carried Forward for Phase 3 Study.

Phase2 Hypotheses Supported Based on Study Results	Phase 3 Study Focus
1. Different dewatering practices impact bioavailable protein differently; some dewatering practices tend to increase odors in the biosolids cake.	Investigated impacts of different dewatering practices on odors, and identified characteristics of centrifuges that affect odors.
2. Based on comparison of results from the one wastewater treatment plant (WWTP) in the Phase 2 study with thermophilic anaerobic digestion and 10 WWTPs with mesophilic digestion, odors from thermophilically digested biosolids cake have different characteristics and patterns of time release than mesophilically digested biosolids cake.	Investigated odor characteristics of different advanced digestion processes such as thermophilic, temperature-phased, and two-phase digestion; and compared them with mesophilic anaerobic digestion.
3. There is evidence from the Phase 2 study that iron in sufficient concentrations binds bioavailable protein in biosolids cake, and thus reduces odor production from dewatered biosolids.	Investigated the impacts of metal cations on cake odors, focusing on iron and aluminum.
Phase 2 Hypotheses Found to be Inconclusive Based on Study Results	**Phase 3 Study Focus**
4. The presence and relative concentration of metal cations in the influent wastewater reduces odors in dewatered biosolids cake. In Phase 2, there was little correlation with either influent cation concentration or influent monovalent to multivalent (M/D) ratio and odor reduction in the dewatered biosolids cakes sampled.	Further investigated the impacts of metal cations in digester feed solids on biosolids odors (also see Hypothesis No. 3 above).
5. Higher concentrations of metal cations such as iron could reduce odors in digested, dewatered biosolids. There appears to be a lack of knowledge about the bioavailability of proteins and other organic material complexed with cations such as iron and aluminum, which caused the project team to remain undecided about this hypothesis.	In addition to investigating influence of metal cations present in sludges (per Hypotheses Nos. 3 & 4), investigate impacts of post-digestion iron and aluminum dosing on biosolids cake odors.
6. A longer solids retention time (SRT) in the digestion process generally produces less bioavailable protein and lower odors in digested, dewatered biosolids.	Conducted a controlled study that correlated digestion SRT with odor levels from digested and dewatered biosolids.
7. Title 40 of the Code of Federal Regulations Part 503 (503 Rule) for vector attraction reduction requires achieving a minimum of 38% volatile solids (VS) destruction during digestion. A clear correlation between biosolids stability per this criterion and reduction in dewatered biosolids odors was not found in this study.	Further investigated and attempted to quantify any relationships between VS destruction during digestion and odor from the digested and dewatered biosolids cakes.
8. Different conveyance practices for biosolids cakes (such as cake pumps, belt conveyors, and screw conveyors), may have impacts on biosolids odors. Cake storage practices and storage times may also have an impact on biosolids odors.	Performed full-scale trials that considered conveyance and storage practices to assess their impacts on biosolids odors.

Research conducted in Phase 3 specifically sought ways to reduce odors from dewatered biosolids cake, thereby reducing negative odor impacts on the public when the cake biosolids or their derivatives are stored, transported, and land applied. The main goal of the Phase 3 study was to provide a general roadmap to WERF subscribers who are seeking ways to optimize biosolids processing and reduce biosolids cake odors. It is important to realize that Phase 3 deals with reduction of odors from biosolids cake as associated with treatment processes, not odor control by containment or absorption and treatment of the odors generated.

Reduced odors also could open the WWTP dewatered biosolids cake to other recycling or disposal opportunities that are currently not used, such as site storage, land application, composting or heat-drying. Additionally, significant cost savings could be realized by minimizing odor controls and other costly options for containment of biosolids.

ES.2 Project Organization

The project team that completed the Phase 3 study is identified in Table ES-2, which shows how each organization was assigned specific tasks within the Phase 3 study.

Table ES-2. Major Responsibilities of WERF Project Team Members.

Team Members	Main Responsibilities
Los Angeles County Sanitation Districts (LACSD)	• WERF member and public outreach activities • Contract administration and project management • Develop and conduct research and develop protocols • Conduct field and pilot testing • Provide final recommendations
CH2M HILL	• Technology management and project management • Develop research and protocols • Develop WWTP Survey • Conduct field and pilot testing • Provide data analysis and final recommendations
Bucknell University and Virginia Polytechnic Institute & State University (Virginia Tech)	• Investigate promising hypotheses from Phase 2 • Develop research protocols • Conduct sample analysis for Phase 3C • Data analysis • Develop Phase 3 recommendations
Technical Advisory Committee	• Provide technical review and expertise during each phase of the project
WERF Project Sub-Committee	• Review sampling and analysis protocols and approaches • Review and comment on research and developments based on Phase 3 findings, recommendations, and conclusions • Review Phase 3 Final Report and periodic progress reports • Attend Phase 3 workshops

The Phase 3 study was completed using three elements: laboratory studies (3A), biosolids processing equipment and management approach surveys (3B), and full-scale field testing (3C). These study elements are not to be considered as study phases, for they were undertaken somewhat concurrently. The study elements were augmented by an addition to the project to study in greater detail the potential impacts of both traditional WWTP chemicals and proprietary chemical, enzymatic, and/or biological agents (CEBAs) on biosolids odors. Many of these CEBA products are claimed to substantially improve solids processing or reduce odor generation. Some are already in use for animal waste mass reduction and breakup of clogging fats, oils, and grease (F.O.G.) in drains. The same organization of core entities (Principal Investigators and contributing organizations) that executed and delivered the Phase 2 project also completed the Phase 3 work.

ES.3 Testing Protocol and Rationale

Throughout Phase 3, the Bucknell University, Virginia Tech, and LACSD laboratories conducted bench scale analyses on biosolids samples collected from various WWTPs using the method for biosolids cake sampling and headspace analysis developed by Novak, et al. (2004) and Glindemann et al. (2006). This sample headspace analysis method was developed and refined during Phase 2 of the project as a faster, simpler, more reproducible, and ultimately less expensive method of odor measurement than flux chamber sampling and odor panel analyses.

When using the headspace analysis method, the headspace storage bottles, which are closed bioactive systems, allow for the measurement of odor consumption by the biosolids cake in the bottle as well as measurement of the odor production-consumption cycles that are ongoing in the closed system. Tests conducted during the development of Phase 2 showed that the odor concentration in the headspace of the bottles is a direct function of the odor concentration of the biosolids cake sample.

In the literature, the generation of volatile organic sulfur compounds (VOSC), reduced inorganic sulfur, and heterocyclic amines has been linked to amino acid metabolism and sulfate reduction. Sulfur-containing functional groups associated with amino acids are removed, and the product either is an odorant itself or undergoes biochemical transformation to form odor-causing compounds. Because of this formation pathway, the WERF team chose to measure total volatile organic sulfur compounds (TVOSC) as a means of quantifying odor levels in a dewatered biosolids cake sample. During Phase 3 of the project, fresh TVOSC readings obtained by analyses of both concentrated sample headspaces and dilute flux chamber gas samples taken from biosolids cakes correlated well (R2 = 0.9824) with results obtained from corresponding odor panel tests of the same headspace and flux chamber gas samples, as shown in Figure ES-1.

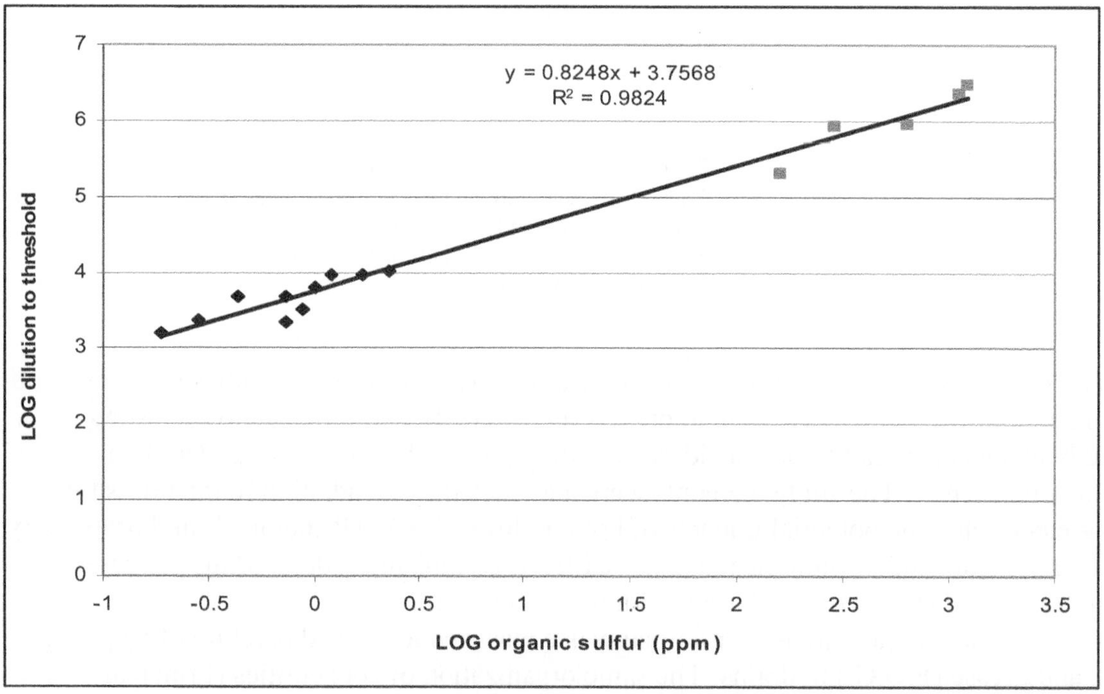

Figure ES-1. Odor Detection Threshold (DT)) versus TVOSC from LACSD (right end of best-fit line). Compared to Biosolids Cake Odors from Flux Chamber Samples (Webster et al, left end of line).

This more recent data has confirmed the validity of the biosolids cake headspace measurement method as a very useful tool for monitoring odor levels from biosolids samples, and for comparing sample odor characteristics and intensities with those of untreated controls.

ES.4 Results, Findings, and Conclusions

In the Phase 3 research, tests were conducted at bench-scale (Phase 3A) and at full-scale (Phase 3C). Phase 3A studies by the Virginia Tech, Bucknell University, and LACSD research laboratories included cation tests, digestion SRT testing, alum addition tests, and bench-scale testing of CEBAs. Phase 3C field studies included full-scale advanced digestion tests, dewatering equipment tests, combined tests on a pre-digestion enhancement process and dewatering equipment, and chemical addition tests.

Findings in Phase 3 of the study are grouped into the following research areas:

- The relationships between odors and TVOSC as shown by example in Figure ES-1 above.
- The effects of the metal cations, such as iron and aluminum, on biosolids cake odors.
- The effects of anaerobic digestion and some enhancements or variations to the anaerobic digestion process on biosolids cake odors.
- The effects of dewatering on biosolids cake odors, with an emphasis on the effects of high-solids centrifuges.

These research findings apply primarily to dewatered biosolids that are stored onsite at WWTPs and transported in bulk to agricultural land application sites, because the handling and land application of dewatered biosolids are known to have generated the most odor complaints. There is evidence both in research and in practice, however, that the reduction of biosolids cake odors also tends to reduce odors from products derived from dewatered biosolids cakes, such as biosolids compost and heat-dried biosolids pellets.

The Phase 3 WERF study produced "process" and "mechanistic" conclusions, where process conclusions are those for which the cause and effect relationships are fully known and understood, and mechanistic conclusions are those that have been indicated by WERF Phase 3 research, but for which the cause and effect relationships are unknown or not fully understood. Following is a summary of process findings and conclusions from the Phase 3 WERF study:

1. TVOSC emissions from digested biosolids dewatered by a belt filter press and a rotary press were both substantially lower than the TVOSC emissions from the same biosolids dewatered by high-solids centrifuge, but dry solids concentration in biosolids dewatered by the belt filter press and rotary press are also typically lower than the centrifuged cake. These results also support Conclusions 2 and 3 below, that reduction of shear during biosolids dewatering will also reduce cake TVOSC and odor emissions.
2. Higher shear forces along the scroll of a high-solids centrifuge are the primary cause of higher TVOSC emissions from centrifugally dewatered cake, and reducing shear during biosolids dewatering results in lower TVOSC emissions from the cake, as proven in side-by-side field trials involving high-solids and low-solids centrifuges.

3. Operational adjustments to high-solids centrifuges, such as reduction of bowl speed or torque, can reduce TVOSC emissions from dewatered cake through reduction of shear forces within the centrifuges. In many cases, significant reductions in TVOSC can be achieved without substantial reductions in cake solids concentrations.
4. Longer solids retention times (SRT) during mesophilic anaerobic digestion result in lower TVOSC emissions from dewatered biosolids cakes. Test data from laboratory-scale digesters indicate that a one-third reduction in peak TVOSC emissions could be achieved by increasing digestion SRT from 15 days to 30 days.
5. Volatile solids (VS) destruction during mesophilic anaerobic digestion increases with longer SRTs, but the increase is not dramatic. The research team found an increase in VS reduction from about 50% at 10 days SRT to 58% at 40 days SRT. The reduction in TVOSC at longer SRTs is more pronounced than the increased VS reduction at the same SRTs, indicating that longer digestion SRTs may have a proportionately greater impact on biosolids odor potential than on VS reduction.
6. Results of trials comparing enhanced and advanced anaerobic digestion processes with conventional anaerobic digestion indicate that pre-digestion enhancements, multi-phase digestion, egg-shaped digesters, and thermophilic digestion could each reduce biosolids TVOSC emissions compared with conventional, mesophilic anaerobic digestion. The mechanisms for TVOSC reduction appear to be a higher destruction rate of odor precursors during digestion, including bioavailable protein, which appears to be the primary odor precursor.
7. Low-level lime addition (up to 5% by weight of lime to dry weight of solids) to anaerobicly digested and dewatered biosolids increased the level of TVOSC emissions from the cake sample headspace.
8. Addition of ferric chloride to anaerobicly-digested biosolids prior to dewatering did not reduce TVOSC emissions from the cake until the ferric chloride dose was at least 8% on a dry mass basis. Another full-scale trial at a different WWTP indicated that iron addition may be effective at lower dosage rates on some digested biosolids. The WERF team concluded that the effectiveness of iron addition in reducing biosolids odors depends on biosolids characteristics and other factors that may be unique to the wastewater and sludge characteristics and processes at each WWTP.
9. Addition of aluminum sulfate (alum) at a laboratory scale to samples of anaerobicly-digested biosolids from two WWTPs prior to dewatering reduced TVOSC emissions from the cake substantially at lower dosage rates than iron, in the range of 0.5% to 1% dry mass of aluminum to mass of dry solids.
10. Results of full-scale alum addition to a high-solids dewatering centrifuge indicate that alum addition can reduce TVOSC emissions from dewatered biosolids. More work needs to be done on the impacts of mixing, residence time, and pH adjustment of the feed sludge in conjunction with polymer dosing for the optimization of alum (or other aluminum compound) addition to biosolids for purposes of reducing cake odors.

Following is a summary of mechanistic findings and conclusions from the Phase 3 WERF study:

1. Methanogenic bacteria play an important role in the cycling of organic sulfur compounds under anaerobic conditions, and hence, TVOSC production from dewatered biosolids. TVOSC emissions from most biosolids tended to increase during the first 7-10 days of sample incubation, and decrease to relatively low TVOSC concentration after 21 days of sample incubation as the methanogen population establishes itself and consumes TVOSC.
2. The relationships between iron, aluminum, and the odor potential of anaerobicly-digested biosolids are complex. Both iron and aluminum form bonds with proteins (the primary odor precursors) in biosolids, but iron-bound protein tends to be released more easily than aluminum-bound protein during dewatering and further processing of digested biosolids.
3. TVOSC emissions from anaerobicly-digested and dewatered biosolids increase with increasing iron content of the feed sludge. This trend is consistent with prior research indicating that biopolymer associated with iron is not completely degraded during anaerobic digestion. Shear during centrifugation subsequently releases the iron-bound protein, which leads to increased odors in the biosolids cake.
4. TVOSC in dewatered biosolids headspace samples decrease as aluminum concentration in the biosolids increases. It appears that aluminum-bound protein is not as readily released during centrifugation as iron-bound protein, leading to lower odors in biosolids cakes with higher aluminum content.
5. None of the CEBAs added to the anaerobic digestion process or downstream of anaerobic digestion showed any significant improvement in TVOSC reduction from dewatered cake without serious consequences such as digester toxicity. However, each CEBA was tested at only a single dosage which was recommended by its vendor or manufacturer. More experimentation of different dosage rates of CEBAs could possibly lead to improved results.

ES.5 Recommendations

Recommendations from this Phase 3 research are grouped into two major focus areas:

1. The research team's recommendations for applying the findings and conclusions of this project to the design and operation of WWTPs (recommendations for practice).
2. The research team's recommendations for further research on reduction of odors from biosolids.

The following sections summarize the research team's specific recommendations in those areas.

ES.5.1 Recommendations for Practice

Potential measures for reducing biosolids odors are grouped into the following two, basic subject areas:

1. Design changes that could be implemented for new WWTPs, new unit processes for existing WWTPs, or engineered modifications to existing processes.

2. Operational changes that could be made at an existing WWTP, either by adjusting operational parameters of processes or equipment, or via chemical addition.

Due to inherent differences in biosolids characteristics, treatment processes, and operating conditions, the Research Team cautions against assuming that similar results to those of this study will be obtained elsewhere. A planned approach is suggested that considers the various odor-reduction strategies summarized in this report, while keeping in mind the degree of cake odor reduction needed. Also, a process change or chemical addition that demonstrates biosolids odor reduction might have side effects on other processes that could outweigh the potential benefits from reduced odors. The selection of available options for odor reduction will therefore be dictated by site-specific conditions at each WWTP.

Figure ES-2 illustrates the potential reduction measures for biosolids odor control that have been explored in the project, as grouped into these two basic subject areas. In many cases, "trial-and-error" approaches will likely be needed to optimize reduction of biosolids cake odors. The required amount of odor reduction should be determined ahead of applying these schemes and then measured after each one is implemented. Odor measurements before and after process modifications provide for more accurate evaluations of a scheme's true odor reduction potential.

In Figure ES-2 "digester pretreatment" refers to application of such emerging waste solids reduction technologies that use temperature, pressure, chemical and biological means (e.g.; sonication, alkaline homogenization, and electrical or chemical lysis) for the purpose of disrupting microbial cell structures in the digester feed to enhance the digestion process. (Note that there is a greater amount of detailed discussion on Figure ES-2 in Chapter 5.0.)

ES.5.2 Recommendations for Further Research

The Phase 3 team sees the specific tasks listed below as some of the high priority research avenues necessary in the future for the resolution of the wastewater treatment industry's biosolids odor problem. These future research tasks could be funded jointly by WERF and targeted collaborative research (TCR) participating utilities, and jointly pursued in active collaboration with other related ongoing biosolids research projects and programs.

1. Investigate the presence of other odorous compounds besides TVOSC, such as nitrogen based odorants, to determine whether TVOSC alone is a reliable indicator for anaerobically digested biosolids.

2. Investigate how the presence or absence of ferric or ferrous iron in biosolids at different points in the wastewater treatment process train may affect biosolids odors.

3. Explore how some of the promising laboratory findings of Phase 3 related to aluminum addition, digestion, and dewatering can be refined and optimized in the field to reduce odors.

4. Identify the fundamental biological and chemicals causative mechanisms for biosolids odor generation and determine how those mechanisms can be neutralized via process changes and chemical addition.

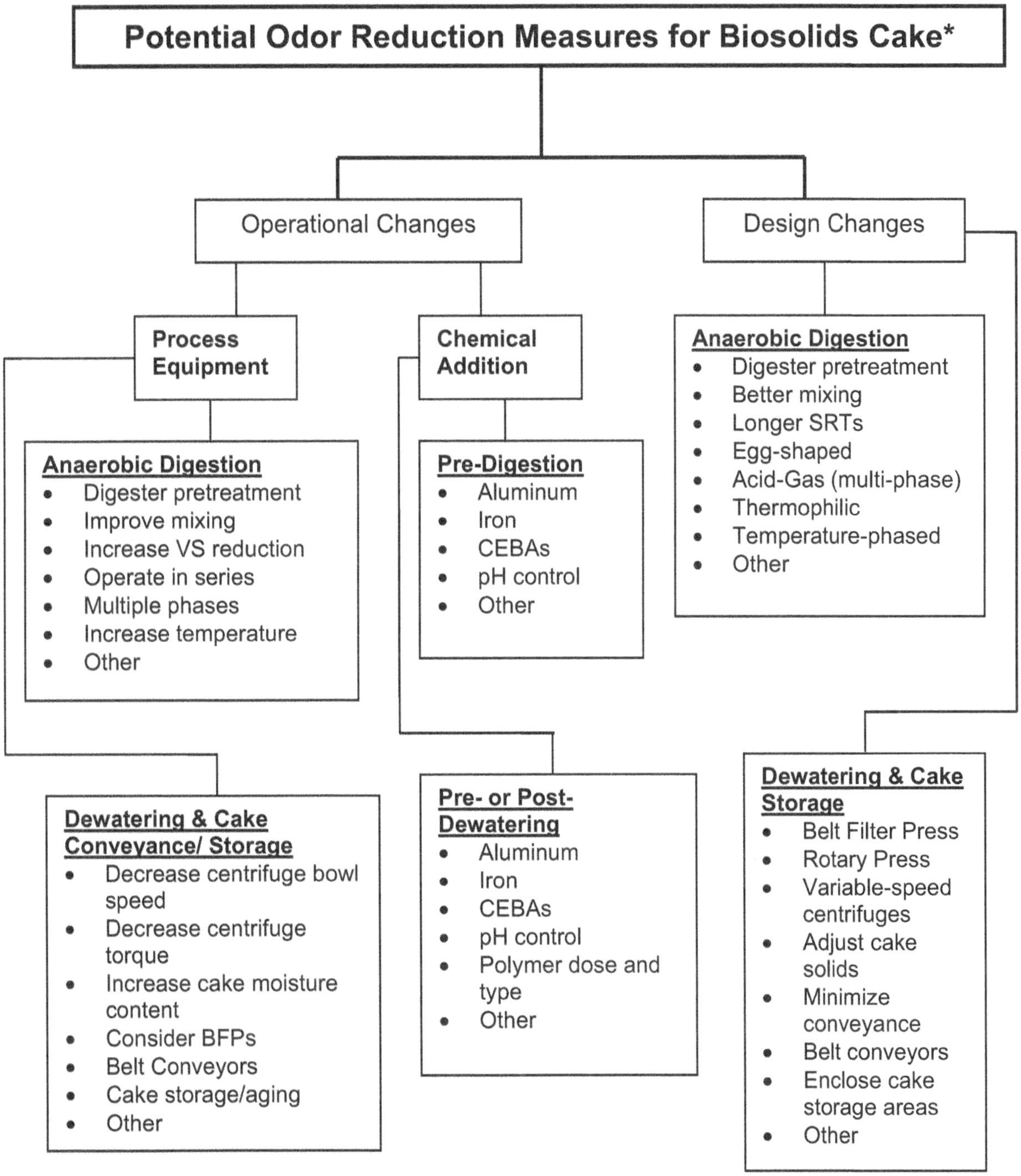

***Note**: None of these options should be considered independently of the others, and odor reductions in one area may impact treatment processes and odors in other areas. Therefore, a consolidated and customized approach is required for each WWTP. Refer to the text of this chapter for additional explanations.

Figure ES-2. Options Summary Chart for Biosolids Odor Reduction.

The research team recommends a three-pronged approach to making further progress in reducing biosolids odors:

1. Demonstrate biosolids odor reduction in full-scale settings at several WWTPs using the Phase 3 approaches, thereby broadening the body of knowledge on this subject.
2. Produce a practice manual for odor reduction in digested and dewatered biosolids, providing results from Phase 3 and some of the recommendations described in this report.
3. Team with the Water Environment Federation, National Biosolids Partnership, or related organization to make results more widely known and applied in the industry.

CHAPTER 1.0

INTRODUCTION

1.1 Background

WERF is completing a multi-phased research project that focuses on the origin and control of odors in the municipal wastewater environment. WERF Project 00HHE5, *Identifying and Controlling the Municipal Wastewater Odor Environment* is divided into three phases. Phase 1 of the research project consisted of an extensive literature search of the current state of knowledge regarding odor characterizations, odor measurements, and odor control for wastewater treatment and collection systems. This phase also looked at what has been successfully implemented in controlling odors at industrial and agricultural sites and whether there is any application to wastewater treatment facilities. A list of potential research agenda items was developed from the information gaps in the Phase 1 literature review. The Phase 1 report was published in 2003 under WERF Project 00HHE5A *Literature Search and Review*.

At a WERF-sponsored specialty workshop at the 2000 Water Environment Federation Technical Exposition and Conference (WEFTEC) in Anaheim, California, the WERF team presented a summary of the Phase 1 literature search and findings. In this meeting, participants voted biosolids odor issues the most important odor problem facing the wastewater industry and the highest priority for future research. Phase 2 of the overall research project was developed as a result of this workshop to focus on the reduction of biosolids odors in the municipal wastewater environment. Specific attention was given to the influence of in-plant processes on biosolids odor quality, including digestion, dewatering, and storage.

Phase 2 involved an in-depth evaluation of biosolids odor generation in 11 wastewater treatment plants (WWTPs) across North America, all of which employ anaerobic digestion of biosolids. The anaerobic digestion process was selected as a Phase 2 focus because anaerobic digestion is used to stabilize the majority of biosolids produced in North America (approximately 70-80% by tonnage). Anaerobic digestion is also the predominant biosolids stabilization process used by WERF utility subscribers. The Phase 2 research aimed at analyzing the means of odor generation in various solids handling processes within the 11 WWTPs surveyed. The Phase 2 project report was published in 2003 under WERF Stock No. 00HHE5T *Impacts of In-Plant Parameters on Biosolids Odor Quality*.

In Phase 2, the project team did not attempt to optimize or adjust process parameters or operating conditions during facility testing. Instead, a comprehensive protocol of wastewater and biosolids sampling and analyses was prepared and performed, in which more than 12,000 data points were gathered from 11 WWTPs during their normal daily operations. The purpose of the Phase 2 field testing and analyses was to develop correlations and observe the statistical trends of the common elements among the WWTPs. These trends were related to plant conditions or process parameters that appeared to affect the generation or reduction of biosolids odors.

The WERF team evaluated a large data set of analytical results from field and laboratory measurements. In addition, detailed WWTP information was collected through Request for Information (RFI) forms filled out by each plant. The resulting correlations between the data points gathered on process and operating conditions at the WWTPs and the measured odor parameters and odor precursors were converted into 17 hypotheses, listed in Table 1-1.

Table 1-1. Summary of Hypotheses Tested in the Phase 2 Study.

Phase 2 Hypotheses	Studied in Phase 3?
Hypotheses Supported Based on Study Results	
1. Higher amounts of bioavailable (labile) protein in biosolids cake create more odors.	No
2. Different dewatering practices impact bioavailable protein differently; some dewatering practices tend to increase odors in the biosolids cake.	Yes
3. Volatile sulfur compounds (VSCs) are the major sources of odors in most digested biosolids. This relationship was shown by a high correlation between odor detection threshold (DT) and concentration of VSCs, indicated by a multiple regression equation having a correlation coefficient of 0.90, which describes this relationship.	No
4. Odor concentrations in mesophilicly digested biosolids cake rise and then decline over time during storage.	No
5. Based on comparison of results from the one wastewater treatment plant (WWTP) in the study with thermophilic anaerobic digestion and 10 WWTPs with mesophilic digestion, odors from thermophilicly digested biosolids cake have different characteristics and patterns of time release than mesophilicly digested biosolids cake.	Yes
6. There is evidence from the Phase 2 study that iron in sufficient concentrations binds bioavailable protein in biosolids cake, and thus reduces odor production from dewatered biosolids.	Yes
Hypotheses Rejected Based on Study Results	
1. There is a positive correlation between high wastewater treatment plant influent sulfate concentrations and odors in biosolids cake.	No
2. Waste activated sludge (WAS) has a higher odor potential downstream of digestion processes than primary sludge.	No
3. Enzyme activity can be used as an indicator of biosolids odor production.	No
Hypotheses Found to be Inconclusive Based on Study Results	
1. High-pressure conveyance of dewatered biosolids leads to increased odors in the biosolids cake end product.	No
2. The presence and relative concentration of metal cations in the influent wastewater reduces odors in dewatered biosolids cake. In Phase 2, there was little correlation with either influent cation concentration or influent monovalent to multivalent (M/D) ratio and odor reduction in the dewatered biosolids cakes sampled.	Yes
3. Mixing of primary and secondary biosolids in pre-digestion processes worsens odors in digested, dewatered biosolids. Some plants in this study showed this correlation, but others did not, hence this relationship cannot be fully supported.	No

Table 1-1. Summary of Hypotheses Tested in the Phase 2 Study.

Phase 2 Hypotheses	Studied in Phase 3?
4. Higher concentrations of metal cations such as iron could reduce odors in digested, dewatered biosolids. There appears to be a lack of knowledge about the bioavailability of proteins and other organic material complexed with cations such as iron and aluminum, which caused the project team to remain undecided about this hypothesis.	Yes
5. Longer primary sludge detention time prior to digestion may increase odors in digested biosolids. This correlation was not supported by the results of the Phase 2 study, though further research is warranted.	Yes
6. A longer WAS retention time in the treatment process, including the activated sludge system plus waste sludge holding prior to digestion, generally produces less odors in digested, dewatered biosolids. This relationship was not seen in the Phase 2 study, however, all plants tested in this study had a very low retention time. Therefore, more plants with longer retention times should be tested in order to confirm or refute this hypothesis.	No
7. Title 40 of the Code of Federal Regulations Part 503 (503 Rule) for vector attraction reduction requires achieving a minimum of 38% volatile solids (VS) destruction during digestion. A clear correlation between biosolids stability per this criterion and reduction in dewatered biosolids odors was not found in this study.	Yes
8. It could not be conclusively shown based on the Phase 2 study results that trimethylamines (TMA) are significant sources of odor in anaerobicly digested, non-limed biosolids.	No

The hypotheses listed in Table 1-1 are a combination of new observations made in the Phase 2 study and traditional hypotheses about biosolids odor generation that were either supported or rejected by the Phase 2 results. The hypotheses were placed into three categories:

- Six hypotheses listed in Table 1-1 were supported by Phase 2 results. Three of these hypotheses appeared to indicate significant relationships between plant operation conditions and biosolids odors, and warranted more rigorous analysis. These three hypotheses were carried over into the Phase 3 study.
- Three industry-based accepted theories were investigated in Phase 2, but were rejected based on the results of Phase 2. These theories were not investigated further in Phase 3.
- Eight trends observed in previous studies or industry-based accepted theories were investigated in Phase 2, but results were inconclusive with respect to biosolids odor reduction. Five of these hypotheses were carried over into the Phase 3 study for more rigorous analysis.

1.2 Purpose

Phase 3 of the overall WERF project was developed to study eight Phase 2 hypotheses (Table 1-1) in more depth, beginning in the laboratory (bench-scale) studies and continuing on to targeted investigations where the WERF team manipulated plant parameters at full scale to identify the best means of reducing biosolids cake odors. The Phase 3 research specifically sought ways to enhance anaerobicly digested and dewatered biosolids to reduce the odor levels in the biosolids end product, thereby reducing negatively perceived impacts on the environment or to the public when beneficially used on land.

The goal of the Phase 3 study was to provide a general application of findings to WERF subscribers who are seeking ways to reduce odors produced by anaerobically digested biosolids. The Phase 3 options summary present a general roadmap for wastewater treatment plant operators seeking to optimize biosolids processing and reduce biosolids cake odors.

Biosolids cakes with minimal odors lead to better public acceptance near biosolids management sites and in neighborhoods adjacent to WWTPs. Reduced odors also could open the WWTP dewatered biosolids cake to other recycling or disposal opportunities that are currently not used due to odor and other concerns (including on-plant site composting or storage). Additionally, significant cost savings could be realized by not requiring extensive odor control or other expensive options for containment and management of biosolids.

1.3 Project Organization

The Phase 3 study was completed using three elements: laboratory studies (Phase 3A), biosolids equipment and management approach surveys (Phase 3B), and full-scale field testing (Phase 3C). These elements were augmented by an addition to the project to investigate the potential impact of commercially available chemical, enzymatic, and biological agents (CEBAs) on biosolids odors. The same organization of entities (principal investigators and contributing organizations) that successfully delivered Phase 2 also completed the Phase 3 work. The Phase 3 elements, project team, and schedule are summarized in this section.

1.3.1 Phase 3 Elements

The scope of the Phase 3 elements was determined following completion of Phase 2, where hypotheses involving biosolids odor reduction were selected for more rigorous analysis. This section discusses these elements, and provides an overview of the CEBA program addition.

<u>Phase 3A: Laboratory Studies</u>

In Phase 3A, laboratory studies at Bucknell University, Virginia Polytechnic Institute, and Los Angeles County Sanitation Districts (LACSD) were completed assessing variables on odor production within the solids processing train. In Phase 3A, the laboratories identified the most promising approaches to reducing dewatered biosolids cake odors, which would then be translated to full-scale applications as part of Phase 3C. The following relationships were tested:

1. <u>The role of metal cations in odor production from anaerobically digested sludges</u>: Trivalent cations, specifically of iron and aluminum, were analyzed for their relationship to the bioavailability of proteins for odor generation following centrifugation. Phase 2 results indicated that biodegradable proteins remain after conventional anaerobic digestion, and it was hypothesized that they are released by shear in centrifuges, thus producing odors. The hypothesis tested in Phase 3A assumed that metal cations bind organic reduced sulfur compounds and cellular protein, which increases digestion efficiency and reduces odors.
2. <u>The effects of digester solids retention time on odors</u>: Phase 2 correlations were inconclusive, as it was unclear if additional digestion time would be helpful in reducing odor-causing proteins. Digester SRT calculations were subject to interpretation. Determination of the relationship between digester SRT and odor

generation would provide important insight into the links between digestion and odors, as if additional solids destruction (and thus protein destruction), is possible at higher SRTs, and if this solids destruction leads to reduced odors, it would suggest that improvements in digestion are likely to reduce odors.

3. The effects of digestion enhancements on odors: Processes that claim to improve solids destruction were tested for their ability to reducing odors from anaerobically digested biosolids. Processes tested include shearing the digester feed, addition of iron to the digester feed, thermophilic digestion, and two-phase digestion.
4. The effects of dewatering on odors: High-speed dewatering centrifuges have been noted in previous studies to increase odor production in the dewatered cake above levels attained with medium- or low-speed centrifuges. The WERF team theorized that the shear imparted to the cake during dewatering contributes to the increased odor production associated with higher speed centrifuges. Therefore, reducing shear could potentially reduce odor production. This part of the investigations examined changes in various operational parameters of centrifuges to determine their impacts on odor production. In addition, comparative tests were conducted on facilities running side-by-side dewatering methods using different technologies, including centrifuges and belt filter presses.
5. The effects of chemical addition on odors: Previous research has shown that the addition of cations (iron and aluminum) could play an important role in binding protein by making the protein less bioavailable, which would thereby reduce production of odorous compounds. Prior investigations have also indicated that low-dose lime addition can reduce odors. These chemical addition alternatives are appealing to utilities since they would provide a relatively low cost, easy to implement solution for controlling odors. The Phase 3A studies tested chemical addition in the laboratories.

Phase 3B: WWTP Surveys

In this portion of the project, the WERF team surveyed WERF members, and others in the wastewater industry, including members of the National Association of Clean Water Agencies (NACWA), the Water Environment Federation (WEF), and the Air and Waste Management Association (AWMA). The survey consisted of questions about the various dewatering and digestion processes that the facilities currently employ. Additionally, the survey asked for any approaches that the WWTPs might be considering using at their facility in order to meet U.S. EPA 40 CFR Part 503 biosolids treatment requirements. Many WWTPs are considering Class A biosolids processing options, new biosolids dewatering approaches, and different management approaches as potential Part 503 compliance mechanisms. Whether or not such approaches simultaneously reduced or accelerated odor levels was a concern of the Phase 3 study.

Under Phase 3B, the project team surveyed WWTPs regarding their biosolids handling, storage, management, and treatment processes. The intent of Phase 3B was to gather information about plants that have processes that would be advantageous to future testing within the overall WERF project, and that may also be amenable to WERF-guided process modifications for reducing biosolids odors.

During this phase of the research, the WERF team requested biosolids samples from various WERF members and survey responders that would be interested in having their biosolids tested and used in the laboratory studies. The choice of individual utilities approached was based on their means of dewatering, enhanced digestion approach, odor quality of their digested biosolids cakes, and potential to have their biosolids tested over a wide range of hypotheses.

Also as part of Phase 3B, two other entities were approached by the WERF research team as part of the Phase 3 study:

- Biosolids dewatering equipment manufacturers and vendors were asked to donate portable and pilot-scale equipment for use in full-scale and side-by-side tests with other dewatering equipment.
- Chemical suppliers were contacted for use in the Phase 3A chemical addition study (iron, lime, and aluminum addition).

Phase 3C: Field Studies

The intent of this third phase of the research was to transfer the most promising approaches identified in the laboratory studies and WWTP survey results to the field, where the hypotheses were tested at full scale, with the goal of reducing actual dewatered biosolids cake odors. While laboratory studies (Phase 3A) provide viable indicators of methods to reduce biosolids cake odors, findings are most substantial when the methods are tested full scale. Also in Phase 3C, the WERF team completed pilot studies at the LACSD Joint Water Pollution Control Plant (JWPCP) and Philadelphia Southwest Water Treatment Plant. At these facilities, side-by-side comparisons of dewatering facilities were conducted.

Phase 3C: CEBA Studies

During the completion of the Phase 3 study, results from other studies indicated that other additives appeared to hold promise for biosolids cake odor reduction. These additives were commercially available chemical, enzymatic, and biological agents, termed CEBAs. However, consistent test protocols and other test parameters were not clearly defined on the other research projects, so the true comparative effectiveness of those CEBAs was not verified. As an amendment to the Phase 3 study, a CEBA addition evaluation was added to Phases 3A and 3C.

The project team contacted various CEBA vendors as well as other related product manufacturers that were known to the team, compiled information on their recommended dosages and application points, developed test protocols that were accepted by the CEBA vendors, and tested the CEBA products according to the endorsed test protocols. The project team used a batch-digestion protocol in which two of the CEBAs were tested concurrently on the same sludge along with an un-dosed control test.

1.3.2 Project Team

The project team that completed the Phase 3 study is depicted in the organization chart contained within Figure 1-1. Table 1-2 shows how each organization was assigned specific tasks within the Phase 3 study.

Table 1-2. Major Responsibilities of WERF Project Team Members.

Team Members	Main Responsibilities
LACSD	• WERF-member and public outreach activities • Contract administration and project management • Develop and conduct research and develop protocols • Conduct field and pilot testing • Provide final recommendations
CH2M HILL	• Technology management and project management • Develop research and protocols • Develop WWTP Survey • Conduct field and pilot testing • Provide data analysis and final recommendations
Bucknell University Virginia Polytechnic Institute	• Investigate promising hypotheses from Phase 2 • Develop research protocols • Conduct sample analysis for Phase 3C • Data analysis • Develop Phase 3 recommendations
Technical Advisory Committee	• Provide technical review and expertise during each phase of the project
WERF Project Sub-Committee	• Review sampling and analysis protocols and approaches • Review and comment on research and developments based on Phase 3 findings, recommendations, and conclusions • Review Phase 3 Report and progress reports • Attend Phase 3 workshops

1.4 Literature and Utility Surveys, Requests for Information (RFIs), and Requests for Proposals (RFPs)

In Phase 3B, the WERF team surveyed WERF subscribers and vendors in the wastewater industry. This section summarizes the results of these surveys, RFIs, and RFPs.

1.4.1 WWTP Agency Surveys and RFIs

The Phase 3 biosolids odor research project (WERF No. 03-CTS-5T) began in mid-2004 with a search for recent papers and journal articles dealing with biosolids odor generation and reduction plus a survey of existing biosolids processes in all WERF-subscribing utilities. The WWTP Request for Information (RFI) form (to be filled out and returned by the respondent), solicitation letter, and a brief one-page research project description were prepared by the project team and sent out by WERF in July 2004. The survey asked for information about each responding WWTP's type of digestion dewatering process, and any possible general interest in joint research with the WERF team for biosolids odor reduction.

In February 2005 a second letter and Request for Interest form was sent out through WERF to all WERF subscribing utilities specifically asking if they were interested in participating in the biosolids odor research project as a TCR (Targeted Collaborative Research) participant. Finally, in December 2005, a follow-up letter was sent by WERF to all survey respondents thanking them for their responses. All survey respondents were later invited to a special closed Phase 3 project seminar held on October 22, 2006 in Dallas, TX, prior to the 2006 Water Environment Federation Technical Exposition and Conference (WEFTEC).

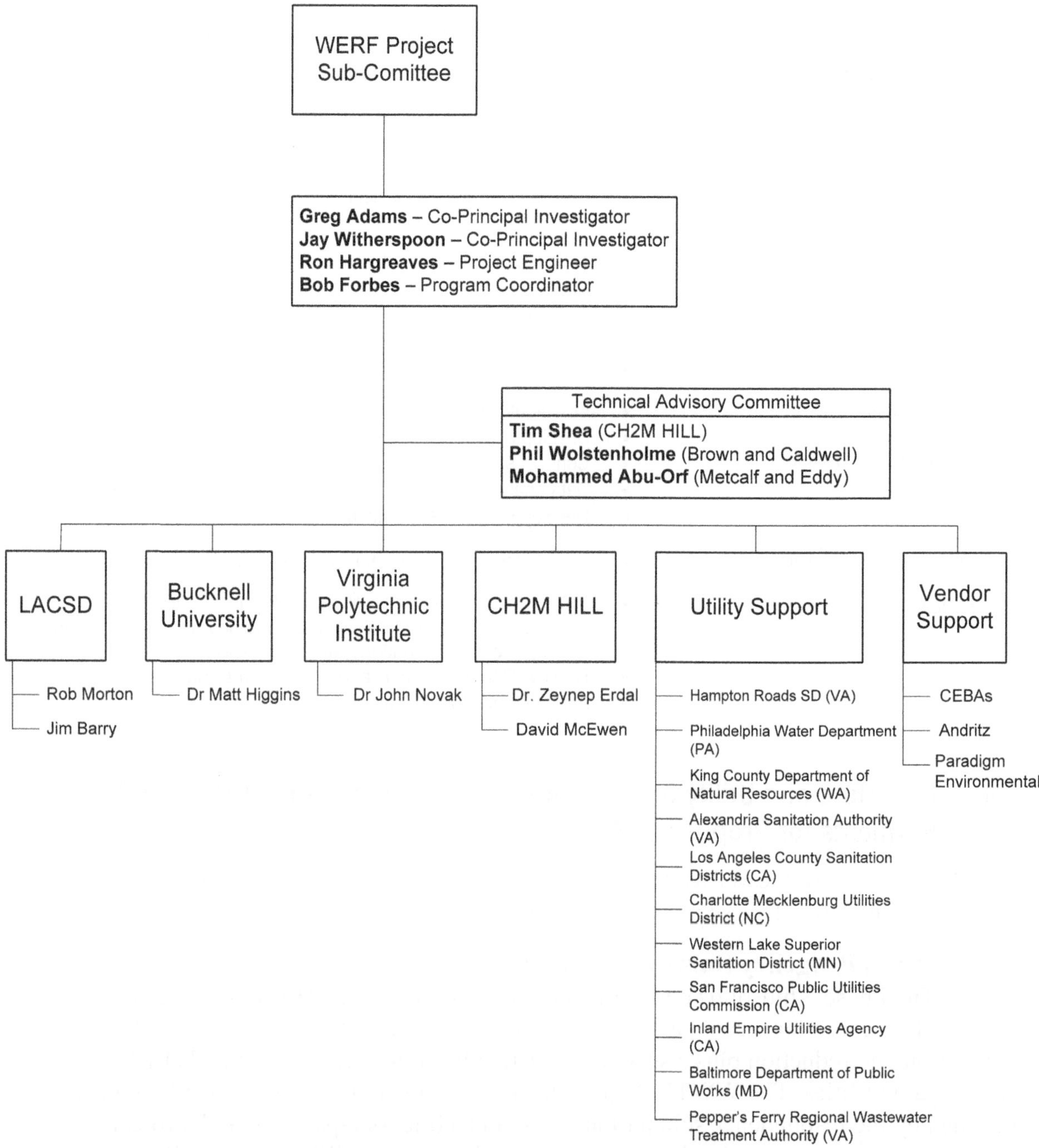

Figure 1-1. WERF Phase 3 Organization Chart.

1.4.2 Dewatering Equipment Vendor RFP

In early February 2005 the WERF project team sent out an RFP to dewatering equipment vendors to check their willingness to participate in the biosolids odor reduction research project. They were specifically asked to propose field testing on sludge dewatering with their pilot or field units to determine the effect on dewatered, stored, biosolids odor generation. All interested vendors were invited to participate in a "jobwalk" teleconference held in late February.

1.4.3 Chemical Enzymatic and/or Biological Agent (CEBA) Vendor RFP

In February 2005 the project team sent out an RFP to chemical, enzymatic, and/or biological agent (CEBA) vendors to survey their interest and willingness to participate in our biosolids odor reduction research project. They were specifically asked to provide information about previous applications and installations and to indicate their willingness to furnish their product samples for our laboratory testing. Interested CEBA vendors were invited to participate in the same "jobwalk" teleconference with the interested equipment vendors in late February.

1.5 Report Organization

The following chapters are contained within this report:

- Chapter 1.0: Introduction
- Chapter 2.0: Sampling and Test Procedures
- Chapter 3.0: Test Results
- Chapter 4.0: Research Findings and Conclusions
- Chapter 5.0: Recommendations

Note that Chapter 3.0 contains a review of all the data associated with this project. Chapter 4.0 provides a summary of findings and conclusions, which may be reviewed much quicker than the time that would be required for reviewing Chapter 3.0. Additional supporting data is provided in the appendix.

CHAPTER 2.0

SAMPLING AND TEST PROCEDURES

2.1 Introduction

The purpose of this section is to present the rationale for laboratory and field analyses and the sampling and analytical test procedures used throughout the Phase 3 biosolids odor research project. Many of the methods used in Phase 3 have proven to be successful during the Phase 2 research and in other research applications. Testing protocols and procedures were followed by the laboratories at Virginia Tech, Bucknell University, and LACSD.

2.2 General Protocols

Prior to start of any laboratory work or sampling, the project team developed comprehensive testing and sampling procedures into a document entitled "General Testing and Analysis Protocol." This document is attached in Appendix A1 of this report, and this section summarizes key elements of the protocol.

2.2.1 Laboratory Analysis Rationale

The Virginia Tech, Bucknell University, and LACSD laboratories conducted bench-scale tests on biosolids samples collected at various WWTPs in Phase 3A and 3C of this project. The means by which the samples were tested was based on a rationale developed in Phase 2 and at the outset of Phase 3. This section presents key elements of this rationale.

2.2.1.1 Headspace Analysis Method

In Phase 3, measurement of emissions from dewatered biosolids cake samples were conducted using the headspace analysis method, which was developed and refined during Phase 2. Method development included comparative tests in the field and laboratory to ensure that laboratory methods simulated the field conditions appropriately. The Phase 2 report and Glindemann, et al. (2006) establish how the headspace analysis method mimics anaerobic incubation conditions in a sludge tank or cake pile.

When using the headspace analysis method, the headspace storage bottles, which are closed bioactive systems, allow for the measurement of odor consumption by the biosolids cake in the bottle as well as measurement of the odor production-consumption cycles that are ongoing in the closed system. Tests conducted during the development of Phase 2 showed that the odor concentration in the headspace of the bottles is a function of the odor concentration of the biosolids cake sample.

The headspace analysis method used in Phases 2 and 3 has several advantages over measurement of biosolids cake emissions using flux chambers, including the following:

- The headspace method is simpler than a comparable test using a flux chamber, with fewer equipment requirements and easy sampling means.
- The storage conditions in headspace bottles are anaerobic and mirror the anaerobic bulk core of large full-scale biosolids cake piles.

- A major advantage over flux chambers, the headspace analysis method characterizes odor production and consumption cycles through long-term storage.
- The headspace analysis method, combined with use of a cryotrap gas chromatograph/mass spectrometer (GC/MS), increases the productivity and speed of odor data production as well as the quality of odor data compared to flux chamber solid-phase microextraction (SPME) methods.
-

In Figure 2-1, the headspace analysis method ("Closed bottle") for biosolids odor emissions measurement is compared to the flux chamber method and to the expected odor generation from a full-scale biosolids cake storage pile. The figure depicts the additional odor consumption element that is included in the headspace analysis method.

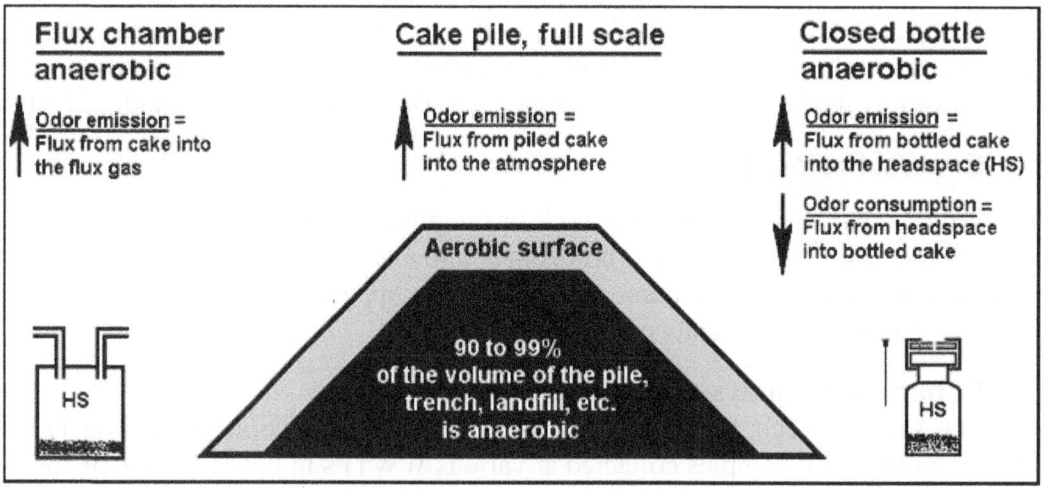

Figure 2-1. Comparison of Headspace Method to a Static Biosolids Pile and a Flux Chamber.

In summation, although both methods have been shown to successfully mimic full-scale conditions, the headspace analysis method has been shown to be representative of the cake storage pile interior, is easier to perform than flux chamber testing, and highly reproducible (Novak et al., 2004; Glindemann et al., 2006). In addition, gas-phase samples can be taken directly from the bottle headspace and analyzed immediately without the need for a concentrating step.

2.2.1.2 Generation and Measurement of Volatile Organic Sulfur Compounds

In the literature, the generation of volatile organic sulfur compounds (VOSCs), reduced inorganic sulfur, and heterocyclic amines has been linked to amino acid metabolism and sulfate reduction. Sulfur-containing functional groups associated with amino acids are removed and the product either is an odorant itself or undergoes biochemical transformation to form odor-causing compounds. These odor-formation mechanisms have been detailed by Higgins et al. (2006) and are depicted in Figure 2-2.

The precursors to odor-causing compounds are proteins. The steps that are taken in the conversion of proteins to volatile organic sulfur compounds are depicted in Figure 2-3.

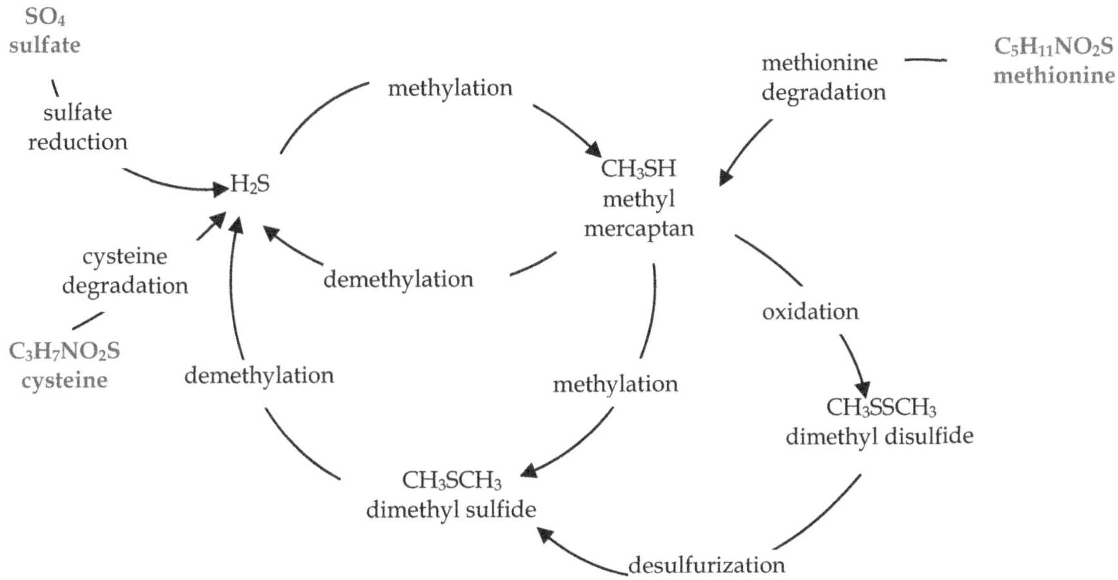

Figure 2-2. Cyclic Pathway Proposed to Describe the Production and Transformations of VOSCs (key substrates for odor production in red bold font).

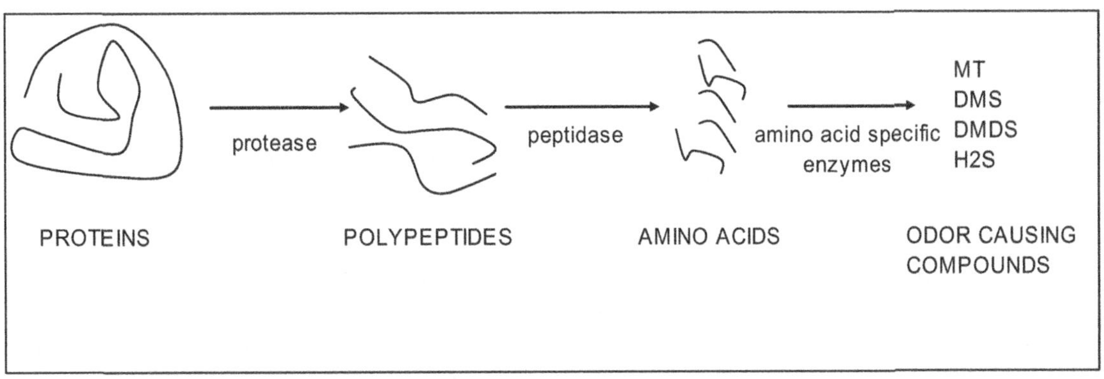

Figure 2-3. Formation Pathway for Odor-causing Sulfur Compounds from the Degradation of Proteins.

The metabolic decomposition of sulfur-based odorants can be both aerobic and anaerobic. Aerobic degradation processes, where reduced sulfur compounds are converted to sulfate, are used in biological air scrubbing units to control exhaust odors from dewatering facilities or other odorous processes. However, most odor problems associated with biosolids come after dewatering during land application and storage, a Phase 2 study finding. The metabolic processes mediating the degradation of compound during application and storage are primarily anaerobic.

The anaerobic degradation of methylated sulfurs and amines is primarily due to methanogenic activity. Under anaerobic conditions, methionine is converted to homoserine and methanethiol, followed by the conversion of methanethiol to methane, carbon dioxide and hydrogen sulfide. Recently, progress has been made into understanding the communities involved in the reduction of methylated sulfur species. These are summarized by Higgins et al. (2006). In these studies, it has been determined that in most sludges, hydrogen sulfide is precipitated as FeS when sufficient iron is present. When iron content is low, hydrogen sulfide accumulates. In the 11 plants studied as part of Phase 2, two plants had hydrogen sulfide levels that were of concern.

As a result of the reduced sulfur compound generation and conversion reactions in sludges, a pattern of increase and then decrease in organic and inorganic sulfur compound concentrations has been observed. Figure 2-4, from Novak et al. (2006), shows a typical pattern of increase and then decrease in concentrations in stored biosolids cake. The plot shows that the major organic sulfur compounds consist of methanethiol (MT) and dimethyl sulfide (DMS) with lesser amounts of dimethyl disulfide (DMDS) and hydrogen sulfide (H2S) being generated. The significance of this trend is that it reflects the growth and decline of the bacterial populations involved in the formation and assimilation of these compounds.

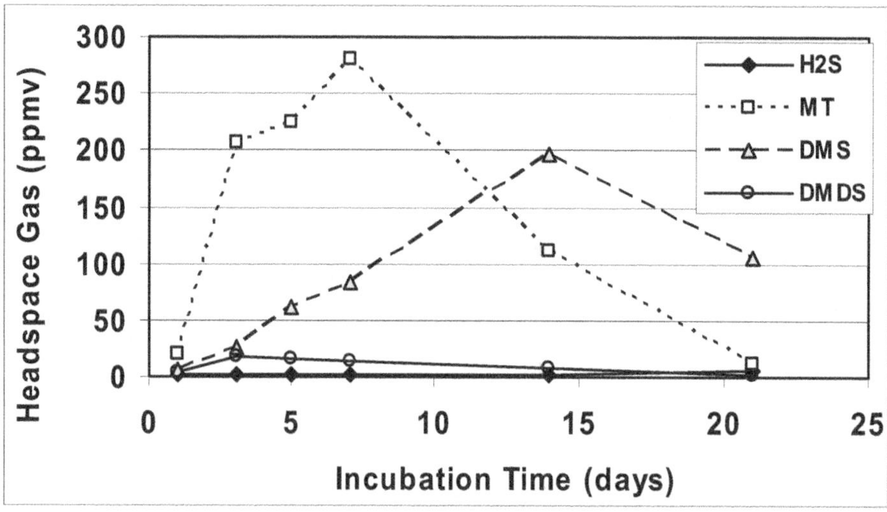

Figure 2-4. Typical Pattern for Production and Loss of Volatile Organic Sulfur Compounds in the Headspace of Incubated Sludge Cake.

Because of the small amount of H2S that is typically generated in biosolids cake storage (see Figure 2-4 as an example), the sum of the VOSCs is typically summarized and presented as total volatile organic sulfur compounds, or TVOSC, where TVOSC is reported as sulfur and includes mostly methane thiol, dimethyl sulfide, and dimethyl disulfide (minimal H2S). The plot of TVOSC concentrations corresponding to the sum of the components in Figure 2-4 over time (referred to as the TVOSC profile) is depicted in Figure 2-5. For this stored biosolids cake, the TVOSC peaks at around 7 days of incubation and then declines. For most mesophilicly digested sludges dewatered by centrifugation, the TVOSC peaks at 3 to 15 days and then declines to near zero after about 30 days.

When reporting odor data using the cake incubation procedure developed by Glindemann, et al. (2006), the peak TVOSC concentration is used to provide a simple representation of the magnitude of the odor-causing volatile organic sulfur compounds. For example, in the TVOSC profile provided in Figure 2-5, the peak TVOSC is approximately 380 parts per million by volume (ppmv). The TVOSC value is reported on an "as sulfur" basis. Therefore, in the TVOSC calculation, the concentration of DMDS is doubled to represent the "as sulfur" basis, due to the presence of two sulfurs in the compound.

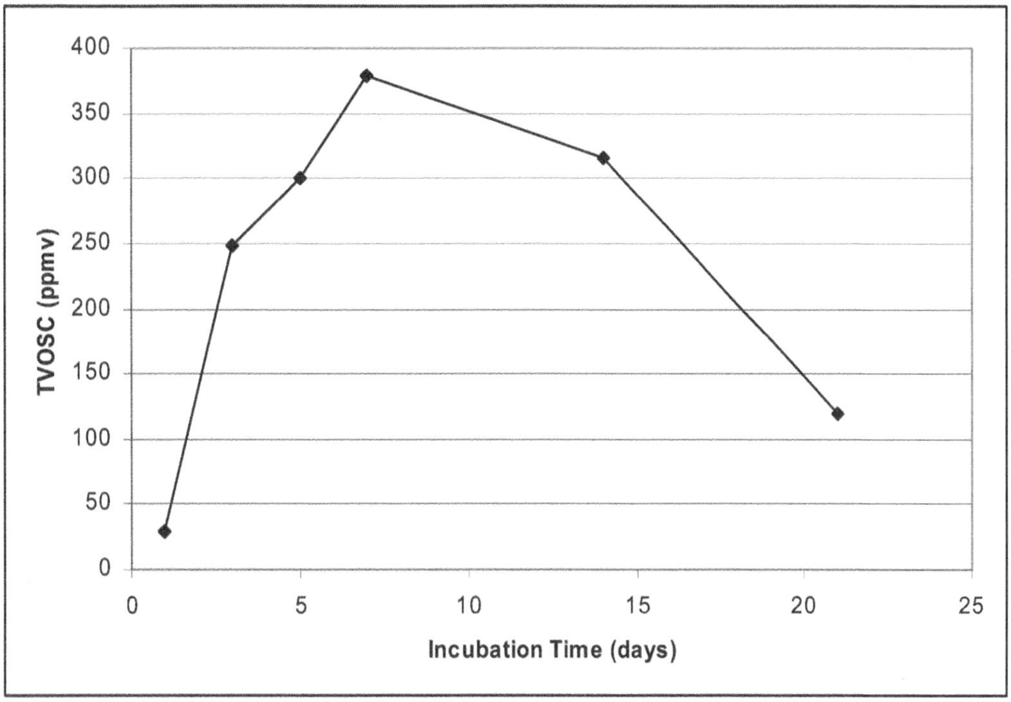

Figure 2-5. Variation in TVOSC with Incubation Time for a Dewatered Sludge Cake (corresponding to Figure 2-4).

In addition to peak TVOSC, the individual compounds that make up the TVOSC concentration (particularly MT and DMS) and the time to reach peak the TVOSC may also be important, so these parameters are often also noted in this report. For example, it was found by Novak et al. (2006) that sludges digested in thermophilic digesters experience a peak TVOSC that occurs much later in the incubation period than the time-to-peak for mesophilicly digested sludges, typically around 40 days. This finding may impact sludge handling management by a POTW that uses thermophilic digestion.

The literature indicates that the measurement of TVOSC for a given headspace sample is strongly correlated with total odor detection threshold (DT), as measured in units of dilutions to threshold (D/T) by an odor panel. Figure 2-6 depicts results from two studies, comparing TVOSC data from the LACSD Joint Water Pollution Control Plant (JWPCP) and TVOSC as measured by a flux chamber (Webster et al, 2006) versus corresponding odor D/T measurements by an odor panel for the same biosolids samples. In both comparisons, the relationship between TVOSC and total odor (as measured by D/T) is consistent over a wide range of TVOSC values, and the correlation between TVOSC and odor detection threshold is very good. Based on the results of this comparison and others like it, the WERF team has concluded that using the headspace analysis method, as described in Section 2.2.1.1, in combination with the laboratory determination of TVOSC, as described above, provides an accurate measurement of odorous emissions of dewatered biosolids, both instantaneously and through days of incubation time.

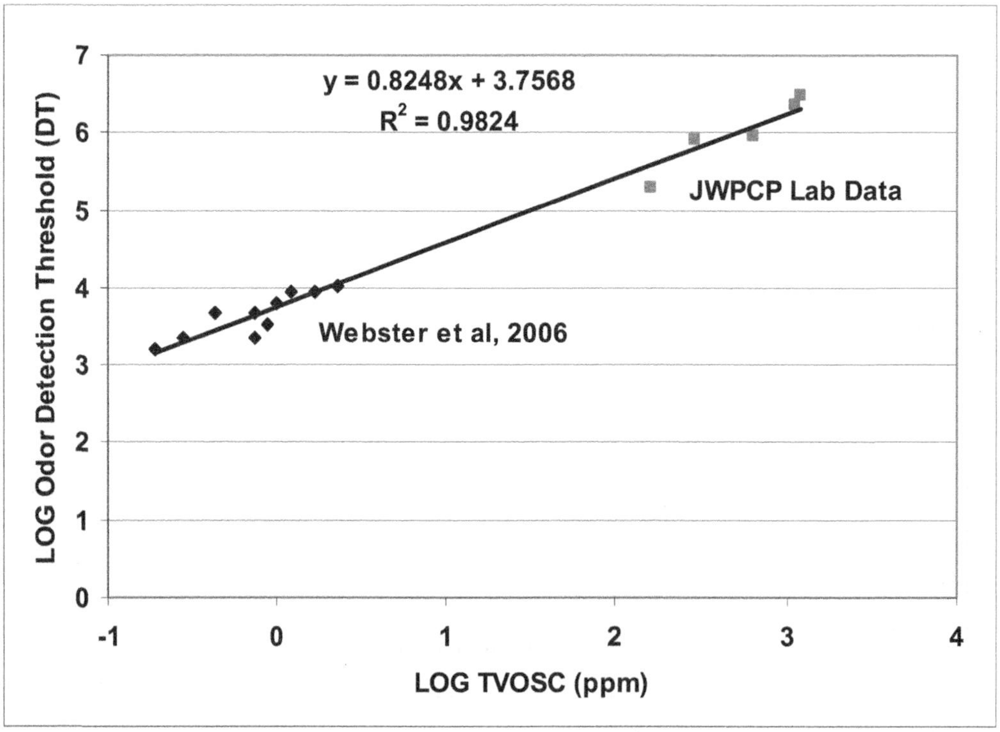

Figure 2-6. Correlation of TVOSC with Odor Detection Threshold in Two Studies.

2.2.1.3 Methane Production and Inhibition

The decrease in TVOSC after day 7 shown in Figure 2-7 is attributed to conversion of organic sulfur compounds to sulfide by methanogenic organisms (Higgins et al., 2006). When methanogens are inhibited, methane generation decreases. This causes organic sulfur compound concentrations to decrease more slowly and subsequently TVOSC usually peaks at a higher concentration. *Therefore, any changes that impact the activity or viability of the methanogenic population will have important effects on TVOSC production and longevity.*

For example, addition of lime immediately following centrifugation can be both beneficial and detrimental to odor generation. If a sufficient amount of lime is available to inhibit *both* the degraders of proteins (organisms that convert protein to MT and DMS) and the methanogens, little odor is generated in the biosolids cake. This trend is shown in Figure 2-7a, in which the TVOSC concentration remains low (near zero) throughout the incubation period given a sufficient lime dose. However, if an amount of lime is added that is large enough to inhibit methanogens, but is insufficient to inhibit the degraders of proteins, the cake TVOSC concentration will rise with incubation time, as depicted in the Figure 2-7a plot, to a level greater than that which is measured without any lime addition. Therefore, Figure 2-7a displays the observation that lime addition can reduce TVOSC or cause it to increase, depending upon the amount added. A companion figure, Figure 2-7b, shows the effect of lime addition on methane production, in that methane production decreases greatly with either level of lime addition. When methane production declines but protein degraders remain active, TVOSC will accumulate in storage biosolids cake.

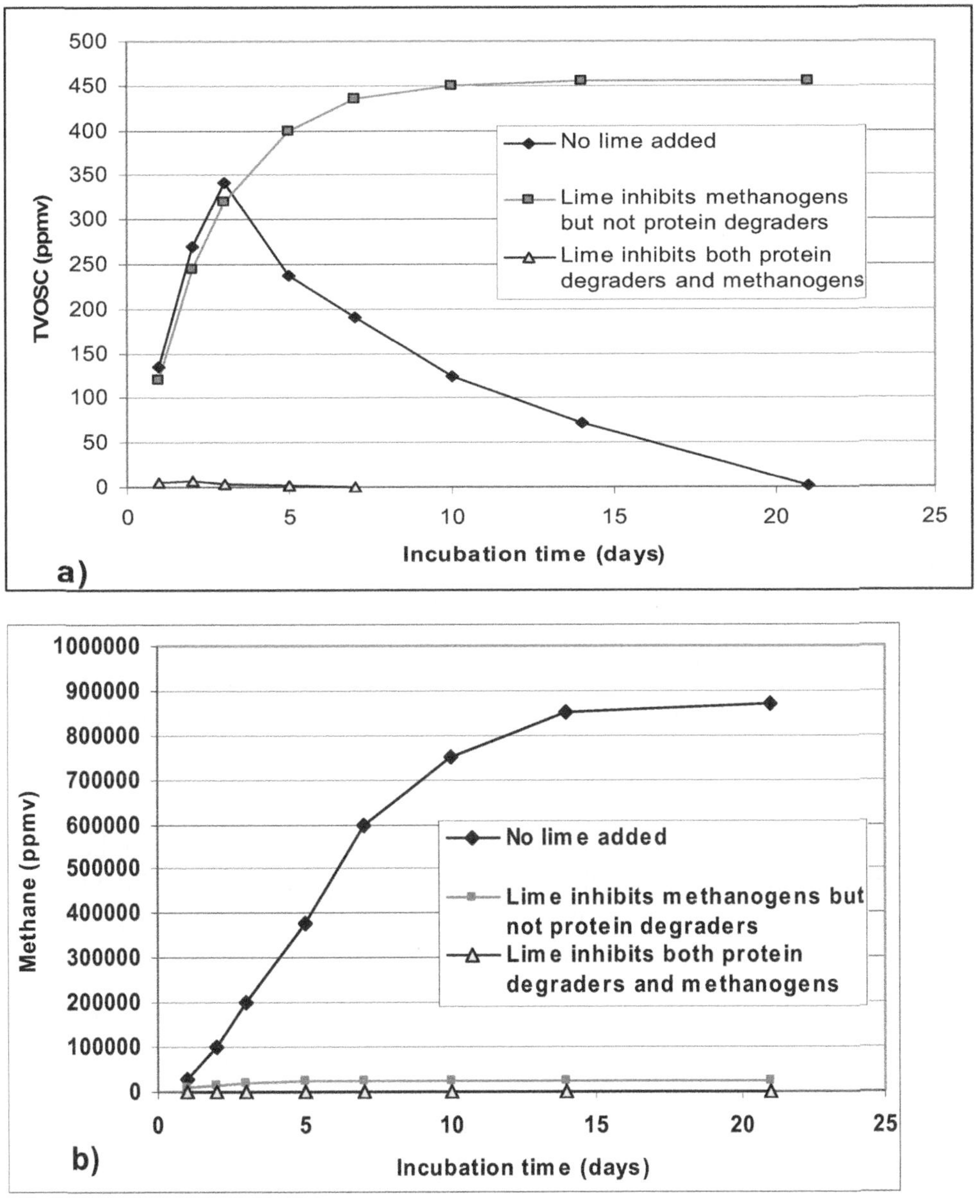

Figure 2-7. Patterns for (a) TVOSC Generation and (b) Methane Production with Different Amounts of Lime Added to the Biosolids.

The conclusion of this analysis is that any change that alters the pH, encourages the growth of methanogens, or inhibits methanogens will change the odor potential of dewatered sludge cakes. Methanogens can also be inhibited in the laboratory intentionally in order to assess the maximum potential for TVOSC production. Figure 2-8, taken from Higgins et al. (2006), illustrates that the addition of bromoethanesulfonic acid (BESA) results in an accumulation of organic sulfur compounds. Therefore, the accumulated peak TVOSC in the presence of BESA can be used as an indication of the "odor potential" for any sludge. The BESA additive can be used to differentiate between processes or chemicals that reduce the proteins that generate odor versus those that increase odor consumption by enhancing methanogenic activity. This differentiation is important when considering the impact of chemical and biological additives on TVOSC production.

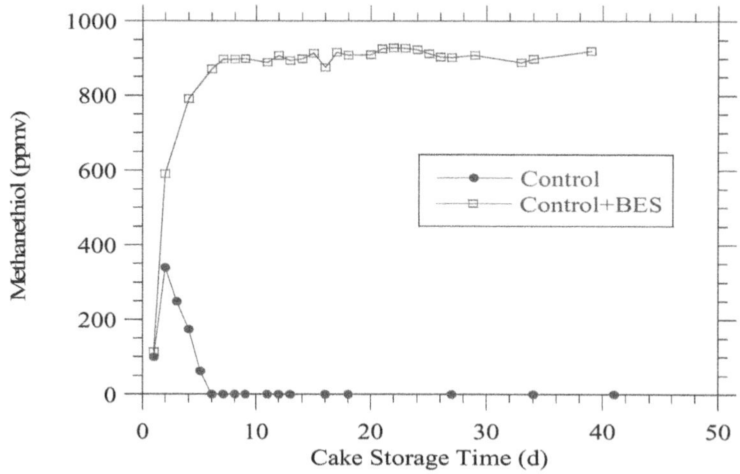

Figure 2-8. The Impact of BESA on Methanethiol Production (from Higgins et al., 2006).

2.2.2 Sampling and Analysis Protocols

The Phase 3A and 3C elements of this research project required sampling from various WWTPs and analysis of various parameters at laboratories. Additionally, the university laboratories conducted bench-scale testing that attempted to simulate full-scale solids process (digestion, dewatering, and storage). This section provides a summary of general protocol used in Phase 3 for sampling and analysis.

2.2.2.1 Sampling Protocols

In all cases during Phase 3, samples were collected under "normal" or baseline operating conditions of the plant, including the anaerobic digesters from which many samples were taken. These "normal" conditions were defined by the average dry weather flows and loads at typical operating conditions of the plant and anaerobic digestion facility.

To ensure that the test results were representative of field conditions and that the samples were shipped to the laboratories in an efficient manner, the project team evaluated the sampling sites and coordinated the sampling event with the facility operations staff in advance. The team also developed site-specific sampling protocols that designated the responsible parties and sample collection, preservation and shipping details. Each site-specific protocol included a sampling protocol in its appendix.

2.2.2.2 Laboratory Headspace Analysis Protocol

In the Virginia Tech and Bucknell University laboratories, dewatered biosolids cake samples were added to closed bottles in which headspace gases were monitored over time. This method mimics conditions present in a biosolids storage pile. The headspace method has been described in detail by Glindemann et al. (2006) and is included in the Phase 3 General Protocol (Appendix A), so will not be considered in detail here. However, a brief review of the method is provided because it is critical to the data collected in Phase 3.

Standards were injected each day prior to the headspace measurements as a means of quality control for the instrumentation. The detection limit for these analyses is on the order of 100 parts per billion by volume (ppbv), and the precision of repeated analysis is on the order of ±5%. The analysis using these standards included tests of concentration of hydrogen sulfide, methane, methanethiol, dimethyl sulfide, dimethyl disulfide, trimethylamine, p-cresol, toluene, ethylbenzene, styrene, indole, and skatole.

In the headspace analysis method, ten grams of wet biosolids cake were placed in 160-mL Wheaton serum bottles after the appropriate treatment was administered (dependent upon the test being conducted). The sample bottles were then sealed using 20-mm Teflon faced butyl rubber septa, tear-off aluminum caps and a crimper. The bottles were stored at a constant temperature (typically 25°C except for in experiments where temperature is a variable) for the duration of the analysis.

The laboratory staff conducted headspace analyses by inserting a needle from a Hamilton gas-tight locking syringe directly through the septa of the serum bottles and removing 1 mL of gas from the headspace. The extracted headspace sample was analyzed using a Hewlett Packard (HP) GC/MS with a 30 m x 32 mm inside diameter x 0.25 m^2 HP 5 column. The carrier gas flow rate was 2 mL/min and the temperature in the cryo-concentration unit was –130°C in the first loop of the GC column.

The duration of each sample run was 18 minutes. During the sample run, the oven temperature is held at 100°C for the first 3 minutes followed by an increase to 220°C at a rate of 15°C per minute for the next 8 minutes. This temperature was then maintained for 7 minutes, completing the sample run.

Volatile organic sulfur compound concentrations and methane concentrations were measured in the headspace of the sample bottles on a regular basis during storage (a separate bottle was sampled each day). VOSCs were identified and quantified by comparing the experimental chromatograms to those of pure standards. Headspace gas chromatography was performed using an HP 5890A Gas Chromatograph equipped with a flame ionization detector (FID). A Restek RT-Sulfur packed column measuring 2 m long with an inside diameter of 32 mm was used. Both the injection port and detector temperatures were 200°C. Zero nitrogen was used as the carrier gas at a flow-rate of 20 mL per minute. Zero air and zero hydrogen were supplied to the FID at flow rates of 450 mL/min and 20 mL/min respectively. All gases were purchased from Airgas Inc.

2.2.2.3 Biosolids Chemical Analysis Protocol

Virginia Tech and Bucknell laboratory staff analyzed biosolids samples for proteins, cations, metals, and residual biological activity (RBA). The samples were fractionated as indicated below prior to measuring the targeted parameters.

<u>Soluble faction extraction</u>:

1. Centrifuge the 100-mL samples at 3,000g for 15 minutes to separate the liquid and solids in the sample.
2. Remove the centrate and filter through a 1.5-μm glass microfiber filter. This filtrate is considered the "soluble" fraction.

<u>Bound or labile fraction extraction</u>:

1. Centrifuge the 50-mL samples at 3,000g for 15 minutes to separate the liquid and solids in the sample.
2. Re-suspend the solids pellets in 100 mL of pH 8 phosphate buffer saline (PBS).
3. Mix at 1,500 revolutions per minute (rpm) for 10 minutes in an unbaffled chamber.
4. Centrifuge at 3,000g for 15 minutes.

5. Remove the centrate and filter through a 1.5-µm glass microfiber filter; the filtrate contains the "bound" or "labile" fraction.

The following list the methods of analysis for various parameters:

Protein analysis was conducted in the same manner for both soluble and bound fractions. Protein was measured using the Hartree (1972) modification of the Lowry et al. (1951) method. Bovine serum albumin was used as the protein standard.

Iron (Fe) and aluminum (Al) were measured in the sludge solids (there is little of either in the liquid phase) by digestion using U.S. EPA Method 3050B. Sludge solids were dried at 105°C and then removed for acid digestion. The centrate from the digestion process was analyzed for Fe and Al using atomic adsorption spectroscopy.

Calcium (Ca), magnesium (Mg), sodium (Na), and potassium (K) in the same samples were measured using a Dionex Ion Chromatograph with a CS12 column and conductivity detection, with self-generating suppression of the eluent. The eluent is 20 mM methanesulfonic acid, introduced at a flow rate of 1mL/min. Soluble Ca, Mg, Na, and K in liquid samples was also measured using the ion chromatography method.

The **total metals content** of samples were determined using nitric acid-heat digestion to dissolve the particulate bound metals followed by analysis using inductively coupled plasma (ICP). Method 3030 E in Standard Methods (1999) was used for the nitric acid digestion procedure. The digest was analyzed by ICP according to the Method 3500.

The **residual biological activity (RBA)** procedure was a modification of the method developed as part of Phase 2. In Phase 3, RBA was determined by measurement of additional volatile solids (VS) destruction, whereby anaerobicly digested biosolids (liquid or cake) samples were batch-digested in a bench-scale unit for *an additional 40 days* at 35°C. In Phase 3, VS destruction was measured by the approximate mass balance method (Appendix C of U.S. EPA, 2003). The final weight of the stored biosolids was determined for wet, dry and inert content after 1 day and after 40 days of digestion (note the time began upon arrival at the laboratory, not upon completion of the sampling, thus "Day 1" refers to the day the samples arrived at the laboratory). Additional VS destruction (RBA) was measured following another 40 days of storage in the sample bottles following the initial 40 days of digestion.

2.2.2.4 Odor Panel Testing

The laboratory at the LACSD Joint Water Pollution Control Plant retains its own certified odor panel, which was used for analysis of the total odor of biosolids samples in Phase 3 in the measurement of odor detection threshold (DT). Results were expressed in standard odor units of dilutions-to-threshold (D/T). The LACSD odor panel consists of nine individuals that are experienced in odor panel testing. The panel can measure a limit of approximately six samples per day (following that number of samples, fatigue of the olfactory sense tends to occur).

In odor panel tests, one bottle containing biosolids cake was used per sample on each analysis day, and the same bottle was not reused on a subsequent day. In each case, a backup bottle was provided for each subsequent days of odor testing. As indicated in Section 2.2.2.3, the day of sample collection was denoted as "day zero" in this and all testing in the Phase 3 work.

2.3 Phase 3A Laboratory Tests

Table 2-1 lists the tests and the analytical work conducted during Phase 3A, with the corresponding laboratories that conducted the tests. CEBA testing was completed by both Virginia Tech and Bucknell, with Virginia Tech testing pre-digestion CEBAs and Bucknell testing post-digestion (pre-dewatering) CEBAs and post-dewatering CEBAs.

Table 2-1. Phase 3A Summary of All Laboratory Analyses and Testing.

	Analytical Work		
Phase 3A Tests	**Virginia Tech**	**Bucknell**	**LACSD**
Cation Tests			
Soluble Proteins and Polysaccharides	X		
Soluble cations (Na, K, Ca, Mg, ammonium)	X		
Cations (Al and Fe)	X		
40-day Headspace Testing	X		
Volatile Fatty Acids (VFAs)	X		
VS Destruction and Residual Biological Activity	X		
Total Solids (TS) and Volatile Solids (VS)	X		
Biosolids Shearing Tests	X		
Digestion SRT Testing			
Soluble cations (Na, K, Ca, Mg, ammonium)	X		
Cations (Al and Fe)	X		
Gas Volume			X
Gas Composition			X
VFAs	X		X
TS and VS			X
Odor Panel Testing			X
Soluble Proteins and Polysaccharides	X		
40-Day Headspace Testing	X		
VS Destruction and Residual Biological Activity	X		
Biosolids Shearing Tests	X		
CEBA Addition Tests			
Digestion Parameters (VFAs, ammonia, pH)	X		
40-Day Headspace Testing	X	X	

Table 2-1. Phase 3A Summary of All Laboratory Analyses and Testing.

Phase 3A Tests	Analytical Work		
	Virginia Tech	Bucknell	LACSD
TS and VS	X	X	
Alum Addition Tests			
40-day Headspace Testing		X	
TS and VS		X	

The details the procedures used for each of these laboratory tests are presented in the following sections.

2.3.1 Cation Testing Procedures and Protocols

Trivalent cations, specifically of iron and aluminum, were analyzed by the Virginia Tech laboratory with respect to their relationship to the bioavailability of proteins for odor generation following centrifugation. Steps taken in these tests include batch anaerobic digestion, sulfur gas analysis, odor profiling and characterization, and cation analysis. The following sections provide detail on each of these steps.

2.3.1.1 Batch Anaerobic Digestion

Batch anaerobic digestion was carried out in the Virginia Tech laboratory using sludges from 12 WWTPs. Ten liters of sludge was batch digested in 20-liter glass containers that were sealed with rubber stoppers. Tedlar bags were connected to each test digester to collect gases generated during anaerobic digestion. All sludges were batch digested at 37°C for 30 days. The contents of the digester were manually mixed once a day for the duration of the 30-day digestion period. The sludges were digested using the same mixture and solids concentration of primary and WAS that was being used at the treatment plant from where the sludges originated.

2.3.1.2 Bench-Scale Sludge Dewatering

Dewatered cakes of anaerobicly digested sludges that were used for odor analysis were prepared using a method that mimics the shearing processes and dewaters cake to a similar solids concentration that occurs in a full-scale high-solids centrifuge. Polymer is also added to the bench-scale centrifuge; the optimum polymer dose for all sludge samples was determined by adding different concentrations of cationic polymer and determining the capillary suction time (CST) (Muller et al., 2004). The optimum polymer dose was assumed to be the one that produced the minimum CST.

Both a Triton Type 304-M and Triton Type 165 CST apparatus were utilized with Whatman 17-CHR chromatography paper. Clarifloc 3275, a liquid cationic high molecular weight polymer, was used to condition all sludges to eliminate the effects of polymer variability on odors. A Warring blender was used to shear the sludge. It is estimated that this blender has a G value of 10,000/s. Liquid sludge was sheared for 30 seconds in 100 mL increments. The optimum dose of cationic polymer for each sample was determined by conducting a polymer dose test after 30 seconds of shearing.

Sludge dewatering using the optimum polymer dose was accomplished using a laboratory centrifuge. The centrifuge was operated at 17,700g for 15 minutes at 25°C. Sludge cakes obtained from the centrifuge could then be further dewatered to increase the solids content using a hydraulic piston press constructed at Virginia Tech, with Whatman 41 paper serving as the filter media.

Multiple sample bottles containing dewatered cake samples with low- and high-solids content were prepared for GC/MS analysis. "Low-solids" cakes refer to biosolids samples that were dewatered using only the laboratory centrifuge and having a solids content similar to dewatered cake achieved at full-scale with a belt filter press. "High-solids" cakes refer to biosolids samples that were further dewatered using the hydraulic piston to achieve solids concentrations similar to those derived from full-scale high-solids centrifuges.

2.3.1.3 Odor Profiling and Characterization

The characterization of odors was accomplished using the headspace analysis method described in Section 2.2 General Protocols.

2.3.1.4 Laboratory Analysis

The total concentration of iron and aluminum cations was measured in sample digester feed sludges on a Perkin Elmer 5100PC Atomic Adsorption Spectrophotometer following acid digestion using U.S. EPA Method 3050B (U.S. EPA, 1996). The resultant concentrations are expressed in units of mg-metal per gram total solids.

Solids and pH tests were conducted according to Standard Methods (APHA, 1999). In order to prepare samples for cation and volatile fatty acid (VFA) analyses, a 500-mL sample of digester biosolids was centrifuged at 17,700g for 15 minutes. Supernatant liquid was filtered through a 0.45-micron syringe filter. From this filtered sample, dilutions were made for cation and VFA analysis. VFA samples were acidified in their individual GC sample bottle by adding concentrated acid at a ratio of 1:10. The VFA samples were analyzed on a GC (Shimadzu Model GC-14A) using a flame ionization detector.

Liquid phase cations were measured on a Dionex D-120 ion chromatograph utilizing a CS-12 column and conductivity detector with self generating suppression of the eluent (Dionex Corp., Sunnyvale, CA). Methanesulfonic acid (20mM) at a flow rate of 1mL/min was used for eluent.

The residual biological activity tests were a modification of the additional digestion test method outlined by U.S. EPA (1993). The tests were conducted by incubating 100mL digested sludge samples in sample bottles at 36.5°C for 40 days. During the incubation period, the bottles were periodically degassed and weighed. The weight lost from each bottle was assumed to be from gasification of the VS in each sample. Additional VS destruction was calculated by dividing the weight lost from each sample at the end of incubation by the weight of its VS content analyzed prior to incubation and as reported as percent VS lost.

Soluble proteins were determined by the modified Lowry method using bovine serum albumin as the protein standard. Soluble polysaccharides were measured by the Dubois method utilizing glucose as the standard as described by Frølund et al. (1996). A laboratory centrifuge was used to separate the solution from majority of solids in sludge. The supernatant after centrifuge was filtered through a 1.5 μm membrane filter and then analyzed separately for proteins and polysaccharides. The values obtained were termed as solution protein and solution polysaccharide.

2.3.2 Digestion SRT Testing Procedures and Protocols

This Phase 3A testing looked into the relationship between digester SRT and odor generation. The Virginia Tech laboratory tested whether additional solids destruction (and thus protein destruction), is possible at higher SRTs, and if this solids destruction leads to reduced odors in the dewatered biosolids cake.

2.3.2.1 Experimental Approach

The digestion SRT study was conducted at LACSD and Virginia Tech using laboratory scale anaerobic digesters that were operated over a range of SRTs (10, 15, 20, 30, and 40 days). Ten-liter volume laboratory digesters were operated by LACSD and Virginia Tech (identical to those used in the cation study). The reactors were fed a 75% primary sludge and 25% thickened WAS blend, which is the same mixture used at full-scale at the LACSD Joint Water Pollution Control Plant, where all sludge for this study was obtained. The reactors were fed and mixed (using a vibration table) once per day.

During the digestion period, feed and digested sludge samples were periodically sent to the Virginia Tech laboratory for TVOSC analysis and extraction of exocellular polymeric substances (EPS). Multiple sample bottles were set up and TVOSC was determined using the centrifuge simulation technique and analysis method described in Section 2.3.1. The extraction methods described below in Section 2.3.2.3 were used to determine specific metal associated floc fractions in the sludge. These fractions were compared to both TVOSC and volatile solids changes with increasing SRT.

2.3.2.2 Bench-Scale Sludge Dewatering

In the digestion SRT study, biosolids cake samples were produced by dewatering polymer conditioned sludges using a laboratory centrifuge identical to what was used in the cation tests (described in Section 2.3.1.2). Samples with and without the addition of BESA were prepared for both low-solids and high solids cakes. Five milliliters of BESA solution (0.127 mmole) was added to about 400 mL of liquid sludge before dewatering it using the lab centrifuge. The use of BESA for methane inhibition is described in Section 2.2.1.3.

2.3.2.3 Selective EPS Extraction

A recent study by Muller et al. (2004) showed that when biosolids in the presence of conditioning polymer are exposed to high shearing forces, proteins are released from sheared sludge particles and rendered bioavailable. Muller et al., (2004) further suggested that degradation of this released material is a cause of odor generation from dewatered biosolids. If this reasoning is correct, a better understanding of floc structure could be usefully linked to the study of odor generation from digested sludges.

It is known that EPS compose a major organic fraction of biosolids floc (Frølund et al., 1996). It has been postulated that there are different pools of EPS that are bound with different metal ions in floc. Degradation of those materials is associated with different digestion conditions (Novak et al., 2003). Important biopolymeric fractions are as follows:

- Divalent cation-associated EPS (calcium and magnesium-bound)
- Iron-bound EPS
- Aluminum-bound EPS

It was observed in previous studies conducted at Virginia Tech that anaerobic and aerobic digestion of waste activated sludge from several different WWTPs resulted in highly variable cation and biopolymer releases and different degrees of VS destruction (Park et al.,

2004). A large amount of protein was recovered from the sludge solution following anaerobic digestion, but the release of divalent cations was not observed in these tests. Furthermore, there was a correlation between the VS reduction by anaerobic digestion and the Fe concentration in feed sludges. These results further suggested that proteins associated with calcium and magnesium are mainly degraded under aerobic conditions, while EPS associated with iron are readily degraded under anaerobic conditions (Park et al., 2004). It was also shown that aluminum plays a significant role in bioflocculation by improving effluent quality with higher aluminum content in floc, but the impact of aluminum and its binding organic matter on sludge digestion was unclear (Park et al., 2004).

Park et al. (2004) showed that three components of floc EPS can be selectively extracted by different chemical extraction methods. These are cation exchange resin (CER) extraction to extract calcium- and magnesium-bound EPS, sulfide extraction of iron-bound EPS, and aluminum-bound EPS extraction, which is achieved using base at pH 10.5. In this study, the CER-extractable fraction (assumed to be associated with calcium and magnesium) and the sulfide-extractable fraction (assumed to be associated with iron) of EPS were extracted over the range of sludge ages. The results were compared to VS and TVOSC changes with digester SRT. Additional details of the Park extraction procedure can be found in Park and Novak (2007).

2.3.2.4 Laboratory Analysis

In the digestion SRT study, the characterization of TVOSC was accomplished using the method described in Section 2.2.1. All other laboratory analysis was conducted as described in Section 2.3.1.4.

2.3.2.5 CEBA Testing Procedures and Protocols

All known CEBA vendors were contacted by e-mail to inform them that we were seeking to bench-scale test CEBA products, strictly from the standpoint of biosolids cake odor reduction, as a part of this research project. Vendors willing to have their products bench tested agreed to furnish a sample of their product along with a recommended single dosage rate for the testing.

Before the CEBA testing started, a detailed, standardized General CEBA Test Protocol (GCTP) was developed by the Bucknell and Virginia Tech laboratories in cooperation with the CEBA manufacturers. The GCTP (attached as Appendix A2) contains all common (non-specific) procedures required for testing three types of CEBA products: pre-digestion, pre-dewatering, and post-dewatering.

A Specific CEBA Test Protocol (SCTP) was developed for each of the eight CEBAs tested. These SCTPs are shown as Appendix A3. The GCTP was used as a common basis for developing each of the SCTPs, after information was obtained from each CEBA vendor on recommended application methods, dosage rates, Material Safety Data Sheets (MSDSs), and any other details suggested by the vendor for testing their CEBA product in the laboratory. Prior to testing, the vendor of each CEBA product tested was required to certify its acceptance of the SCTP for its product by signing the bottom of the SCTP.

CEBA addition was performed according to the dosages and manner mandated in the SCTP developed from the individual CEBA vendors. Individual dosage variations for each CEBA rate and frequency are included in the results section (Chapter 3.0) of this report. The following sections summarize the laboratory test protocols conducted using these protocols, for each type of CEBA product.

2.3.2.6 Pre-Digestion CEBAs

Four pre-digestion CEBAs (CEBAs 1, 4, 5, and 6) were tested at the Virginia Tech laboratory. The laboratory characterized the impact of CEBAs that were added to the solids stream prior to or during digestion at Feed Points 1 and 2, respectively, as shown in the schematic diagram in Figure 2-9.

The Pepper's Ferry WWTP in Pepper's Ferry, VA provided primary and WAS samples for use in this CEBA testing. The WERF team stated that the digester feed sludges at the Pepper's Ferry WWTP have characteristics typical of a municipal WWTP that employs anaerobic digestion of biosolids. The primary sludge and WAS were mixed in the laboratory and fed into the laboratory digester at a typical 50:50 ratio (by dry mass of solids) to simulate a typical digester feed sludge. Digested biosolids (approximately 20% by volume) were also provided as initial seed for the laboratory digester. Inclusion of seed sludge insured that an anaerobic microbial population was available to digest the sludge solids.

After the first seven days of operation, half of the contents of each test digester were removed and replaced with fresh primary sludge or WAS, utilizing the same 50:50 ratio of sludge used at the outset of the test. The same procedure was repeated after another seven days. The CEBA product was added to the feed sludge at its prescribed dosage rate each time the digestion reactor contents were changed. This staged batch digestion procedure allowed for acclimation of the digestion system to the CEBA product. After these two cycles, the digester was left alone to digest the sludge for an additional 21 days. The total digestion time, including the two seven-day "acclimation periods", was 35 days.

The laboratory digesters were maintained at a constant temperature of 35°C (95 °F) with an average SRT of 26.25 days. The TS, VS, pH, and ammonia nitrogen concentration in the inlet feeds and the effluent liquid sludge from the test digesters were measured. After 35 days, the sludge from each CEBA-dosed digester and the non-dosed control digesters was sheared, dewatered, tested using the sample headspace analysis method described in Section 2.2 General Protocols.

2.3.2.7 Post-Digestion CEBAs

Two CEBA products (number 7 and 8) that were to be added following anaerobic digestion but prior to dewatering were evaluated at Bucknell University. The addition point in the solids train is Feed Point 3 as depicted in Figure 2-10.

A protocol for conditioning, dewatering, and processing solids was developed that when enacted in the laboratory generated a cake with a similar odor production profile as a high-solids centrifuge. This process is termed the high-solids centrifuge simulation process (HSCSP). The HSCSP was used for testing all post-digestion CEBA products. The step-by-step procedure for conditioning and dewatering is summarized below. Control samples with no CEBA products added were prepared using the same process.

Conditioning: The CEBA products were added according to the suppliers' direction for dosage rate and holding time of the liquid digested solids. After the CEBA addition, the samples were conditioned and dewatered. The polymer used was provided by the plant for which the study was being performed, identical to the polymer used at full scale. One liter of solids was placed in a baffled reactor, and polymer was added to the solution. The sludge was mixed at a predefined mixing intensity and time to achieve a given energy input or shear, which is was measured by its *Gt* value. The typical *Gt* value used for these experiments was 100,000. After mixing, the capillary suction time (CST) was measured. This process was

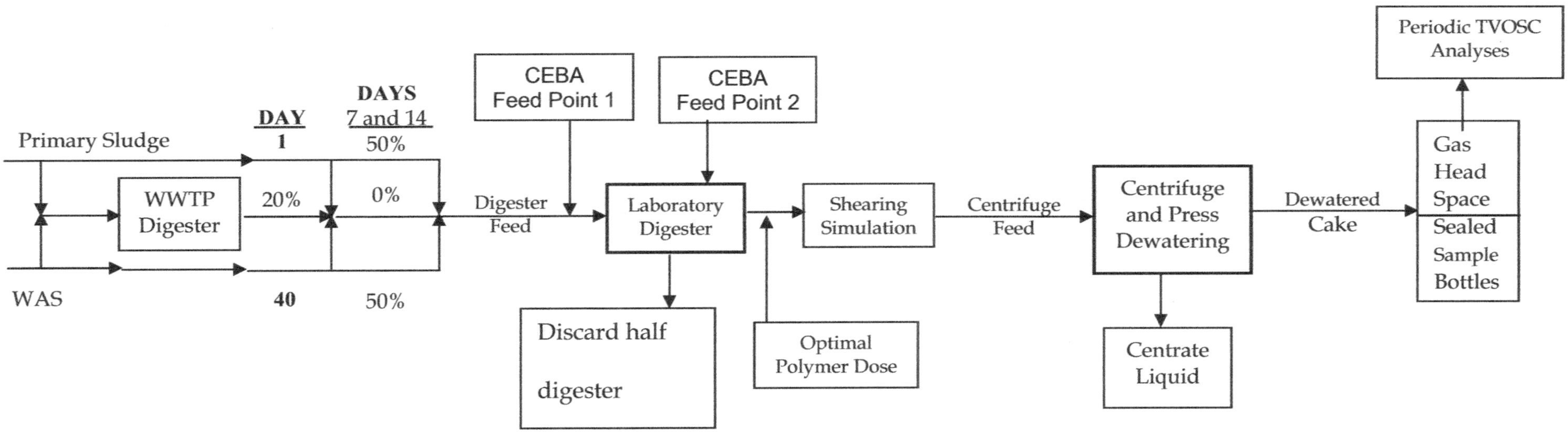

Figure 2-9. CEBA Feed Point Locations Prior to or During Digestion.

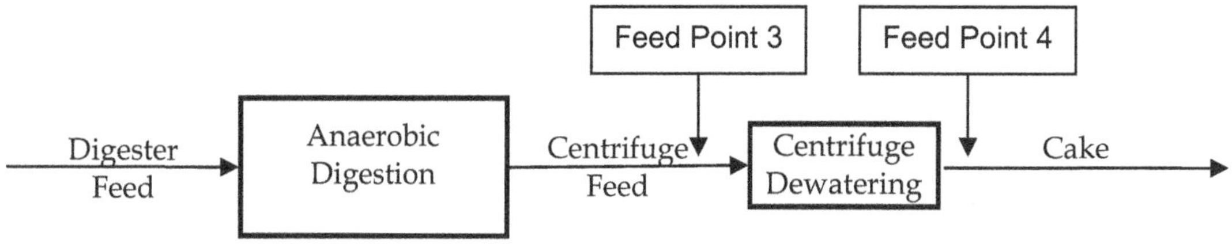

Figure 2-10. CEBA Feed Point Locations Prior to and after Dewatering.

repeated with varying polymer dosages until the optimum polymer dose (OPD) was obtained, as measured by the polymer dose that produced the minimum CST.

Dewatering: The sludge conditioned with the OPD was then placed in 250-mL sample bottles, and dewatered using a laboratory centrifuge operating at 3,000g for 10 minutes. After centrifuging, the supernatant was removed (the centrate was analyzed for different constituents as well), and the cake was combined from the different tubes. Typically, about 200-300 g of wet cake in total was generated, containing a solids content of around 25%. Higher cake solids could also be obtained with an additional dewatering unit such as a belt filter press simulator (as was done at the Virginia Tech laboratory).

Processing: The wet cake samples were processed to simulate the high shear experienced in a full-scale centrifuge. To achieve this, the cake was processed through a meat-grinder apparatus that pushes the cake forward using a scroll-conveyor, followed by extrusion through a small opening at the end of the conveyor.

Headspace Analysis: After processing, the cake samples were placed in sampling bottles for headspace analysis using the protocols detailed in Section 2.2 General Protocols.

2.3.2.8 Post-Dewatering CEBAs

CEBAs number 2 and 3 were added directly to the cake samples (after dewatering and shearing) using the dosage rates recommended by their respective vendors and indicated in their respective SCTPs. Cake samples were generated using the method described in Section 2.3.3.2. Once the cake samples were generated, each CEBA was added to the cake using a spray misting bottle at the correct dosage, and then mixed into the cake per the direction of the vendor. The dose point is indicated in Figure 2-10 as Feed Point 4. Control samples with no CEBA products added were prepared using the same process.

2.3.2.9 Alum Addition Testing Procedures and Protocols

In the Phase 3A research, Bucknell University conducted several laboratory-scale trials with alum addition to the dewatering process sludge feed. These trials used the high solids centrifuge simulation process (HSCSP) described in the General CEBA Test Protocol to produce cake that had similar solids content as cake samples from full-scale high solids centrifuges. In the laboratory testing, alum is added to digested sludge prior to polymer addition, and the samples are mixed and then dewatered using a laboratory centrifuge.

Bucknell University completed alum testing in four trials using sludge from three different WWTPs: two trials were performed using the Philadelphia Water Department (PWD) Biosolids Recycling Center (BRC) sludge, one trial was performed using the San Francisco Southeast Wastewater Treatment Plant sludge, and one trial was performed using the LACSD Joint Water Pollution Control Plant sludge.

A summary of the testing protocol for alum addition is as follows:

1. The optimum polymer dose (OPD) for the laboratory centrifuge and the test sludge was determined using a defined mixing regime and capillary suction time (CST) as the measure of dewaterability. The *Gt* used for mixing was 100,000. A more detailed explanation of the dewatering experimental setup can be found in Higgins et al. (2004). Once the OPD was determined for the sample, this polymer dose was used for all doses of alum tested, with the one exception being for the Philadelphia Water Department BRC tests, in which a second trial was conducted where the OPD was determined for each alum dose.
2. During each test, alum was added ranging in dose from 0 to 4% as Al (in terms of aluminum mass to dry solids mass), followed by the polymer dose.
3. The sample was mixed for 100 seconds at a velocity gradient (*G*) equal to 1,000/s.
4. After conditioning, the samples were dewatered using a laboratory centrifuge, and the cakes were then sheared.
5. The cake samples were placed in sample bottles, and the headspace analysis procedure was conducted as indicated in Section 2.2 General Protocols.

Both liquid digested sludge (pre-dewatering) and dewatered cake samples were shipped from the plants to Bucknell for testing. Bucknell compared the TVOSC profile produced for various alum doses using the laboratory dewatering protocol versus the TVOSC profile produced by the dewatered cake sample shipped from the plant (the "field cake" biosolids dewatered by the full-scale centrifuges), which had no alum addition.

2.4 Phase 3C Full-Scale Experiments and Test Procedures

Table 2-2 lists the tests and the analytical work conducted during Phase 3C, with the corresponding laboratories that conducted the tests.

Table 2-2. Phase 3C Summary of All Full-Scale Analyses and Testing.

	Analytical Work		
Phase 3C Tests	Virginia Tech	Bucknell	LACSD
Advanced Digestion Studies			
Soluble proteins and polysaccharides	X		
Soluble cations (Na, K, Ca, Mg, ammonium)	X		
Cations (Al and Fe)	X		
40-day Headspace Testing	X		
Volatile Fatty Acids (VFAs)	X		
VS Destruction and Residual Biological Activity	X		
Dewatered Cake TS and VS	X		
Biosolids Shearing Tests	X		

Table 2-2. Phase 3C Summary of All Full-Scale Analyses and Testing.

Phase 3C Tests	Analytical Work		
	Virginia Tech	Bucknell	LACSD
Dewatering Equipment Tests			
Soluble proteins and polysaccharides		X	
Soluble cations (Na, K, Ca, Mg, ammonium)		X	
Cations (Al and Fe)		X	
40-day Headspace Testing		X	
Dewatered Cake TS and VS		X	
Combined Microsludge and Dewatering Pilot Tests			
40-day Headspace Testing	X	X	
Dewatered Cake TS and VS	X	X	
Odor Panel Testing			X
Centrate and Filtrate TS and VS	X	X	
Biosolids Shearing Tests	X	X	
Chemical Addition Tests (Lime, Ferric Chloride)			
Soluble proteins and polysaccharides		X	
Cations (Al and Fe)		X	
40-day Headspace Testing		X	
Dewatered Cake pH		X	
Dewatered Cake TS		X	
Alum Addition Tests			
Soluble proteins and polysaccharides		X	
Cations (Al and Fe)		X	
40-day Headspace Testing		X	
Centrate Solids %		X	
Dewatered Cake TS		X	

2.4.1 Advanced Digestion Studies

Several wastewater treatment plants that had operational advanced digestion processes, either pilot or full-scale, were studied to determine how advanced digestion processes influenced odor potential from centrifugally dewatered sludges. Most of these advanced digestion processes were multi-stage processes, some with a thermophilic stage. Procedures followed at the three facilities used in Phase 3 are described below.

2.4.1.1 Inland Empire Utilities Agency Phased Digestion

The flow scheme for the Inland Empire Utilities Agency (IEUA) WWTP is depicted in Appendix A4. At IEUA, all sludge is initially digested for 2 days in a mesophilic acid phase digester. The solids train then splits into two digestion trains: a second stage mesophilic digester and a second stage thermophilic digester. Both of these digesters have a detention time of approximately 16 to 17 days, making the total two-stage digestion period 19 to 20 days. In addition, for the field testing, a portion of the sludge from the meso-thermo train was further digested for an additional 7 days in a mesophilic digester for a total of 26 days. Therefore, the three digestion trains considered in these experiments provide the following treatment processes for the IEUA solids:

1. Acid-gas mesophilic-thermophilic (AGMT)
2. Acid-gas mesophilic-mesophilic (AGMM)
3. Acid-gas mesophilic-thermophilic-mesophilic (AGMTM)

Prior to the start of the field sampling, WERF project team developed an "IEUA Regional Plant (RP)-1 Advanced Digestion Facilities Biosolids Cake Odors Evaluation Sampling Plan." This plan is included in Appendix A4. Samples were collected from all three digestion trains as well as from the thickened primary sludge/WAS feed line to the acid phase digester, and shipped to Virginia Tech overnight for analysis the next day. The sludge samples were dewatered and analyzed for TVOSC using the protocols detailed in Section 2.2 General Protocols.

2.4.1.2 Baltimore Back River WWTP Phased Digestion

The test protocol for the Baltimore Back River WWTP is described on pp. A1-14 to A1-16 of Appendix A1. During the field testing, the solids digestion train consisted of the following digestion systems:

1. One single-stage mesophilic pancake digester (providing 15 days or greater detention time)
2. One single-stage mesophilic egg-shaped digester (with 14 days detention time)
3. One two-stage acid/gas pilot mesophilic digester (acid stage detention time was 2 days, gas stage detention time was 14 days)

Samples were collected, shipped to the Virginia Tech laboratory overnight, and analyzed the next day. The sludge samples were dewatered and analyzed for TVOSC using the protocols detailed in Section 2.2 General Protocol.

2.4.1.3 Western Lake Superior Sanitary District Temperature Phased Digestion

The Western Lake Superior Sanitary District (WLSSD) plant in Duluth, MN does not have primary treatment of wastewater, so it only has WAS available for digestion, not primary sludge. The test protocol for WLSSD is described on pp. A1-14 to A1-16 of Appendix A1. WLSSD operates a two-stage thermophilic/mesophilic system. Each stage has 6 days of detention time. The interest the WERF team had in the WLSSD system for the Phase 3 research was that the second stage (the mesophilic stage) was not cooled and as a result, had temperatures ranging from 42-48°C. The WERF team theorized prior to the field study that because the second digestion stage of the WLSSD plant was not actually operating in the mesophilic range (which typically has temperatures in the range of 33-39°C), poor digestion and high odors would result.

Therefore, the WLSSD plant was studied to determine if digestion at the "normal" temperature instead of at the elevated temperature found in the field units would reduce TVOSC in the dewatered cake. A sample of sludge digested by both stages was shipped to Virginia Tech along with sludge that was digested by the first stage only. The sludge from the first stage was then batch digested in the Virginia Tech laboratory (using the protocol contained in Section 2.2 General Protocols) for an additional 30 days, and both sludges were tested for TVOSC production.

2.4.2 Dewatering Equipment Tests

Several full-scale tests were performed to examine the impact of dewatering processes and operation on odor production from dewatered bisolids cake. The general approach to this testing was to collect samples including digester effluent/centrifuge feed, dewatered biosolids cake, and centrate for the given conditions being studied. For example, at the King County WWTP, samples from side-by-side belt filter press and high-solids centrifuge samples were collected to compare odor production from these two dewatering devices. Specific details on the testing conditions for each full-scale study are provided in the following sections.

2.4.2.1 Philadelphia Water Department Biosolids Recycling Center

The Philadelphia Water Department (PWD) Biosolids Recycling Center (BRC) was tested in fall 2004 to compare odor emissions from various solids processing units. Goals of the PWD sampling were to accomplish the following:

1. Compare the effect of different centrifuge types on cake odor production: dewatered biosolids cake samples were collected from six different centrifuges in operation at the BRC on the day of the sampling event. Each centrifuge was dewatering the same digester sludge effluent from the PWD Northeast WWTP.
2. Examine the effect of centrifuge torque settings on cake odor production: this effect was investigated using the same Humbolt high-solids centrifuge, on which the torque was adjusted to three different settings. After each change in torque, the centrifuge was operated until steady-state conditions were achieved, and then samples were collected.
3. Examine the effect of centrifuge bowl speed on cake odor production: similarly, the bowl speed was adjusted on the same centrifuge to examine the effect of three speeds on cake characteristics and odor production.

Samples were shipped to Bucknell University where the standard headspace analysis was performed, as described in Section 2.2 General Protocols.

2.4.2.2 LACSD Joint Water Pollution Control Plant

The LACSD Joint Water Pollution Control Plant (JWPCP) was tested in spring 2005 to compare odor emissions from various solids processing units in a full-scale study similar to the PWD sampling. Goals of the LACSD sampling were to accomplish the following:

1. Compare the effect of different centrifuge types on cake odor production
2. Examine the effect of centrifuge torque on cake odor production
3. Examine the effect of two different polymer types on cake odor production
4. Examine the effect of centrifuge bowl speed on cake odor production
5. Compare cake quality and odor production from a side-by-side comparison of centrifuge and a fournier rotary press

After changing the settings on a single centrifuge unit, such as the torque setting, the WERF field team allowed a sufficient amount of time to pass to assure that the process was in a steady-state condition.

Samples were shipped to Bucknell University where the standard headspace analysis was performed, as described in Section 2.2 General Protocols.

2.4.2.3 King County West Lake WWTP

The goal of the King County sampling was to compare the dewatered biosolids cake quality in terms of odor production from a side-by-side centrifuge and belt filter press that were dewatering the same anaerobicly digested sludge. Similar to the other dewatering full-scale tests, samples were collected from each dewatering device when they were operating under steady-state conditions. Samples were shipped to Bucknell University where the standard headspace analysis was performed, as described in Section 2.2 General Protocols.

2.4.3 Combined MicroSludge™ and Dewatering Equipment Pilot Testing

In March 2006, the WERF team conducted pilot testing at the LACSD JWPCP to study the impact of the Microsludge™ technology digestion enhancement on biosolids odors, in addition to performing additional dewatering comparison tests. The Microsludge™ technology includes two primary processes: 1) Caustic chemical pre-treatment to weaken cell membranes and significantly lower the viscosity of WAS and 2) An industrial-scale high-pressure homogenizer or "cell disrupter" to provide a very large and sudden pressure drop to lyse the bacterial cells in the WAS. In this process, the WAS in the cell disruption valve is accelerated to a velocity of 300 meters per second in about 2 microseconds. This high velocity increase causes fluid shear and cavitation. The WAS then impinges on an impact ring, disrupting the cell membranes and liquefying the components of the WAS.

Following cell disruption, within the Microsludge™ unit the processed WAS is liquefied and then added to primary sludge before anaerobic digestion (note the Microsludge™ technology is not used on primary sludge).

The purpose of these pilot tests was to assess the odor potential of biosolids generated with and without the Microsludge™ technology digestion enhancement, and following dewatering using two types of equipment. The following test elements were included:

- Enhanced digestion testing using MicroSludge™ technology for cell lysis and VS destruction enhancement of WAS. The equipment was provided by Paradigm Environmental Technologies, Inc. Impacts on biosolids odors were studied.
- Dewatering equipment (centrifuge and belt filter press) operation. The equipment was provided by the Andritz Group. Impacts on biosolids odors were studied.

Prior to the start of this field sampling, the WERF team developed a specific test protocol entitled "Field Testing Activities at JWPCP of LACSD: Dewatering Equipment and Enhanced Digestion Tests." The protocol is attached as Appendix A5. Tests were conducted as outlined in the protocol and are summarized in the sections below. Liquid and solid biosolids samples analysis was conducted at the Bucknell, Virginia Tech, and LACSD laboratories as indicated in Table 2-2. In addition, field data recording sheets were kept to monitor field activities and ambient conditions.

2.4.3.1 Centrifuge Adjustment Tests

Throughout the dewatering test program, a series of dewatering equipment control and operations parameters were varied in a stepwise manner to determine the impacts of each on odor generation potential of the final biosolids cake product. Table 2-3 lists the parameters that were varied to produce the centrifuge adjustments testing conducted at the JWPCP. The testing was completed using an Andritz pilot scale D4L model centrifuge. All the adjustments to the centrifuge were performed by Andritz field personnel.

Table 2-3. Sampling Matrix for LACSD JWPCP Microsludge and Centrifuge Adjustments Tests.

	System Parameters for Sample			
Sample Date	Digester	Polymer Type	Dewatering Unit	Centrifuge Speed
3/8/06	Control	None	None	N/A
3/8/06	Microsludge	Mannich	None	N/A
3/9/06	Control	Mannich	Field Centrifuge	Normal
3/9/06	Control	Mannich	Full-Scale Centrifuge	Match full-scale
3/10/06	Microsludge	Mannich	Field Centrifuge	Normal
3/10/06	Microsludge	Emulsion	Full-Scale Centrifuge	1,500g
3/13/06	Control	Emulsion	Field Centrifuge	2,500g
3/13/06	Control	Mannich	Field Centrifuge	1,500g
3/14/06	Control	Mannich	Field Centrifuge	2,500g
3/14/06	Control	Emulsion	Field Centrifuge	1,500g
3/15/06	Microsludge	Emulsion	Field Centrifuge	2,500g
3/15/06	Microsludge	Mannich	Field Centrifuge	1,500g
3/16/06	Microsludge	Mannich	Field Centrifuge	2,500g
3/16/06	Microsludge	Emulsion	Field Centrifuge	2,500g
3/20/06	Control	Emulsion	Field Centrifuge (High Load)	2,500g
3/20/06	Control	Emulsion	Field Centrifuge (Low Load)	2,500g
3/21/06	Microsludge	Emulsion	Field Centrifuge (High Load)	2,500g
3/21/06	Microsludge	Emulsion	Field Centrifuge (Low Load)	2,500g

2.4.3.2 Centrifuge Scroll Test

At the LACSD JWPCP, tests on the centrifuge scroll were conducted in order to attempt to establish the odor generation potential at various locations within the centrifuge scroll as biosolids are conveyed through a centrifuge. Figure 2-11 illustrates a typical centrifuge scroll and the sampling points that were used in this test event.

Table 2-4 lists the system parameters and the conditions that were varied to conduct the centrifuge scroll testing at the LACSD JWPCP. The testing was completed using an Andritz pilot scale D4L model centrifuge. All the adjustments to the centrifuge were performed by Andritz field personnel.

2.4.3.3 Belt Filter Press Comparison

Table 2-5 lists the parameters and the test conditions for the belt filter press dewatering tests conducted at the JWPCP of LACSD using Andritz pilot scale SMX S8 (1.0-

m) model belt filter press. All the adjustments to the field belt filter press unit were performed by Andritz field personnel.

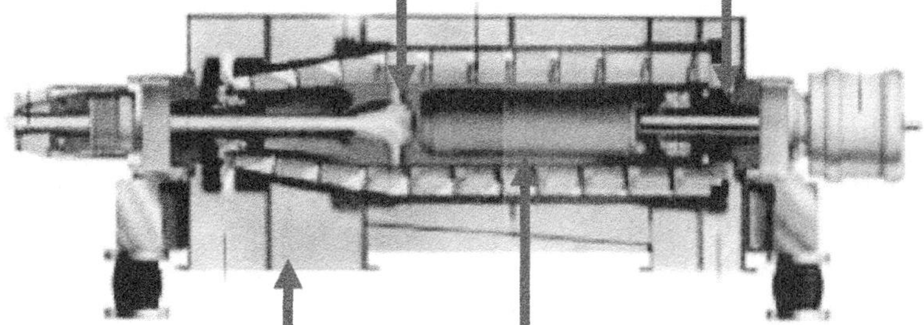

Figure 2-11. Centrifuge Scroll and the Sample Locations.

Table 2-4. Sampling Matrix for JWPCP Centrifuge Scroll Tests.

	System Parameters for Sample			
Sample Date	Digester	Polymer Type	Dewatering Unit	Centrifuge Speed
3/22/06	Control	Emulsion	Field Centrifuge Scroll 1	2,500g
3/22/06	Control	Emulsion	Field Centrifuge Scroll 2	2,500g
3/22/06	Control	Emulsion	Field Centrifuge Scroll 3	2,500g
3/22/06	Control	Emulsion	Field Centrifuge Scroll 4	2,500g

Table 2-5. Sampling Matrix for Belt Filter Press Comparison Tests.

	System Parameters for Sample			
Sample Date	Digester	Polymer Type	Dewatering Unit	Centrifuge Speed
3/23/06	Control	Emulsion	Belt Filter Press	N/A
3/23/06	Control	Emulsion	Belt Filter Press	N/A
3/23/06	Control	Emulsion	Belt Filter Press + Lab Press	N/A
3/27/06	Microsludge	Emulsion	Belt Filter Press	N/A
3/27/06	Microsludge	Emulsion	Belt Filter Press	N/A
3/27/06	Microsludge	Emulsion	Belt Filter Press + Lab Press	N/A

2.4.4 Chemical Addition Tests (Ferric Chloride and Lime)

The WERF team studied the effect of full-scale chemical addition (ferric chloride, lime, and alum) on odor production in dewatered biosolids cake at two WWTPs: the Charlotte-Mecklenburg (CMU) McAlpine Creek Wastewater Management Facility (WWMF) and the Philadelphia Water Department BRC. Procedures followed at the two facilities used in Phase 3 are described below.

At the Charlotte-Mecklenburg Utilities McAlpine Creek WWMF, the WERF team field tested the effect of ferric chloride addition prior to dewatering and lime addition after dewatering, mixed along the centrifuge scroll conveyance system. The chemical addition tests were conducted in two field trials. During the first trial, eight combinations of iron and lime dosages were studied, as indicated in Table 2-6.

Table 2-6. Summary of Lime and Ferric Chloride Dosages for CMU McAlpine Creek WWMF Field Trial 1.

Sample Number	Lime Dose (%)	Ferric Chloride Dose (%)
1	0	0
2	0	3
3	0	6.6
4	0	9
5	5	0
6	5	3
7	5	8
8	5	9

Following ferric chloride addition, the field testing team allowed time for steady-state conditions to be achieved prior to collecting dewatered cake samples. The lime-amended dewatered cake samples were collected downstream of the conveyance system on the solids processing train. Samples were shipped overnight to Bucknell University for headspace TVOSC analysis the next day, per the procedures detailed in Section 2.2 General Protocols.

A second trial was then conducted at the CMU McAlpine Creek WWMF to test additional combinations of ferric chloride and lime. The identical sampling and laboratory analysis procedure was conducted in the second trial, with ferric chloride and lime doses as indicated in Table 2-7.

Table 2-7. Summary of Lime and Ferric Chloride Dosages for CMU McAlpine Creek WWMF Field Trial 2.

Sample #	Lime Dose (%)	Ferric Chloride Dose (lb Fe/lb dry solids)
1	0	0
2	0	0.016
3	1	0
4	3	0.016
5	4	0.016
6	5	0
7	5	0.016
8	5	0.02
9	5	0.03

2.4.5 Alum Addition Tests

Full-scale testing of alum (aluminum sulfate) addition at the Philadelphia Water Department BRC took place on July 26, 2006. This field testing was conducted following positive results from laboratory tests in which alum addition correlated with a significant decrease in odor production in dewatered cake. The goal of the field testing was to examine the effect of alum addition prior to dewatering on dewatering performance and on dewatering biosolids cake quality.

2.4.5.1 Alum Addition Sampling Protocol

The WERF team produced a specific test protocol for use in the full-scale alum addition testing that detailed the sampling locations and protocol. The alum feed point for the testing was upstream of a high-solids dewatering centrifuge (Centrifuge No. 1) at the BRC, as depicted in the process schematic in Figure 2-12. Centrifuge No. 1 ran in parallel with an identical high-solids centrifuge that was not receiving alum feed (Centrifuge No. 2). This centrifuge also served as a backup in case Centrifuge No. 1 was not operational during the field testing. The alum addition was performed in conjunction with a dewatering polymer dose that was optimized for each corresponding alum dose.

The WERF field team took three samples: the centrifuge feed (prior to alum addition), the centrate, and the dewatered cake (see Figure 2-12 for these three sampling locations). Aluminum recovery was measured in both the centrate and dewatered cake. Control samples were taken from Centrifuge No. 1 at the beginning and end of the testing period. Centrifuge No. 2 also served as a source of parallel control samples if necessary.

2.4.5.2 Trial Run Testing Procedure

During the afternoon of the day prior to the sampling event, Tuesday July 25, 2006, a trial run was conducted to assure that sampling procedures would be suitable for the testing that would occur on the following day. Samples collected in the trial run were not shipped to Bucknell University for laboratory analysis. In the morning of the trial run, Delta Chemical

Company shipped a tanker containing more than 2,000 gallons of alum solution (see Appendix A for the alum dosage calculation sheet) to the BRC loading dock along with hose

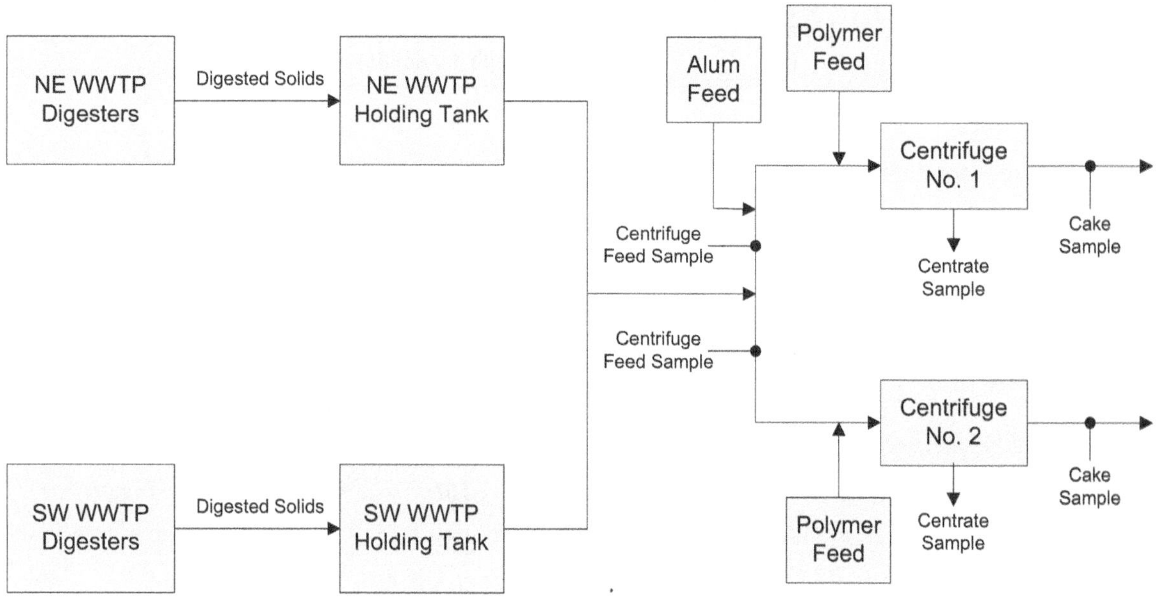

Figure 2-12. BRC Alum Full-Scale Test Schematic, Showing Test and Control Centrifuges.

and an alum feed pump. The Delta representative then made the necessary connections for pumping alum to an injection port in the feed line of Centrifuge No. 1. Two test trials of alum doses with polymer were applied in the following manner:

1. Alum was pumped into the feed line at the first prescribed feed rate of 0.7 gallons per minute (gpm), as indicated below in Table 2-8.
2. The polymer feed was subsequently reduced until cake solids dryness and centrate clarity from the test centrifuge had been optimized.
3. The same procedure was then followed at the 1.4 gpm alum feed rate.

Once trial runs had been completed for the first two alum feed rates, all testing equipment was left in place for sampling the next day.

2.4.5.3 Alum Addition Testing Procedure

The WERF field team tested alum addition as described above on Wednesday, July 26, 2006. In the laboratory analysis protocol, this is referred to as "Day 0" (Day 1 and subsequent days refer to the analysis days). The field team established an optimal polymer feed rate (without alum addition) for both test centrifuges at the beginning of the operating shift, which was 6:00 AM. This is the standard time that the centrifuge feed sludge is changed over from the Southwest (SW) WWTP digested solids holding tank to the Northeast (NE) WWTP digested solids holding tank (see Figure 2-12). The NE WWTP digested solids was the preferred feed sludge for testing, since its iron content is typically lower than the SW WWTP digested solids.

In the testing, each sample set consisted of a centrifuge feed sample, followed a few minutes later by a dewatered cake and centrate sample, which were taken simultaneously. The first set of samples was taken from the sludge feed to Centrifuge No. 1 before commencing the alum dosing. This constituted the control sample, and corresponds to Sample #1 on the sample matrix shown in Table 2-8.

Table 2-8. Philadelphia Water Department BRC Sample Matrix: Test Run 1 (NE WWTP Sludge).

Sample #	Alum Feed Rate (gpm)	Alum Dose (% Al)
1	0	0
2	0.7	0.5
3	1.4	1.0
4	2.8	2.0
5	5.6	4.0
6	0	0

After the control samples were taken, the alum feed was increased to the first dose rate of 0.5% Al, which corresponds to 0.7 gpm on the alum feed pump, as indicated on the dosage calculation sheet (Appendix A6). The polymer dose was then decreased slowly until the optimum dose corresponding with the alum feed rate was found. When this was achieved, the centrifuge was run for more than 30 minutes to achieve steady-state conditions, at which point Sample No. 2 was taken.

The alum feed rate was then increased again to the second test dose rate of 1% Al, or about 1.4 gpm on the alum feed pump. The optimal polymer dose was determined using the same procedure described above. The centrifuge was then run for 60 minutes, after which Sample No. 3 was taken along with a corresponding control sample from Centrifuge No. 2.

The same procedure was followed for Samples No. 4 and No. 5, corresponding to Al dose rates of 2% and 4% respectively. After Sample No. 5 was taken at the field testing maximum Al dose rate of 4%, the alum feed was stopped and the polymer feed rate was increased to optimal conditions, similar to the polymer feed rate from Sample No. 1. Sample No. 6 was then taken from the Centrifuge No. 1 as a final control sample. This sample was taken to compare analysis results with Sample No. 1, as a means of gauging any differences in centrifuge feed sludge quality during the test run. Collection of Sample No. 6 resulted in the end of the test run.

Since there was sufficient time left in the day after Sample No. 6 was taken, another test run was conducted on digested sludge form the SW WWTP sludge holding tank. The second test run (referred to in this report as Test Run 2) was conducted in an identical manner as the first test run that used sludge from the NE WWTP holding tank as centrifuge feed. This constituted a modified test run to see if there were any similar trends in odor comparisons for that sludge stream.

2.4.5.4 Sampling and Analysis Protocol

The following samples were taken during the alum addition field testing at the Philadelphia Water Department BRC:

- Sampling volumes of 500 mL of digested biosolids (centrifuge feed solids) and 250 mL of centrate were taken for each control and test sample.
- Two 1-gallon sealed plastic bags of dewatered centrifuge cake, double-wrapped for leak and odor control were taken for each control and test sample. One plastic bag was shipped to the Bucknell University laboratory for TVOSC analysis and other tests, and the other was shipped to the LACSD laboratory for odor panel testing.
- Sample bottles for headspace analysis by Bucknell University were filled on-site within a few minutes of collecting the sample. These bottles were then placed in a cooler with no ice (at ambient temperature) and shipped to the Bucknell University laboratory for analysis.

Laboratory Analysis of the collected samples for the alum addition field testing was conducted as indicated in Section 2.2 General Protocols.

2.5 Quality Assurance / Quality Control Procedures

During the course of the project, it was important for the WERF team to assure the quality of data that was being collected, especially the headspace TVOSC measurements by the Bucknell University and Virginia Tech laboratories. In addition, both Bucknell University and Virginia Tech developed centrifuge simulation procedures that used different approaches. These methods needed to be verified to determine if they produce cakes with similar odor profiles as full-scale centrifuges and similar odor profiles to each other.

A Quality Assurance / Quality Control (QA/QC) study was performed with the following objectives:

1. Perform and inter-laboratory study to compare TVOSC analytical results using the same cake sample. The three laboratories that were tested were Bucknell University, Virginia Tech, and LACSD. These three laboratories conducted TVOSC analysis over the course of the Phase 3 work.
2. Compare the Bucknell University and Virginia Tech centrifuge simulation processes to results from full-scale centrifuge cakes.
3. Compare the Bucknell University and Virginia Tech centrifuge simulation processes to each other.

In order to complete these tests to reflect the consistency of the data in the Phase 3 research, the WERF team compiled a "QA/QC Protocol," which is attached in Appendix A7 of this report. The protocol includes the methods of sampling and analysis used by the three Phase 3 laboratories and how the QA/QC comparative testing was carried out. Liquid sludge and solid biosolids samples were collected from two WWTPs for this QA/QC testing: the San Francisco Southeast WWTP and the LACSD Joint Water Pollution Control Plant.

For the comparison of TVOSC analysis methods at the three laboratories, a dewatered cake sample was collected from a high-solids centrifuge at the LACSD Joint Water Pollution Control Plant (JWPCP), split up into even quantities, and sent to three laboratories: 1) to Bucknell University via overnight shipping in a cooler with no ice, 2) to Virginia Tech via overnight shipping in a cooler with no ice, and 3) to the LACSD laboratory located at the JWPCP (no shipping, as the sample remained on site). Upon arrival at each laboratory, cake samples were placed in headspace bottles and analyzed for TVOSC concentration during cake storage. Dewatered cake TVOSC profiles were compared for the three laboratories. Results from the QA/QC testing are provided in Section 3.4.1.

To test the laboratory methods used by Bucknell University and Virginia Tech that produce a cake that has similar properties to full-scale high-solids centrifuges, a set of experiments were also designed in the QA/QC protocol. For these experiments, samples of JWPCP digested sludge was shipped to the Bucknell and Virginia Tech laboratories for bench-scale dewatering. A field cake sample was also shipped to each lab in order to compare the laboratory procedure with the cake generated by the JWPCP centrifuge. For each sludge sample, a comparison was made between the TVOSC profiles generated by the following means (results from the QA/QC testing are provided in Section 3.4.2):

- The Bucknell University laboratory bench-scale dewatering unit
- The Virginia Tech laboratory bench-scale dewatering unit
- The field cake sample (the TVOSC profile generated by the cake dewatered at full-scale at the JWPCP).

CHAPTER 3.0

TEST RESULTS

3.1 Introduction

The purpose of this section is to present the test results for the Phase 3A and Phase 3C activities conducted under this project. The sections that follow include the results collected for the laboratory and field tests summarized in Section 2. The data sets used in compiling these results are included in Appendix B of this report.

3.2 Phase 3A Laboratory Testing Results

Laboratory dewatered biosolids odor tests that were conducted as part of Phase 3A included experiments on cation testing, digestion SRT, CEBA addition, ferric chloride addition, lime addition, and alum addition. The following sections present the test results for these portions of the Phase 3 research and provide conclusions based on those results.

3.2.1 Results of Cation Tests

In the Phase 3 research, to study the influence of cations on odor production from anaerobicly digested biosolids, raw sludges from 12 different WWTPs were anaerobically batch digested in the Virginia Tech laboratory for 30 days at 37°C. The following sections discuss the test results with respect to cation content in the sample sludges before and after digestion, and effects on odor production in the dewatered biosolids cake.

3.2.1.1 Effects of Anaerobic Digestion on Biosolids Odors

The 12 feed sludges were analyzed for total and volatile solids (TS and VS), solution cations, total iron (Fe) and aluminum (Al), and solution biopolymers (proteins and polysaccharides). The compositions of the raw feed sludge samples are listed in Table 3-1 (proteins and polysaccharides measurements are not shown).

The mix of primary sludge and WAS in the samples was generally the same as generated at the treatment plants from where the sludge originated. Feed sludges from all plants consisted of a blend of primary and WAS with the exception of Plants 11W and 12W, whose samples consisted only of thickened WAS.

Table 3-1. Composition of Feed Sludge Samples (Cation Tests).

Plant	TS (g/L)	VS (g/L)	Na^+ (mg/L)	NH_4^+ (mg/L)	K^+ (mg/L)	Mg^{2+} (mg/L)	Ca^{2+} (mg/L)	Fe (mg/g TS)	Al (mg/g TS)
1	14.4	10.7	53.1	10.2	20.3	13.6	41.9	16.7	5.3
2	6.9	5.5	60.1	13.0	11.3	5.4	27.7	7.6	6.9
3	39.2	29.2	72.8	618.4	227.3	75.4	156.2	35.0	2.9
4	40.9	25.7	56.5	414.0	53.4	65.8	179.7	96.9	5.1
5	41.9	32.6	67.5	280.0	83.9	54.0	115.7	27.6	5.6

Table 3-1. Composition of Feed Sludge Samples (Cation Tests).

Plant	TS (g/L)	VS (g/L)	Na^+ (mg/L)	NH_4^+ (mg/L)	K^+ (mg/L)	Mg^{2+} (mg/L)	Ca^{2+} (mg/L)	Fe (mg/g TS)	Al (mg/g TS)
6	42.9	36.7	158.0	85.4	98.6	68.7	205.1	10.9	3.1
7	47.8	38.2	98.0	146.3	438.6	121.8	211.4	25.9	6.3
8	33.9	25.1	~	~	~	~	~	49.1	3.8
9	37.5	31.7	67.7	105.1	151.3	51.0	137.9	28.5	5.5
10	39.5	31.8	58.1	278.6	199.1	85.1	129.4	15.3	8.5
11W	10.4	8.0	35.74	19.75	35.82	22.33	48.01	13.7	9.8
12W	52.0	41.0	25.8	367.0	221.2	15.9	48.5	11.8	8.6

~: did not measure
W refers to sludge samples containing only WAS

Recent studies have shown that cations play an important role in biofloc formation and destruction during anaerobic digestion (Novak et al., 2003; Park et al., 2006). WERF team members have postulated that there are different pools of EPS that are bound with different metal ions in floc, and degradation of those materials occurs under different digestion conditions, with some fractions degrading anaerobically and some aerobically (Novak et al., 2003). These EPS fractions include divalent cation-associated EPS that degrades aerobically, iron-associated EPS that degrades anaerobically, and aluminum bound EPS that appears to be resistant to degradation.

After 30 days of batch anaerobic digestion in the Virginia Tech laboratory, each of the 12 sludge samples was analyzed for pH, TS, VS, solution cations, total Fe and Al, solution biopolymers (proteins and polysaccharides), optimal polymer dose and odors. Table 3-2 depicts some of the parameters measured in the analysis of the 12 sludge samples, following anaerobic digestion. Data in this table is directly comparable to data in Table 3-1.

Table 3-3 depicts a comparison of protein and polysaccharide concentration before and after laboratory anaerobic digestion of the sludge samples. The data indicates that both protein and polysaccharide concentrations increase in sludge following anaerobic digestion, but on average, proteins increase much more than polysaccharides (note in Table 3-3 that protein concentration increased to levels over 1,000 mg/L in several samples, while polysaccharide concentration for the most part stayed roughly in the same order of magnitude following digestion).

Table 3-2. Chemical Content of Sludge Samples Following Laboratory Anaerobic Digestion for 30 Days.

Plant	pH	TS (g/L)	VS (g/L)	VSR* (%)	Na+ (mg/L)	NH4+ (mg/L)	K+ (mg/L)	Mg+2 (mg/L)	Ca+2 (mg/L)	Fe (mg/gTS)	Al (mg/gTS)
1	~	10.4	7.2	33.1	54.4	229.9	37.9	16.9	14.9	12.4	4.0
2	6.7	4.5	3.0	44.5	61.9	210.7	40.4	13.8	16.5	12.1	16.8
3	7.6	25.0	15.0	48.7	71.1	1516.4	209.6	64.9	87.1	61.3	4.0
4	7.3	21.6	10.1	60.5	59.4	716.0	62.9	46.3	70.8	134.0	5.5
5	6.5	31.4	21.9	32.8	64.7	981.6	171.8	36.4	78.9	43.0	13.6

Table 3-2. Chemical Content of Sludge Samples Following Laboratory Anaerobic Digestion for 30 Days.

Plant	pH	TS (g/L)	VS (g/L)	VSR* (%)	Na+ (mg/L)	NH4+ (mg/L)	K+ (mg/L)	Mg+2 (mg/L)	Ca+2 (mg/L)	Fe (mg/gTS)	Al (mg/gTS)
6	6.4	27.6	21.5	41.5	141.5	962.5	296.4	83.7	137.6	19.1	10.2
7	6.8	29.7	20.9	45.3	97.2	1217.9	433.2	118.9	109.7	41.7	10.1
8	6.9	23.6	15.0	40.2	253.4	895.2	157.1	60.1	28.9	70.6	5.5
9	7.2	29.4	17.5	44.6	3047.5	950.0	108.0	15.3	18.1	38.2	9.0
10	6.9	30.8	20.7	35.0	1912.7	1361.1	352.2	21.0	nd	13.8	7.8
11W	7.4	7.2	4.8	39.4	37.9	300.5	76.3	22.7	43.8	20.0	14.3
12W	7.7	36.8	25.4	38.0	46.0	1844.0	482.1	15.5	8.6	15.2	8.6

*VSR – Volatile solids reduction, nd: not detected, ~: did not measure
W refers to sludge samples containing only waste activated sludge.

Table 3-3. Solution Biopolymer Concentration in Feed and Digested Sludge Samples

Plant	Feed Sludge Protein Concentration (mg/L)	Digested Sludge Protein Concentration (mg/L)	Feed Sludge Polysaccharide Concentration (mg/L)	Digested Sludge Polysaccharide Concentration (mg/L)
1	56.6	413.1	10.0	41.0
2	43.5	192.3	5.4	15.7
3	904.4	1,407.2	160.6	168.3
4	693.6	906.2	97.1	59.4
5	604.0	2,411.3	145.6	435.8
6	728.8	2,315.4	134.4	372.9
7	1,934.4	2,995.9	341.5	430.2
8	~	1,189.6	~	115.6
9	634.5	1,303.8	106.7	44.4
10	745.6	1,734.2	200.1	41.8
11W	51.2		~	23.9
12W	968.8	2,848.7	130.3	47.1

~: did not measure
W refers to sludge samples containing only WAS

Since VS reduction is one of the key parameters used to assess the performance of an anaerobic digester, in these tests, VS reduction was compared to the chemical composition of the sludges after anaerobic digestion. The data plotted in Figure 3-1 shows that the percent VS reduction was found to generally increase with iron content of the digested sludge, especially in sludges with iron content higher than about 20 mg/kg. These results are similar to those found by Park et al. (2006) for waste activated sludges.

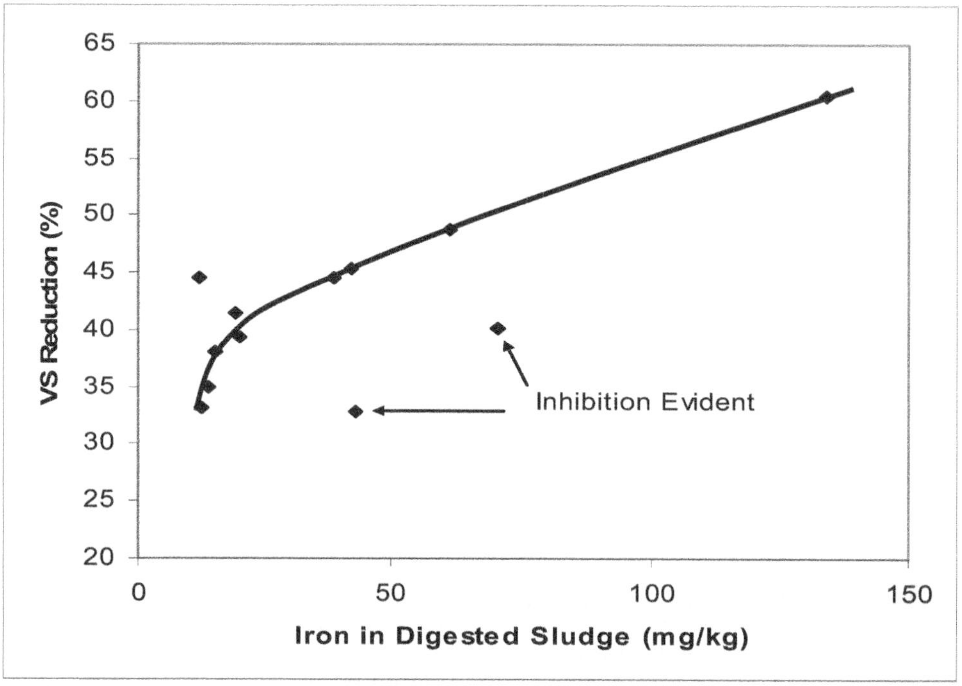

Figure 3-1. Relationship of Iron Content and Volatile Solids Destruction in 12 Sludge Samples.

Two of the 12 sample sludges had VS destruction much less than expected (indicated in Figure 3-1 as "Inhibition Evident"). Both of these sludges are from plants that receive large quantities of industrial wastes. Both of these sample sludges also contained indications of microbial inhibition which in theory caused the VS destruction to be less than expected based on the iron content of the sludge. Excepting those two outlying points, the results of this correlation are in accordance with previous observations that VS destruction increases with an increase in the iron concentration measured in the digested sludge.

Protein has been reported as one of the major organic compounds degraded during anaerobic digestion (Novak et al, 2003). While some of the released protein remains in solution, most of the protein is degraded, accounting for much of the VS destruction in anaerobic digestion. Figure 3-2 depicts the plot of solution proteins in digested sludge as a function of VS reduction. From analysis of the data it can be seen that higher solution proteins were generally measured in sludge samples that had higher VS destruction.

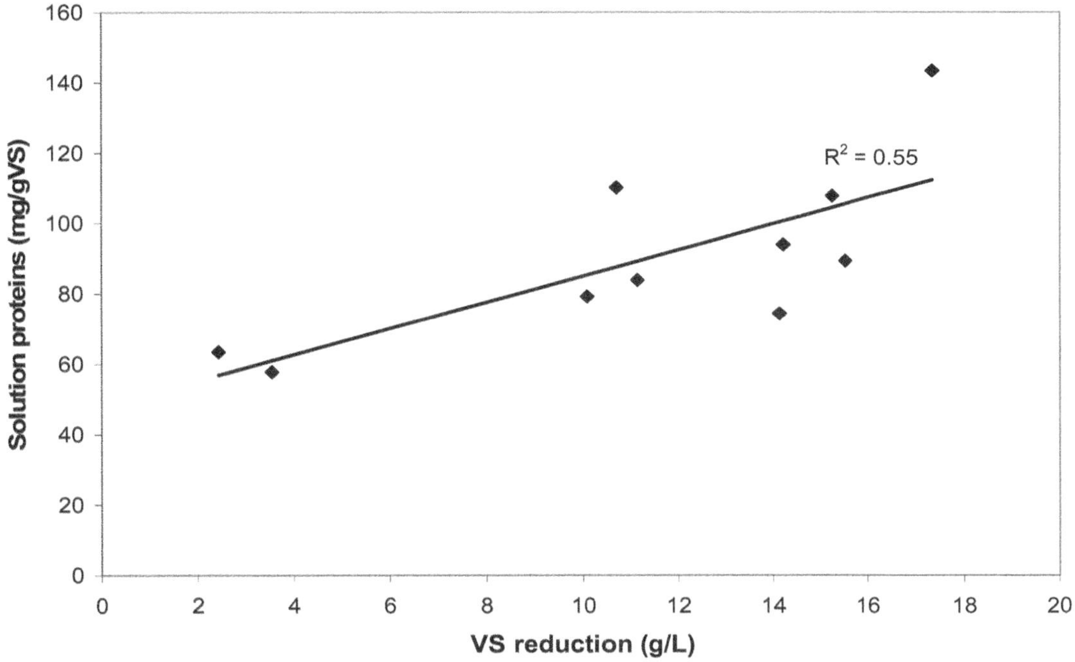

Figure 3-2. Impact of VS Reduction on Solution Proteins.

Figure 3-3 (produced using the same 12 sludge samples) indicates that digested sludge samples that experience a higher VS reduction result in increasing ammonium ion concentrations. This observation serves as an indication of the degradation of nitrogen-containing organic matter that occurs during anaerobic digestion, which primarily consists of protein degradation. Data in Figure 3-3 indicates that ammonia production is closely associated with VS reduction during anaerobic digestion. The data show the same trend as was observed by (Park et al., 2006), though the correlation in the Phase 3 research was not as strong. The weaker correlation in these tests is likely due to the use of a blend of primary sludge and WAS instead of only WAS, as was the case in the previous research (Park et al., 2006). Studies have shown that primary sludge is not homogeneous and its characteristics typically differ substantially from one facility to another. This observation could explain the variability in the data in Figure 3-3.

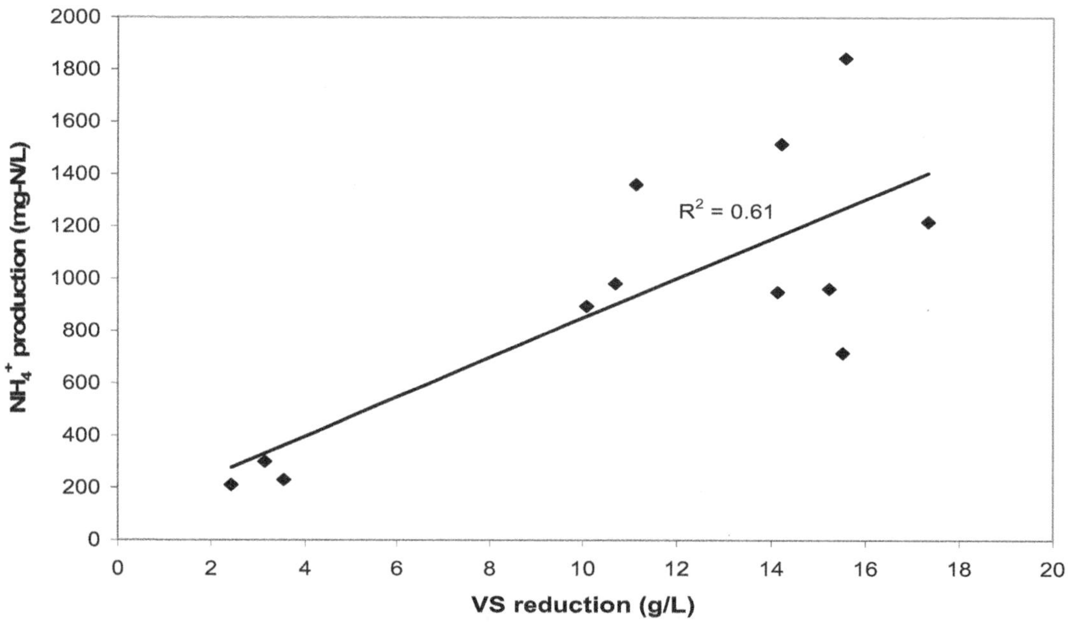

Figure 3-3. Measured Ammonium Ion Production for Various VS Reduction Levels on 12 Sludge Samples.

3.2.1.2 Role of Methanogens in Biosolids Odors

Research has shown that methanogenic bacteria are able to degrade volatile organic sulfur compounds (VOSCs) (Higgins et al, 2006). In fresh water sediments, a balance typically exists between the production of VOSCs and their degradation, resulting in little emission of these compounds unless the system is disturbed (Lomans et al., 2001). A similar balance is theorized to exist in anaerobic digesters since VOSCs are typically not emitted except under conditions that cause stress to methanogenic bacteria. Higgins et al. (2005) have shown that methanogens play an important role in cycling of VOSCs from biosolids. Higgins et al. (2005) also showed that near stoichiometric amounts of VOSCs are produced relative to the amount of substrate available when methanogens are inhibited.

Headspace gas in bottles containing dewatered cake from the 12 WWTPs was periodically analyzed at the Virginia Tech laboratory using a GC/MS to generate TVOSC concentration versus time plots (TVOSC profiles). As shown in Figure 3-4, dewatered cake from WWTP No. 1 that was amended with BESA to reduce the methanogenic activity produced higher TVOSC concentrations than dewatered cake that was not treated with BESA. This supports the assertion by the WERF team that BESA-amended cakes will provide the maximum "odor potential" of a dewatered cake with regard to its headspace TVOSC profile.

The TVOSC concentration from cakes which were not amended with BESA peaked earlier and did not peak as high as the BESA-amended cakes. Additionally, the TVOSC concentration decreased rapidly with time following the peak due to degradation by methanogens. The Figure 3-4 plot is in agreement with those reported by Higgins et al. (2004), suggesting that methanogens play an important role in cycling of VOSCs.

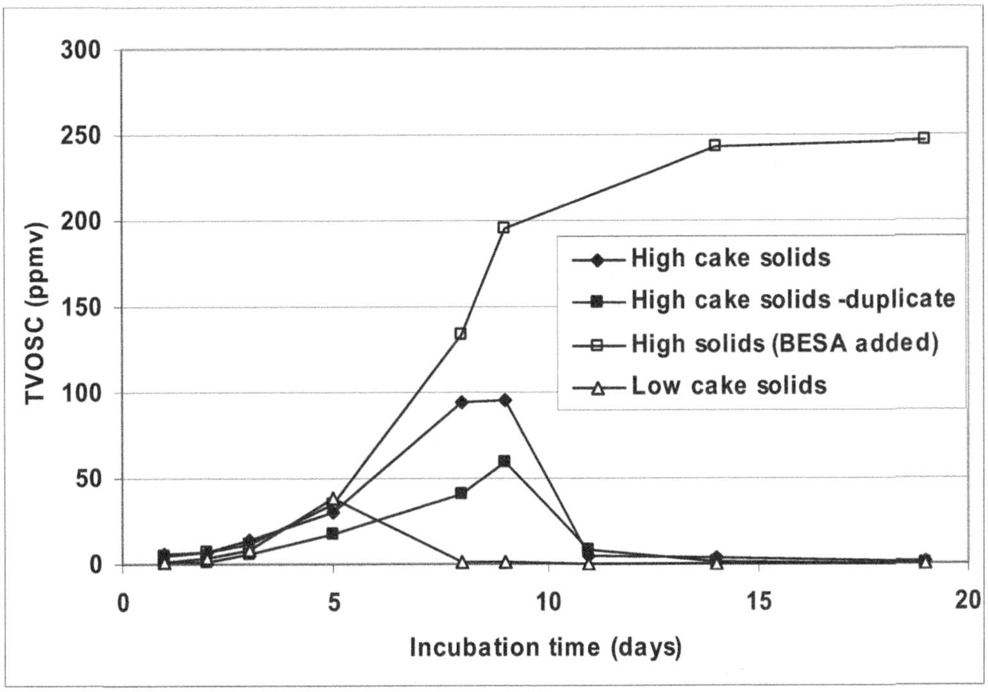

Figure 3-4. Headspace TVOSC Concentration of WWTP No. 1 with Varying Cake Solids.

To show the same data another way, the headspace sulfur concentrations were normalized to the initial quantity of volatile solids added to the incubation vial. The normalized concentrations were plotted as a function of the incubation time in Figure 3-5. The general TVOSC increase observed in the BESA-amended cake was found as well.

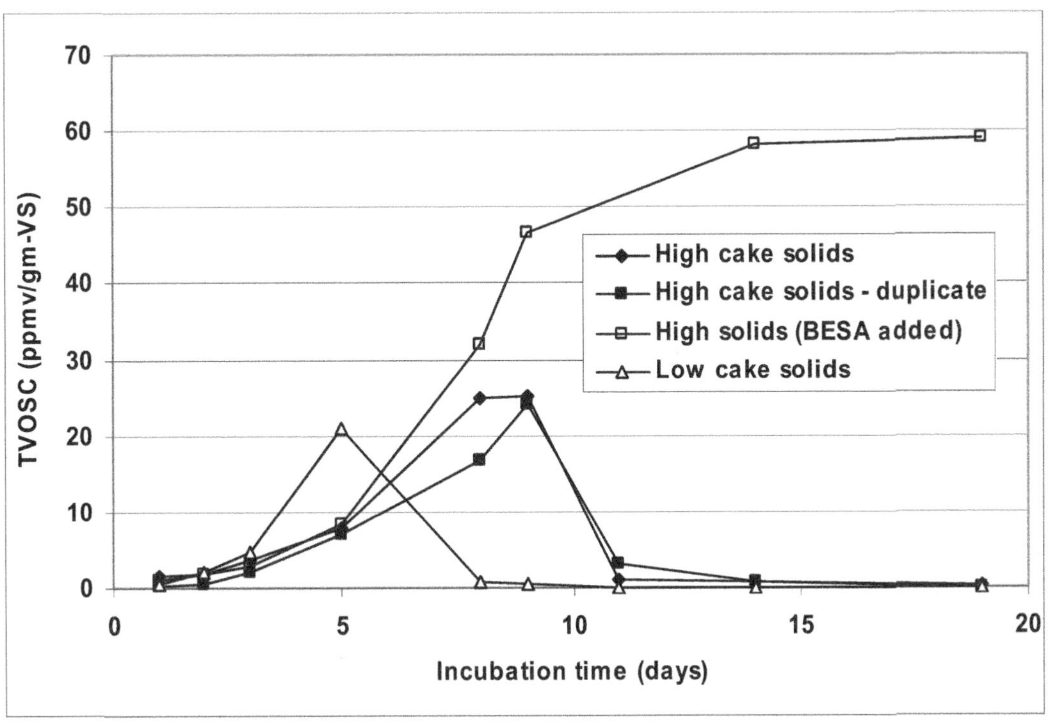

Figure 3-5. Headspace TVOSC Concentration of WWTP No. 1 (Cake Solids Normalized to Initial VS Content).

The data in Figures 3-4 and 3-5 show that the concentration of TVOSC in the headspace of dewatered cakes with a low-solids concentration was lower than that measured for high-solids cake. This observation suggests that the solids concentration of dewatered cake affects the cycling of VOSCs. This observation is in agreement with recent results reported by Muller et al. (2004). It is theorized that as the cake moisture is reduced, methanogens are inhibited, either by exposure to oxygen or lack of moisture. However, although the methanogens are inhibited, other organisms in the sludge cake (many of which are facultative) continue to degrade protein to produce sulfur gases. As a result, the syntrophy between microbial reactions that lead to VOSC production and consumption is lost, leading to accumulation of VOSCs in the biosolids sample headspace.

The WERF team also compared the odor generation potential of all 12 sludge samples by comparing the peak TVOSC concentrations for both samples without BESA and BESA-amended (methanogens inhibited) samples. Table 3-4 lists the peak TVOSC concentrations measured in headspace incubation bottles for each plant sample, with peak TVOSC values indicated for high- and low-solids samples. The table indicates that BESA-amended samples contained a higher peak TVOSC than the comparable sample that was not amended with BESA for 11 of the 12 sample sludges that were dewatered to arrive at a high-solids cake. The low-solids samples showed this trend for all sludges tested. Therefore, the trend shown for WWTP No. 1 (Figures 3-4 and 3-5) held true for almost all plants tested: BESA-amended cakes result in higher peak TVOSC concentrations.

Table 3-4. Peak TVOSC Concentration Comparison for Dewatered Cake Samples (Varying Solids Content).

	High-Solids Samples			Low-Solids Samples		
		Peak TVOSC Concentration (mg/m^3)			Peak TVOSC Concentration (mg/m^3)	
Plant	TS %	No BESA	With BESA	TS %	No BESA	With BESA
1	22.0%	101	323	11.9%	50	-
2	11.8%	5	59	10.6%	2	33
3	24.3%	179	331	20.6%	~	294
4	36.4%	274	1,015	24.3%	~	589
5	20.4%	997	1,000	17.8%	~	925
6	24.0%	11	19	19.2%	~	47
7	17.9%	758	873	16.7%	~	817
8	27.4%	922	1,149	19.6%	852	1,098
9	27.6%	160	178	19.0%	~	156
10	15.8%	201	306	13.4%	~	290
11W	19.0%	5	277	~	~	~
12W	19.2%	513	493	15.2%	59	451

~: did not measure.
W refers to sludge samples containing only WAS

A similar trend was also observed for most of the 12 WWTPs with respect to solids content as was seen in WWTP No. 1, in that lower-solids content cakes had lower peak TVOSC than the corresponding high-solids cake samples. Table 3-4 also displays this trend.

3.2.1.3 Feed Sludge Cation Content Impact on Biosolids Odors

The WERF team also compared the odor generation potential of sludges with different metal concentrations in the feed sludge. Headspace peak TVOSC concentrations were determined for sludges amended with BESA and those without BESA.

Impact of Iron in Feed Sludge on Biosolids Odors

Data in Figure 3-6 shows the peak TVOSC concentrations generated from high-solids cake samples from the tested WWTPs as a function of iron (Fe) concentration in the feed sludge. Peak TVOSC concentration values fit reasonably well with iron concentration in the sludge solids except those from Plants 3 and 4 (indicated as hollow boxes in the), which receive water plant sludge containing high amounts of iron. The WERF team theorizes that iron in these sludges was not able to bind with the proteins in the same way as iron in the sewage and therefore did not contribute to TVOSC in the same way as other iron in the wastewater. These two points were not considered in the Figure 3-6 correlation.

The WERF team additionally noted that sludge samples from plants 11W and 12W were different from sludge samples from the other 10 plants because they consisted only of thickened WAS. Because of this, data from Plants 11W and 12W were not included in the Figure 3-6 plot.

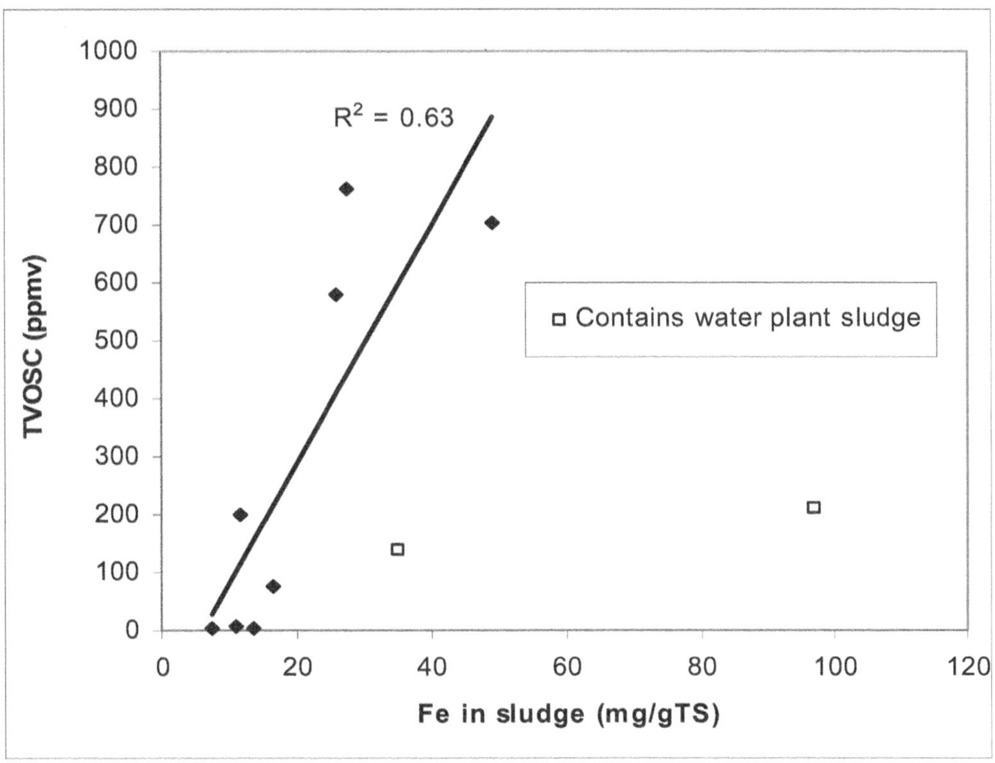

Figure 3-6. Impact of Iron in Feed Sludge on High-Solids Cake Peak TVOSC Concentration (No BESA Added).

The data in Figure 3-7 shows that when BESA is added, the correlation between iron concentration and TVOSC is better. This was expected because when methanogenic activity is suppressed, the total potential TVOSC is realized. When methanogens are active, some of the VOSCs are converted to hydrogen sulfide, making the peak TVOSC more variable.

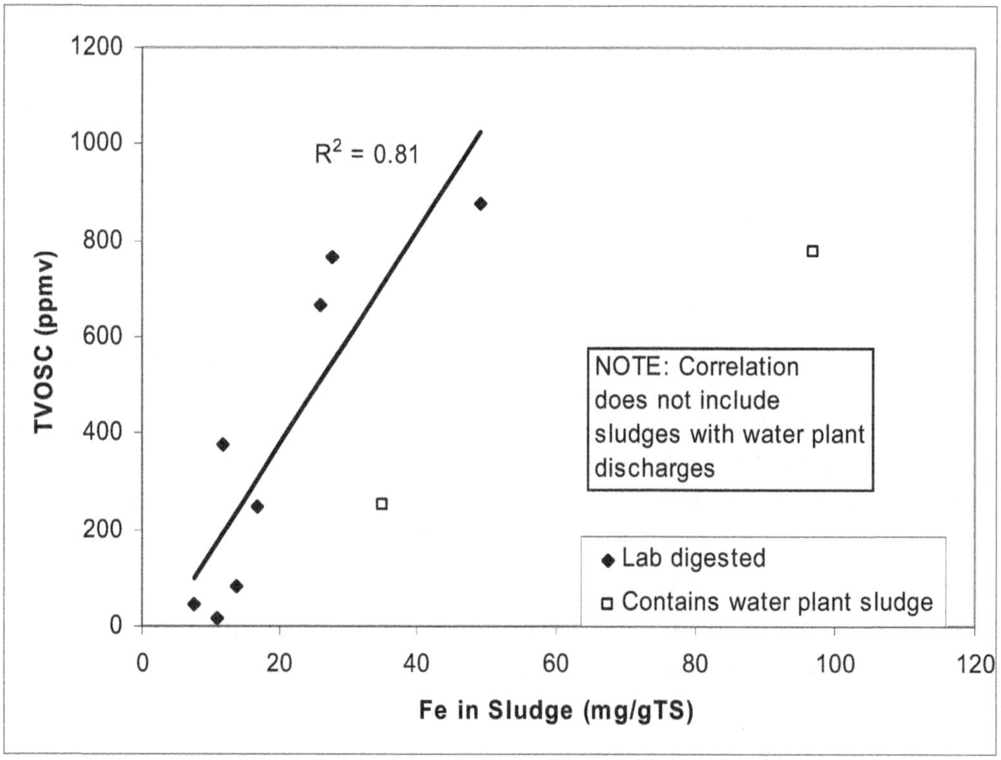

Figure 3-7. Impact of Iron in Feed Sludge on High-solids Cake Peak TVOSC Concentration (BESA Added).

It is clear from data shown in Figures 3-6 and 3-7 that for high-solids dewatered cake samples, peak TVOSC headspace concentration increases with increasing iron content of the feed sludge. The data obtained in this study is consistent with two previous research findings: 1) Biopolymers associated with iron are not completely degraded during anaerobic digestion (Adams et al., 2003; Muller et al., 2004), and 2) When these biopolymers, primarily proteins, are made bioavailable by shear in the presence of dewatering polymer they undergo degradation and generate VOSCs.

Peak TVOSC concentration generated from dewatered cake with low solids was also plotted as a function of iron content in the feed sludge (figure not shown). A trend similar to that observed for high-solids cake samples (Figures 3-6 and 3-7) was observed, with the peak TVOSC concentrations increasing with increasing iron content of the feed sludge.

Iron and TVOSC testing was also performed on two cake samples that were digested in the field at full-scale. Figure 3-8 depicts a composite of both field and laboratory TVOSC data measurements versus feed sludge iron content for BESA-amended high-solids cake samples. The figure indicates that the relationship between iron and TVOSC is valid for both field and laboratory digested samples.

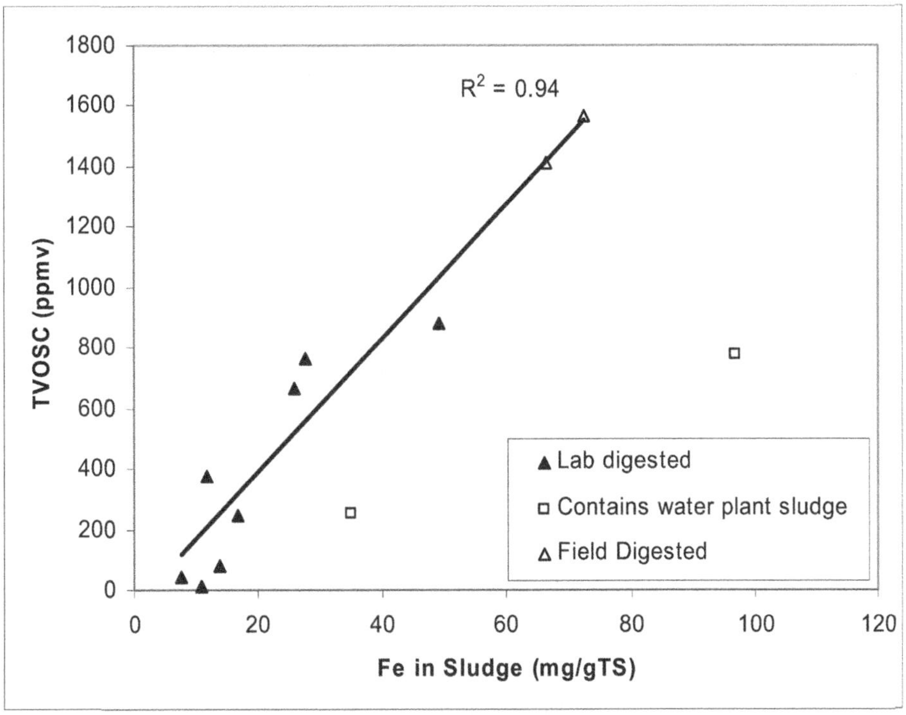

Figure 3-8. Impact of Iron in Feed Sludge on High-solids Cake Peak TVOSC Concentration (BESA Added).

Impact of Aluminum in Feed Sludge on Biosolids Odors

A plot of TVOSC in the headspace versus aluminum concentration in the feed sludge is depicted in Figure 3-9. The plot indicates that for feed sludge samples with relatively high iron content, increasing amounts of aluminum result in *lower* peak TVOSC in the headspace of the dewatered biosolids product. Note that the hollow boxes in the plot represent WWTP samples that contain low iron content. Low iron sludges will also have low quantities of proteins associated with iron, so there would be little potential for VOSC generation. Because of this, these two points are not considered in the correlation. In addition, the two plant samples that contained water plant sludge (with high iron content) and the two plants that consisted only of thickened WAS are not included in the correlation.

Impact of Feed Sludge Monovalent to Divalent Ratio on Biosolids Odors

Prior to the Phase 3 study, the WERF team theorized that a high monovalent to divalent ratio (M/D) in a feed sludge would result in lower odors in the dewatered biosolids cake. This theory was put forth because it has been observed in previous studies that monovalent cations can cause poor flocculation and poor binding of proteins to iron. This relationship was tested with the 12 WWTPs samples provided in this cation study.

The relationship between M/D ratio and peak TVOSC concentration in the headspace of bottles containing high-solids cake (amended with BESA) is shown in Figure 3-10. The concentrations of cations for this plot are reported in milliequivalent/milliequivalent (meq/meq) basis. The monovalent cations considered in this test were Na, K and NH_4 and the divalent cations considered were Mg and Ca. As indicated in the plot, no apparent relationship of peak TVOSC with M/D ratio was observed.

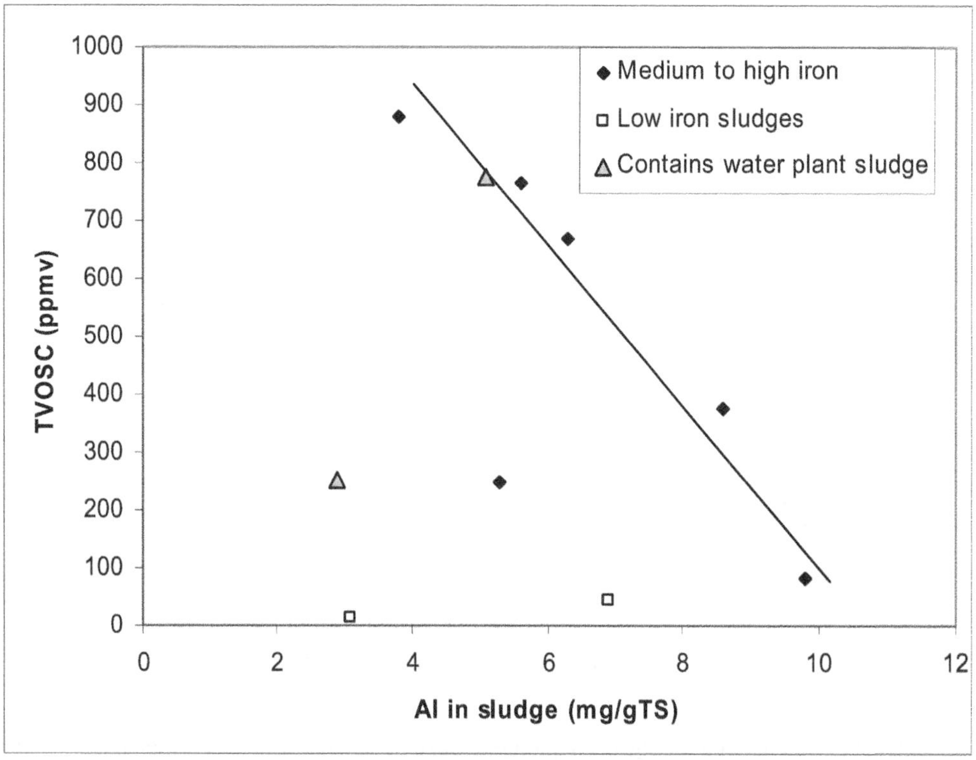

Figure 3-9. Impact of Aluminum Content in Feed Sludge on High-solids Cake Peak TVOSC (BESA Added).

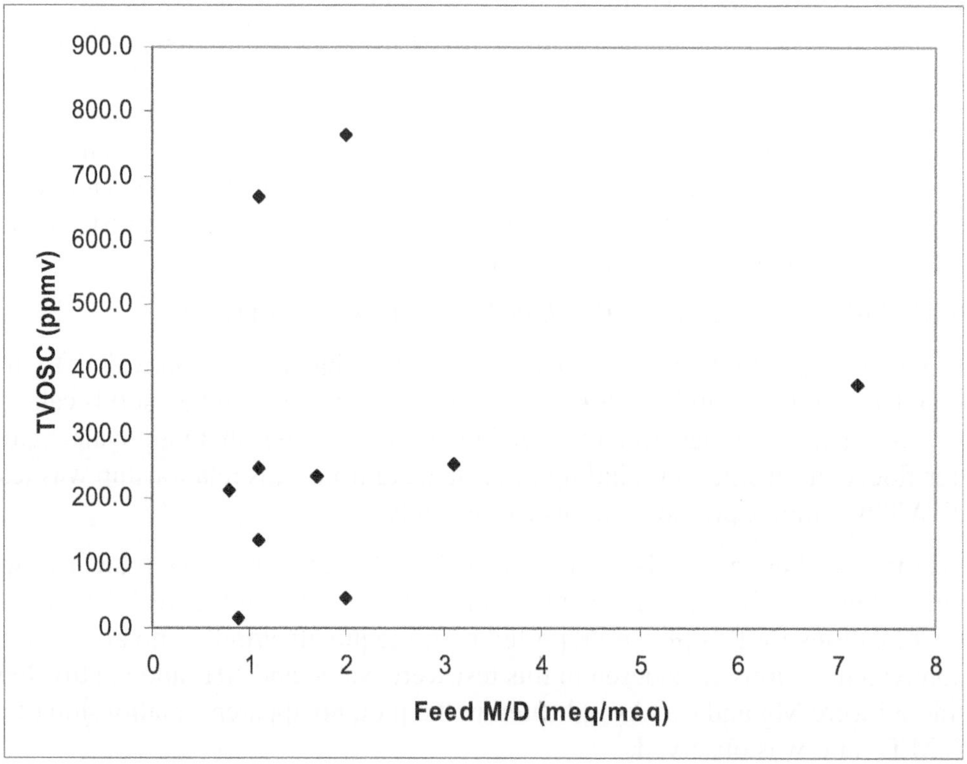

Figure 3-10. Impact of M/D Ratio in Feed Sludge on High-solids Cake Peak TVOSC Concentration (BESA Added).

Observed Relationships between Feed Sludge Cations and Biosolids Odors

The plots in Figures 3-6 through 3-9 indicate that the iron and aluminum content of feed sludge plays a crucial role in odor generation from dewatered cake. The relationship between generation of VOSCs and iron provides a method for predicting (by analysis of the feed sludge iron content prior to anaerobic digestion) if a sludge is likely to generate odors upon being dewatered by a centrifuge. This observation provides a tool to help a POTW in selection of dewatering equipment and in making decisions regarding its overall sludge handling management.

The data also suggest that higher aluminum concentration in feed sludge can indicate that there will be lower odors in the dewatered biosolids product. This information is also potentially helpful to a POTW that is concerned about odors in biosolids cake following anaerobic digestion, dewatering, and storage.

3.2.1.4 Volatile Solids Destruction and Biosolids Odors

VS destruction in anaerobic digestion is one of the key parameters for regulatory compliance with the Vector Attraction Reduction (VAR) requirements of the 40 CFR Part 503 rule. Based on previous research it was anticipated that plants that achieve a higher percentage of VS reduction would benefit from this larger percentage reduction with a lower odor level in their dewatered biosolids cake. This study was included in the Phase 3 research to determine the correlation between the percentage of VS reduction achieved by an anaerobic digester and biosolids odors.

The same sludge samples collected from 12 different facilities were batch digested for 20-days, and the percentage of VS destruction achieved after batch digestion of these samples was plotted against the corresponding dewatered cake peak TVOSC (Figure 3-11). The lack of any correlation in Figure 3-11 is in line with the Phase II findings. As Figures 3-6 through 3-9 indicate, the correlation between the varying iron concentrations in these sludges and the dewatered biosolids cake TVOSC is greater than a possible correlation between VS reduction and cake TVOSC.

This further emphasizes the conclusion that VS reduction is not a good parameter for comparing relative TVOSC (or odor) potential for biosolids collected from different facilities. However, as research conducted using sludge from the same facility has shown, a correlation between VS reduction and odor potential exists (see Section 3.2.2 and Figures 3-13 through 3-15) and that higher VS reduction does result in lower TVOSC when the same sludge is evaluated. Moreover, Figure 3-12, which contains a plot of data from eight (8) different plant biosolids, shows that TVOSC increases with RBA as well.

In addition, the WERF team noted that the cake headspace TVOSC for digested sludges that achieved greater than the 38% VS destruction level ranged from near 0 to 900 ppmv (the highest TVOSC measured). These results indicate that the 38% VS destruction requirement for VAR compliance is not a useful parameter for predicting the odor quality of dewatered biosolids. In addition, the percentage of VS reduction achieved in a digester, in general, does not correlate with TVOSC when comparing sludges from different facilities.

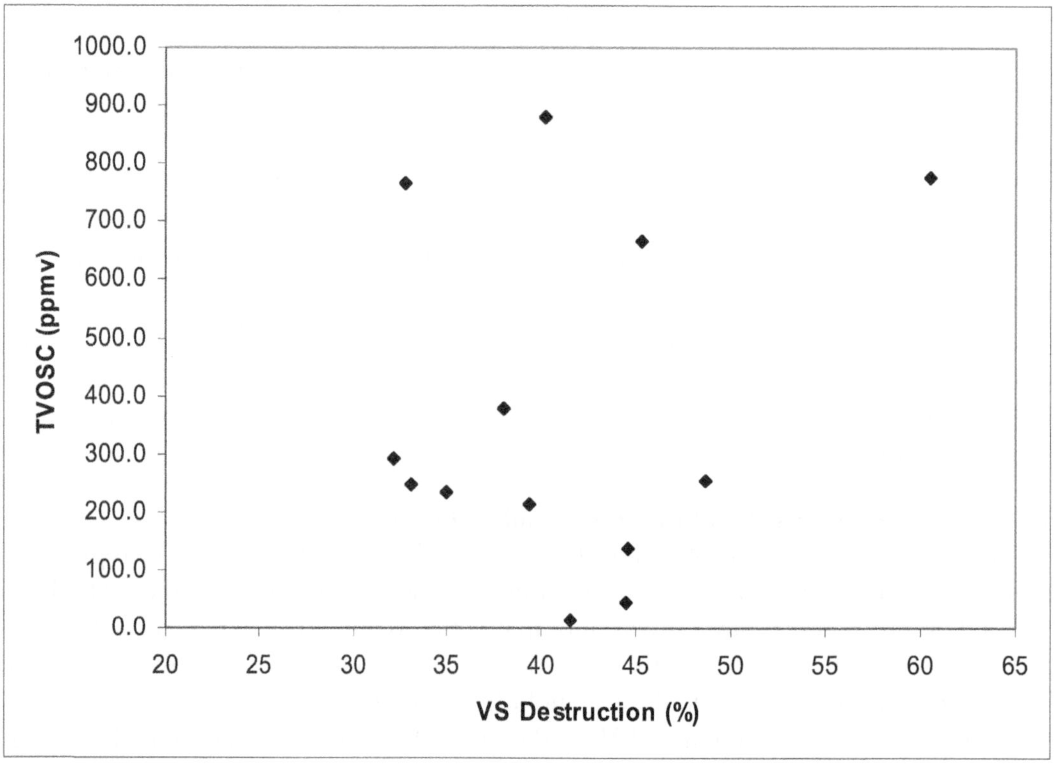

Figure 3-11. Correlation of VS Destruction with Peak Headspace TVOSC in High-solids Cake (BESA Added).

3.2.1.5 Residual Biological Activity and Biosolids Odors

Residual biological activity (RBA) is considered to be a measure of the biological stability of digested biosolids, since RBA represents the potential for additional biological activity (additional VS destruction) in digested biosolids. Figure 3-12 shows the relationship between dewatered biosolids RBA (measured as a percentage of additional VS destruction after 40 additional days of digestion) and peak TVOSC in the headspace of dewatered cake using the WWTP sludge samples sent to Virginia Tech for analysis.

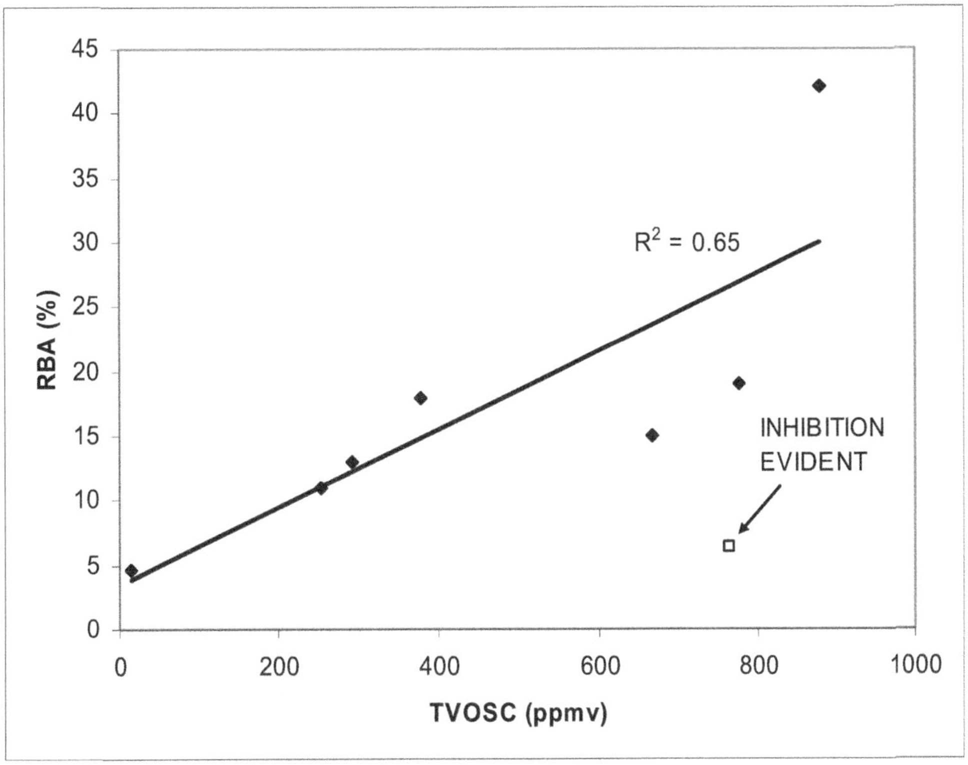

Figure 3-12. Impact of Additional VS Destruction (40 Additional Days of Digestion) on Peak TVOSC from High-solids Cake Samples (BESA Added).

In this testing, a similar plot was produced for the samples that were not amended with BESA. In general, for both BESA-amended samples and those without BESA, dewatered cake samples with higher peak TVOSC concentrations after the original 40 days of digestion had higher RBA values (higher additional VS destruction was achieved with an additional 40 days of digestion).

The data point that varies most from the plot (indicated as a hollow square in Figure 3-12) is from one of the plants where inhibition or toxicity was evident. Based on this study, the WERF team concluded that RBA might be a useful tool for predicting the odor potential of dewatered biosolids. Although the data contain considerably variability, the RBA relationship with TVOSC is much better than the relationship between TVOSC and percentage of VS destruction, as indicated see Section 3.2.1.4.

3.2.1.6 Cation Study Conclusions

The results from this cation study can be used to explain differences in biosolids odor emissions from different wastewater treatment plants. In general, this research has shown that factors that either 1) increase the bioavailability of protein or 2) decrease VOSC degradation through methanogen inhibition result in greater net VOSC production from the final dewatered biosolids cake product.

As indicated above, results from this study show that anaerobic digester feed sludges with a higher iron content correlated with higher amounts of TVOSC in the dewatered biosolids cake. This suggests that iron-bound material in sludge is highly sensitive to shear imparted in dewatering units. The WERF team theorizes that in sludge that is relatively higher in iron content, the iron binds a greater amount of protein that sludges with lower iron content. Some of iron-bound protein is released into solution due to reduction of Fe^{3+} to Fe^{2+} and degraded during anaerobic digestion. *However, complete degradation of the iron-associated protein is not achieved in anaerobic digestion.* Therefore, when this high iron, anaerobicly digested sludge is exposed to the high-shear conditions of a centrifuge in the presence of dewatering polymer, the iron-bound proteins are made bioavailable. Degradation of these released materials causes generation of VOSC.

Parameters commonly used to assess the performance of anaerobic digesters like volatile solids reduction, volatile fatty acid content of feed sludge and residual biological activity were also measured by Virginia Tech and plotted against TVOSC of the dewatered biosolids cakes. Results showed that VS destruction increased with an increase in the iron content of sludge. In general, greater VS destruction and higher TVOSC concentration was observed for digested sludge with high iron content.

The conclusions that can be drawn from this study are as follows:

1. Higher VOSC concentrations in the dewatered biosolids cake of samples from 12 WWTPs correlated with relatively higher iron content in the (anaerobically digested) centrifuge feed sludge. It is thought that when iron is reduced in the anaerobic digester, the binding of proteins associated with iron are weakened and can be released in the centrifuge due to shear, where they can then be degraded to organic sulfur compounds.
2. Methanogenic microorganisms play an important role in the cycling of organic sulfur compounds. Cake samples that were amended with a methanogenic inhibitor (BESA) produced higher TVOSC concentrations, which persisted over several weeks of incubation in the laboratory. It has been shown that methanogenic organisms can convert organic sulfur compounds to sulfide (Higgins, et al., 2006). Sulfide precipitates as FeS as long as a sufficient quantity of iron is present in the biosolids.
3. The percentage of volatile solids reduction achieved for different sludges showed no apparent correlation with TVOSC in the dewatered cake. It appears that some fractions of volatile solids contain more sulfur-containing proteins than other fractions. Therefore, a reduction in volatile solids will not directly correlate with destruction of organic sulfur precursors in biosolids.
4. Residual biological activity (RBA), as measured by additional VS destruction, was weakly correlated with TVOSC. RBA may be a useful parameter for determining if a digested sludge is likely to generate odors following dewatering. The additional volatile solids that are degraded by an additional 40 days of digestion appear to contain organic sulfur precursors, so when these solids are degraded, TVOSC is typically reduced.

3.2.2 Results of Digestion SRT Study

This aspect of the Phase 3 study was conducted to determine how extended sludge retention times (SRTs) would influence TVOSC generation in dewatered biosolids cake. In the WERF Phase 2 research, it was theorized that increasing the SRT of sludge in an anaerobic digester would result in a decrease in TVOSC of the dewatered biosolids cake. Data from the WERF Phase 2 study was deemed to be inconclusive with regard to this relationship, because the samples were collected from facilities with other prevailing conditions dominating, and a conclusion was not apparent.

However, determining the *degree* of TVOSC reduction possible with increasing SRT was an important part of this Phase 3 study. If extending the SRT can produce a low-odor biosolids product compared to a standard SRT that otherwise would produce a high odor level, then extending the SRT in an anaerobic digester could be used as an additional tool for utilities for biosolids odor control. Alternatively, if the observed odor reduction under an extended digester SRT is minimal, then it can be concluded that the amount of digester SRT is not an important consideration in biosolids odor control.

Sludge from the Los Angeles County Joint Water Pollution Control Plant (JWPCP) was used for the study because it is a relatively high odor-generating sludge. Laboratory digesters were run at the JWPCP and sludge samples were shipped on four occasions to the Virginia Tech laboratory for processing and analysis. The methods used for the study are presented in detail in Section 2.3.2.

3.2.2.1 Anaerobic Digester SRT Effect on Odors

The headspace peak TVOSC concentrations of JWPCP dewatered bisolids cake (high-solids) are plotted as a function of digester SRT in Figure 3-13 (the plot includes sludge samples with and without BESA addition). The TVOSC concentration from the dewatered cake of the digester feed sludge is shown at an SRT of 0 days. As was observed in Adams et al. (2002), the headspace peak TVOSC decreased with increasing SRT. In this testing, for both BESA-amended samples and those without BESA, the correlation between these two parameters is strong.

In this testing, the measured TVOSC from non-BESA amended samples differed little from high-solids cake samples amended with BESA, indicating that BESA addition did not have much effect on TVOSC reduction. The data suggests that some type of compound inhibitory to methanogenic organisms is present in the dewatered cake samples. Further study is required to verify this and improve our knowledge of toxicity effects on odor generation from dewatered cakes.

In addition, note the headspace TVOSC concentration of the feed sludge was over 2,000 ppmv and the 40-day SRT sludge had a TVOSC of between 800 and 1,000 ppmv, approximately. These concentrations are much higher than those reported by Adams et al. (2003) for the 11 anaerobicly digested sludges tested in the WERF Phase 2 study. This indicates that the sludge obtained from the JWPCP had a very high odor potential, and even at a very long digestion SRT (40 days), the odor potential remained high.

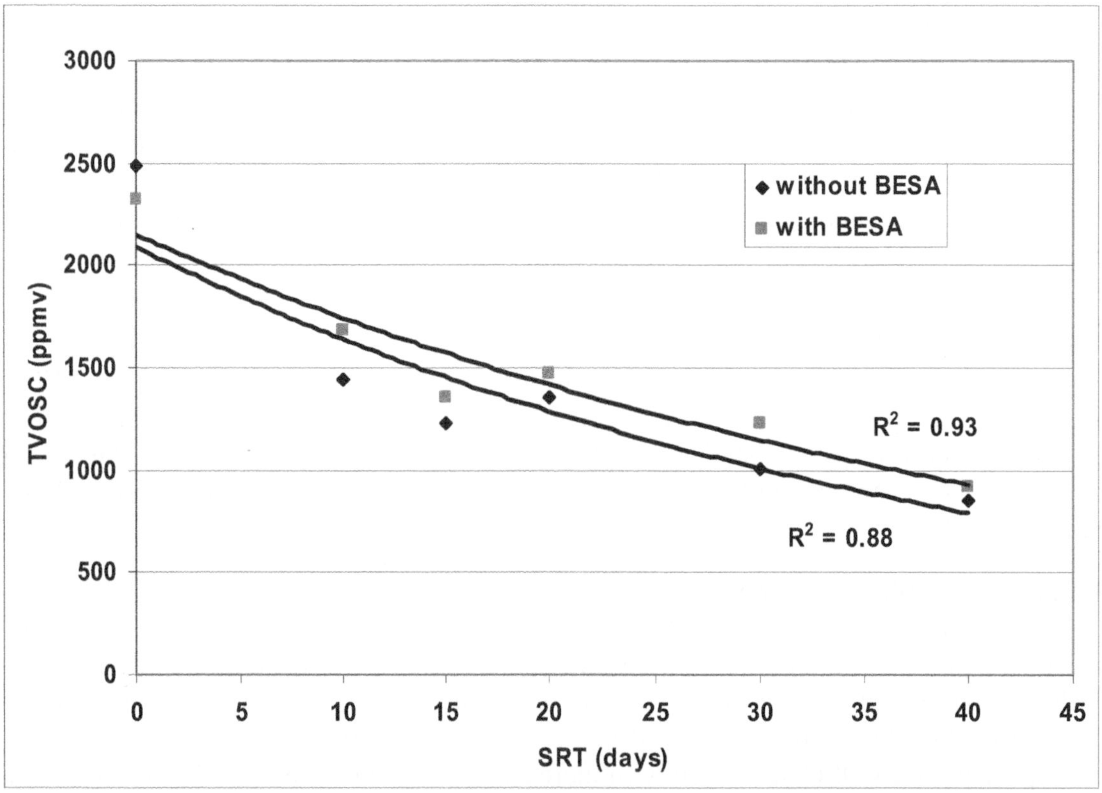

Figure 3-13. Impact of Anaerobic Digester SRT on Peak TVOSC Concentration in the Dewatered Cake (High Solids).

Figure 3-14 shows a comparison of the headspace TVOSC concentrations in bottles containing samples of dewatered cake sample with low solids content (TS = 17.0%) and high-solids content (TS = 28.3%). All values for this plot were generated from samples that were amended with BESA (greater reductions in TVOSC concentration were observed in samples that were not amended with BESA). The "High Solids" data in the plot are identical to the "with BESA" data in Figure 3-13.

Like Figure 3-13, this plot also shows that TVOSC concentration decreased with increasing SRT. Moreover, the low-solids cake produced lower TVOSC concentrations than high-solids cakes at the same SRT. This observation agrees with results recently reported by Muller et al. (2004). Note that as the headspace gas represents only a portion of the total TVOSC (the rest is in the liquid phase), the difference in TVOSC between the high-solids and low-solids samples could be due to the difference in the sludge cake water content. These plots clearly show the potential for odor problems after switching from low-solids to high-solids centrifuges. Some research has indicated that methanogenic activity may be inhibited due to the dryness of cakes as well (Lomans et al., 2001).

In addition, sludge dewatered using high solids centrifuges may be stripping more substrate due to greater shear imparted on the sludge (e.g., protein) that can serve as the precursor for VOSC formation. Research presented later in this chapter also showed that if a dewatered cake from a belt filter press at 20% solids is pressed further to get a drier cake, the TVOSC will increase (see Figure 3-117).

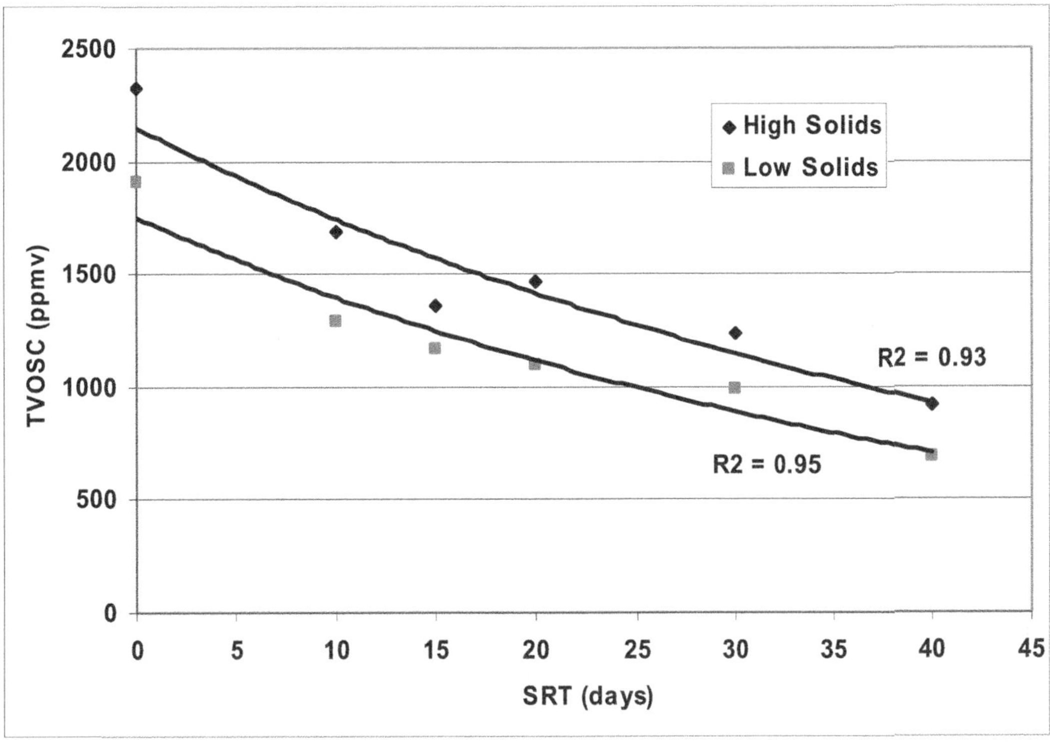

Figure 3-14. Impact of Laboratory Anaerobic Digester SRT on Peak TVOSC Concentration in the Dewatered Cake (BESA Added to all Samples).

3.2.2.2 Volatile Solids Reduction Effects on Odors

Volatile solids (VS) reduction is used as a design parameter for wastewater treatment and is also used for classification of biosolids as Class A or B per 40 Code of Federal Regulations (CFR) Part 503. Achieving at least 38% VS reduction is one of the major parameters for regulatory compliance with the Vector Attraction Reduction requirement of the 40 CFR Part 503 rule. It has been theorized that higher VS destruction in the digester will have a beneficial impact on digested biosolids quality and dewatered biosolids odors. In the Phase 3 research SRT laboratory testing, the WERF team observed the relationship between SRT, VS reduction, and biosolids cake odors using JWPCP sludge samples.

The plot in Figure 3-15 shows the VS reduction versus digester SRT for four sets of sludge samples from the JWPCP. The data show that VS reduction increases slightly with detention in a laboratory anaerobic digester, and that each set of samples was relatively consistent with regard to achieving a similar VS reduction over periods of time during digestion (at 10, 15, 20, 30, and 40 days).

Note that for all four sludge sample sets, about 50% VS reduction was observed after 10 days in the laboratory anaerobic digesters, which is significantly greater than the 38% VS reduction VAR requirement of the 40 CFR Part 503 rule. After the 10-day incubation and measurement, additional VS reduction occurred in all sample sets: on average, the laboratory digesters gained 4 % additional VS reduction between 10 and 20 days, an additional 2% VS reduction between 20 and 30 days, and an additional 2% between 30 and 40 days (a total of approximately 58% for the entire 40-day digestion period, on average for all sample sets).

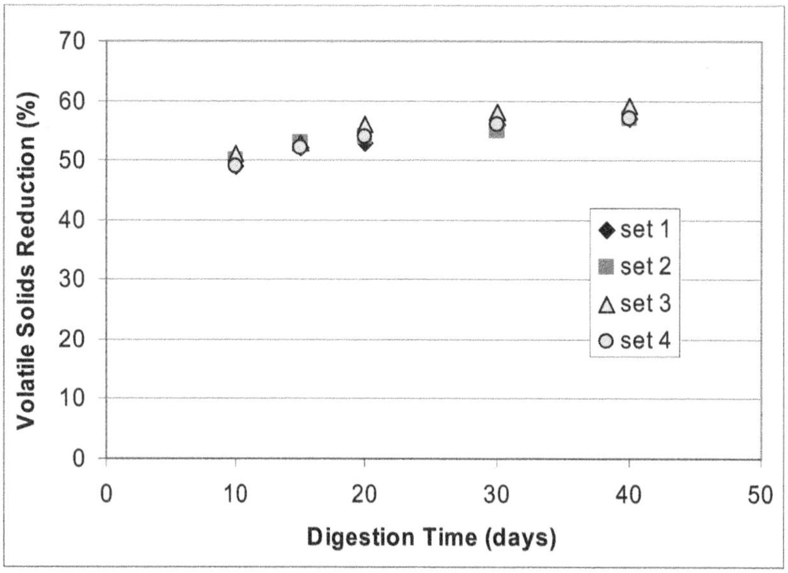

Figure 3-15. VS Reduction as a Function of Digester SRT (JWPCP Samples).

The change in TVOSC with increasing digester SRT for the same four JWPCP sample sets is shown in Figure 3-16. The headspace TVOSC concentration of the samples varied from approximately 2,500 to 700 ppmv over the 40-day digestion period, which corresponded to a measured 50 to 58% VS reduction range for the same sludge sample sets. In general, lower headspace TVOSC concentrations were measured in cake samples from digesters with higher VS destruction, which suggests that VS destruction results in the specific degradation of odor generating organic matter.

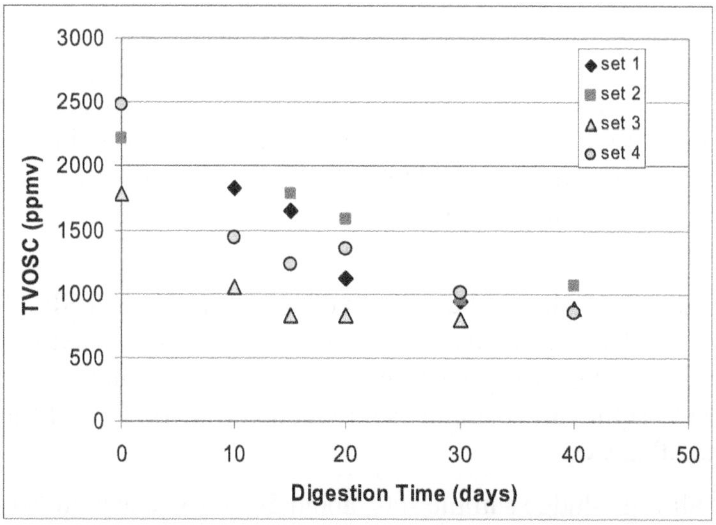

Figure 3-16. Impact of Digestion Time on TVOSC (JWPCP Samples).

Note that although there was minimal additional VS destruction after 10 days of digestion (Figure 3-15), TVOSC was reduced significantly in all sample sets between 10 and 40 days digestion. This indicates that the extra VS in the sludge that is destroyed at longer SRTs is primarily comprised of odor-generating organic material. Therefore, it can be concluded for the JWPCP sludge that although additional VS destruction generated between the 10 and 40 day point in the incubation period correlates with lower TVOSC (and therefore, lower odors), the relationship observed was not linear. This is because the majority of the VS reduction took place early in the digestion period (in the first 10 days), and TVOSC reduction occurred throughout the 40-day digestion period.

The WERF team noted that although TVOSC concentrations did generally decrease with increasing SRT for the individual sludge samples, the reduction was not due to approaching and exceeding the 38% VS reduction VAR requirement of the 40 CFR Part 503 rule. In fact, the 38% requirement was exceeded after only 10 days of detention in the laboratory digesters. This result in combination with the high TVOSC measurements in Figure 3-16 (greater than 1,000 ppmv at 10 days of incubation, which corresponded to approximately 50% VS reduction) indicates that achieving the required 38% VS reduction does not guarantee a low-odor final biosolids product.

3.2.2.3 Residual Biological Activity Relationship with Odors

Residual biological activity (RBA) is considered to be a measure of the biological stability of digested biosolids, since it represents the potential for further biological activity through endogenous decay and consumption of the substrate available in the form of volatile acids and proteins. In the Phase 3 research, the WERF team conducted RBA tests using the LACSD JWPCP sludge as measured by additional VS destruction achieved following an additional 40-day digestion period in the laboratory anaerobic digesters. In this study, sludge samples from the JWPCP were initially digested in separate laboratory digesters at Virginia Tech for periods ranging from 10 to 40 days. RBA was measured using an additional 40-day digestion period following that initial digestion (thus the total digestion period ranged from 50 to 80 days).

Figure 3-17 depicts the correlation between VS reduction in the digester achieved during the initial digestion period (ranging from 10 to 40 days, as labeled in the figure) and the corresponding RBA for each data point, measured as the additional VS reduction percentage achieved following the additional 40 days of digestion of each sample. JWPCP sludge sample set 4 was used for this plot, though each of the four sample sets showed a similar RBA trend with respect to the VS reduction pattern observed: sludge samples that had a greater initial SRT (and VS reduction as shown in Figure 3-15) had a lower RBA (less additional VS destruction was achieved when digested an additional 40 days).

Figure 3-18 depicts a similar plot of the measured dewatered cake peak headspace TVOSC concentration versus RBA (additional VS destruction achieved in 40 days additional digestion), also for JWPCP sludge sample set 4. The plot indicates that the samples with higher original TVOSC (and in general the lowest SRT) experienced the highest amount of additional VS destruction in the additional 40 days of digestion.

These results suggest that RBA could be a useful indicator of the odor potential of anaerobicly digested sludge, as samples that were able to achieve greater VS destruction had higher TVOSC in the original digestion period.

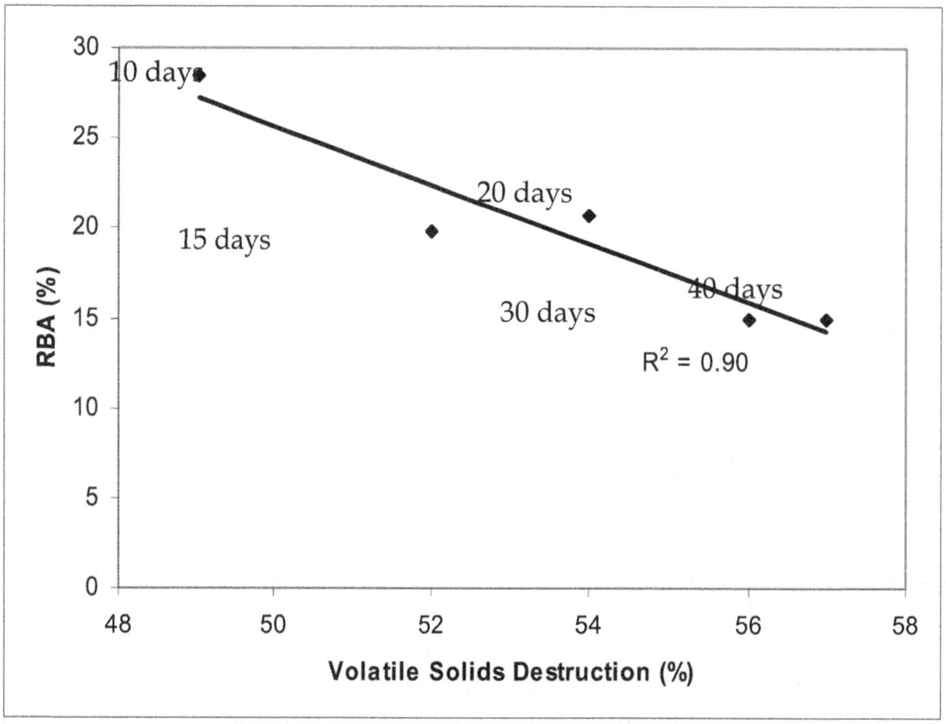

Figure 3-17. Relationship of Initial VS Destruction with Residual Biological Activity (Measured as Additional VS Destruction Achieved with 40 Additional Days of Digestion) for Biosolids from Digesters Operated at Different SRT (Initial SRT Shown for Each Point in Days). (Data for Sample Set 4).

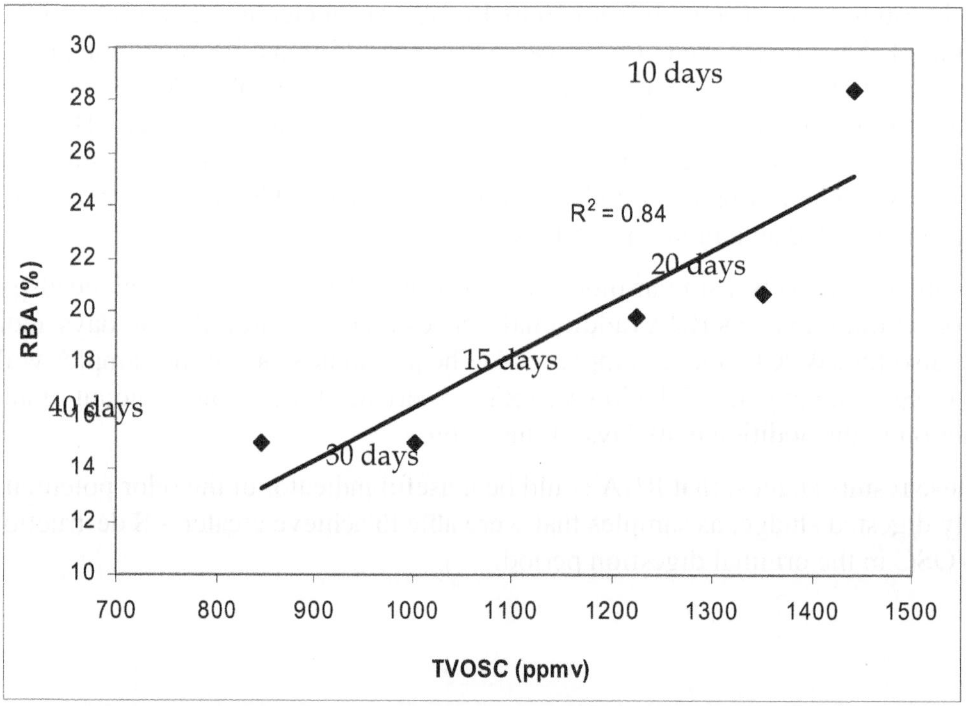

Figure 3-18. Relationship of Initial TVOSC with Residual Biological Activity (Measured as Additional VS Destruction Achieved with 40 Additional Days of Digestion) for Biosolids from Digesters Operated at Different SRT (Initial SRT Shown for Each Point in Days). (Data for Sample Set 4).

3.2.2.4 SRT Effects on Dewatering Rate and Solution Biopolymer Concentration

In this study, Virginia Tech also conducted SRT experiments that studied the impact of varying digester SRT on dewatering rate and protein and polysaccharide concentration, which is related to dewatering rate. The LACSD JWPCP sludge was used to conduct these experiments. Figure 3-19 depicts the impact of varying anaerobic digester SRT on the dewatering rate of digested sludge, measured as the capillary suction time (CST). The CST for the feed sludge was highest, measured at 80,000 seconds (data point not included in Figure 3-19). Measured CST then decreased with increasing digester SRT, leveling off at around 20 days of SRT.

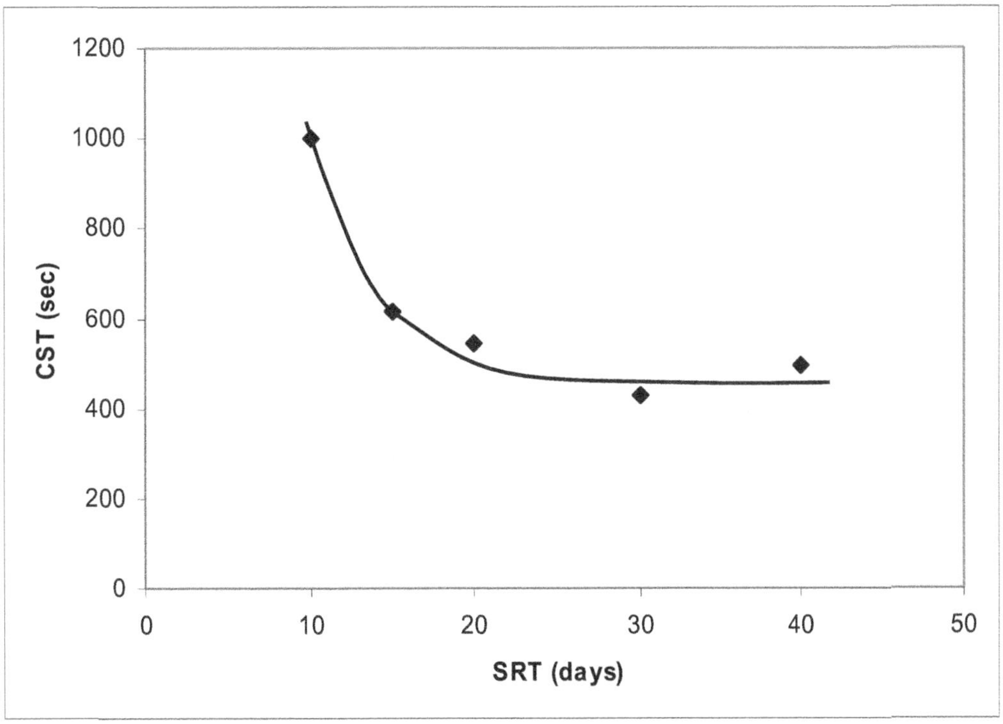

Figure 3-19. Digester SRT Impact on Dewatering Rate (Measured as CST). (Data for Sample Set 4).

Figures 3-20 and 3-21 depict the variation in the concentration of solution proteins and polysaccharides, respectively, in the digested sludge as a function of digester SRT. Both solution protein and polysaccharide concentrations decreased with an increase in SRT, also leveling off after about 20 days of SRT. Since some of these constituents are colloidal and interfere with sludge filtration, there is also a resulting higher CST (as is observed Figure 3-19). During days 10 to 40 of the digestion period, much of the soluble and colloidal material (and a smaller amount of solids, as depicted in Figure 3-15) are degraded. As a result of this degradation, the CST decreases (Figure 3-19).

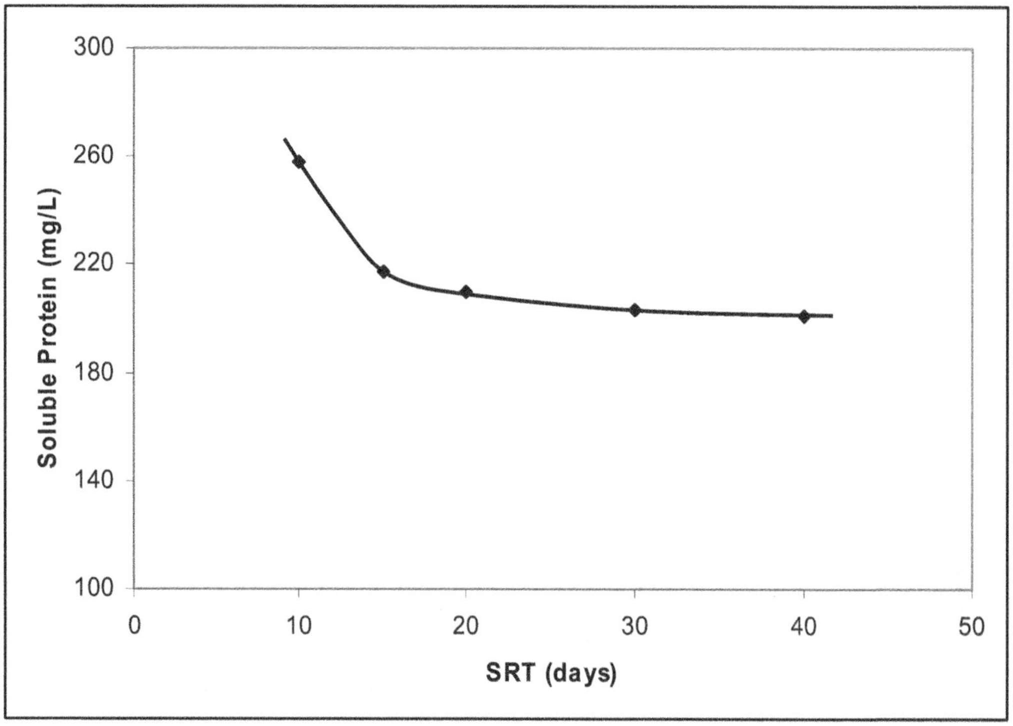

Figure 3-20. Protein Concentration in Solution Variation with Digester SRT. (Data for JWPCP Sample Set 4).

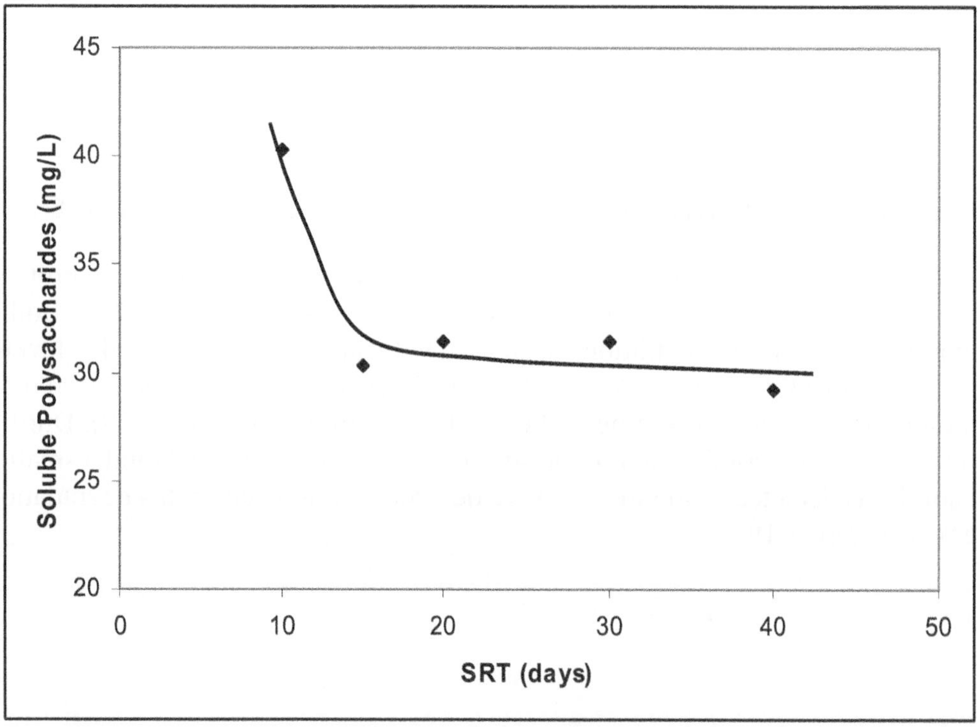

Figure 3-21. Polysaccharide Concentration in Solution Variation with Digester SRT. (Data for JWPCP sample set 4).

3.2.2.5 Exocellular Polymeric Substances Effect on Odors

Virginia Tech also conducted an exocellular polymeric substances (EPS) extraction study that studied the relationship of varying digester SRT on extractable proteins in JWPCP sludge samples. Three types of extraction of proteins were used in this study:

- The cation exchange resin (CER) method extracts calcium and magnesium-bound material
- The sulfide extraction method extracts iron-bound material using sulfide addition
- The BASE method extracts aluminum bound material using a strong base.

A plot of the measured data from JWPCP sample set 1 is depicted in Figure 3-22. The figure indicates that proteins extracted by the sulfide extraction method (iron-bound material) decreased with increasing SRT, while the CER-extractable protein content (calcium and magnesium-bound material) did not vary with SRT. The figure indicates that along with the measured decrease in odor potential with increasing SRT (shown by the reduction in TVOSC in Figure 3-16), there is also a corresponding degradation of iron-bound protein.

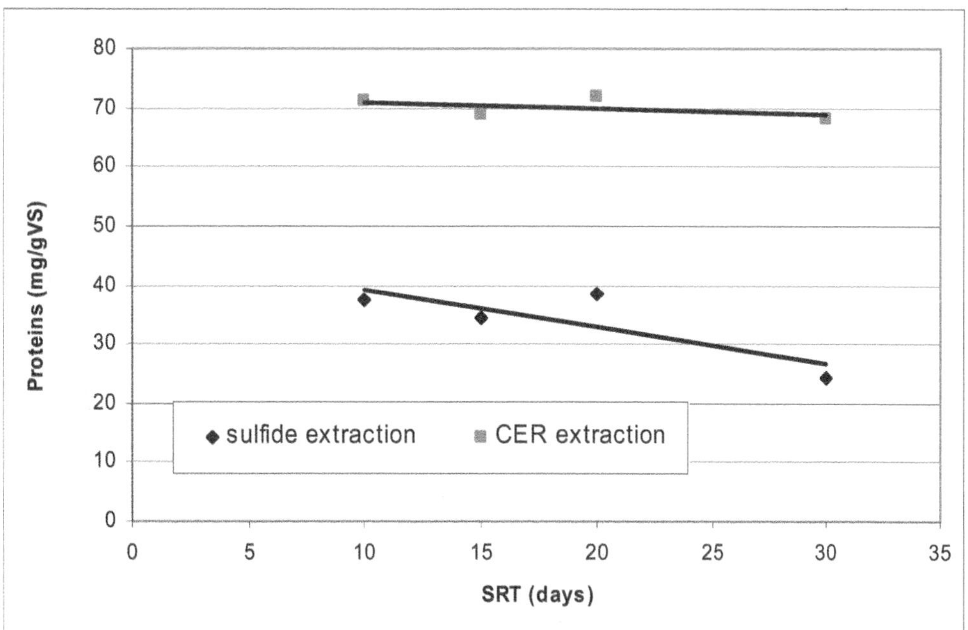

Figure 3-22. Variation of Measured Proteins with Digester SRT. (Data for sample set 1).

The importance of iron-bound protein in anaerobic digestion observed in the Phase 3 research is in accordance with a previous study (Novak et al., 2003). This study showed that release of iron-bound protein by reduction of iron under anaerobic conditions and its further degradation is related to the digestibility of sludge in an anaerobic digester. The sulfide extraction data from the Phase 3 observations (Figure 3-22) further implies that there are additional proteins bound with iron in anaerobicly digested sludges, and that this material leads to the generation of odor in dewatered cakes.

Figure 3-23 shows the protein concentrations extracted from the feed sludge and from the sludge following 40 days of digestion, using the three extraction techniques. The data in the figure illustrates that anaerobic digestion was closely related to the degradation of the iron-bound protein fraction (the sulfide-extractable protein).

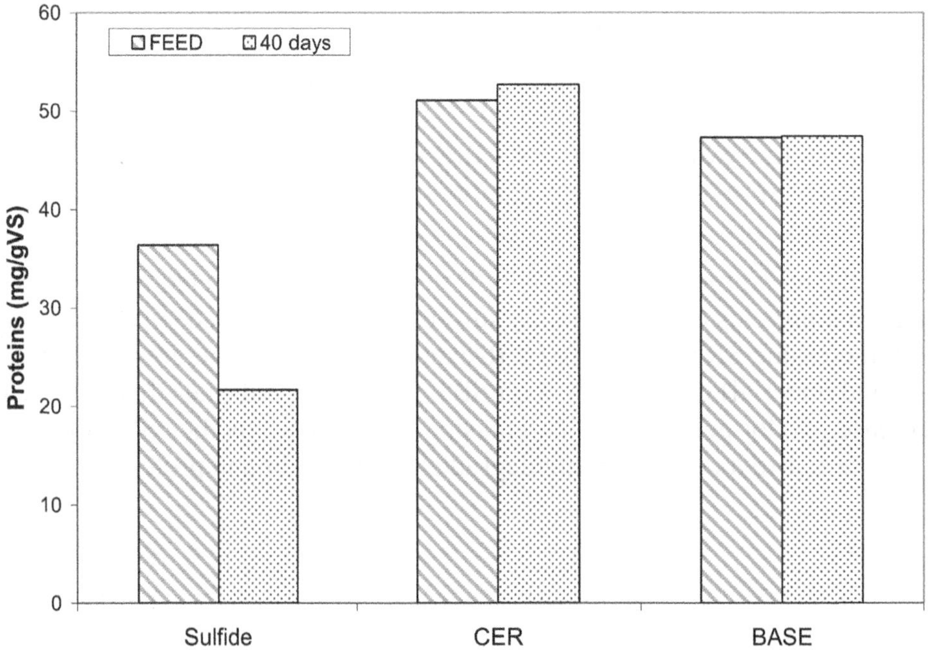

Figure 3-23. Proteins Extracted from the JWPCP Feed Sludge and After Digestion by Different Extraction Methods.

While proteins extracted using the CER and base extraction methods did not change before and after digestion, almost 45% of the Fe-associated protein was removed after 40 days of anaerobic digestion. This suggests that iron-linked proteins were selectively and continuously degraded under anaerobic conditions. This result is also in accordance with earlier studies that iron and the iron-associated proteins play an important role in determining the digestibility of sludge under anaerobic conditions (Park et al., 2006). The WERF team believes that this is most likely because Fe^{3+} undergoes reduction under anaerobic conditions, reducing the strength of the binding between iron and proteins. This increases the bioavailability of proteins and also makes them more susceptible to dispersion by shear.

Figure 3-24 depicts a plot of peak TVOSC against sulfide-extracted protein concentrations for two sets of JWPCP samples. For sludge from both data sets, the extracted protein concentrations correlated well with the peak TVOSC headspace concentrations: TVOSC concentration increased with an increase in the amount of proteins bound to iron. This suggests that the sulfide extraction test, hence the iron-bound protein content, may be a useful tool for determining if a sludge is likely to generate odors following dewatering. Sulfide-extracted protein in each sample also correlated with the digestion SRT, or VS reduction (samples extracted from digesters operating at different SRTs, feed sludge being the same).

Figure 3-25 contains a plot of sulfide-extracted protein concentration versus RBA for JWPCP set 4 sludge samples. The plot indicates that sulfide-extraction protein concentrations in the sludge correlated well with the RBA data. This observation suggests that the materials associated with iron that are left un-degraded after anaerobic digestion are degraded during the extended digestion period of 40 days.

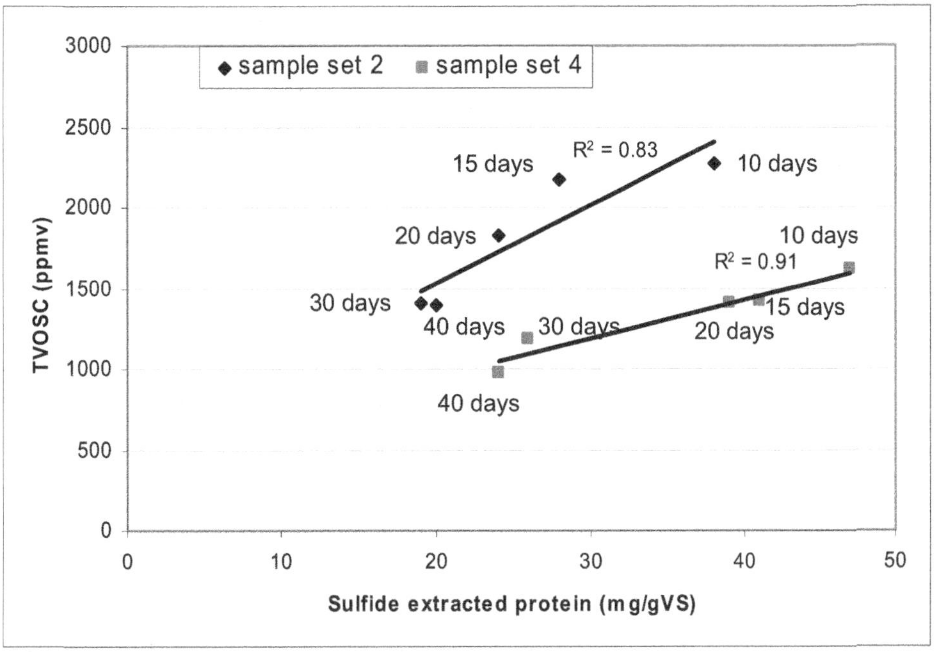

Figure 3-24. Sulfide-Extracted Protein Relationship with Peak TVOSC (JWPCP Sludge Samples Dewatered Using Laboratory Scale High Solids Centrifuge Simulation Technique).

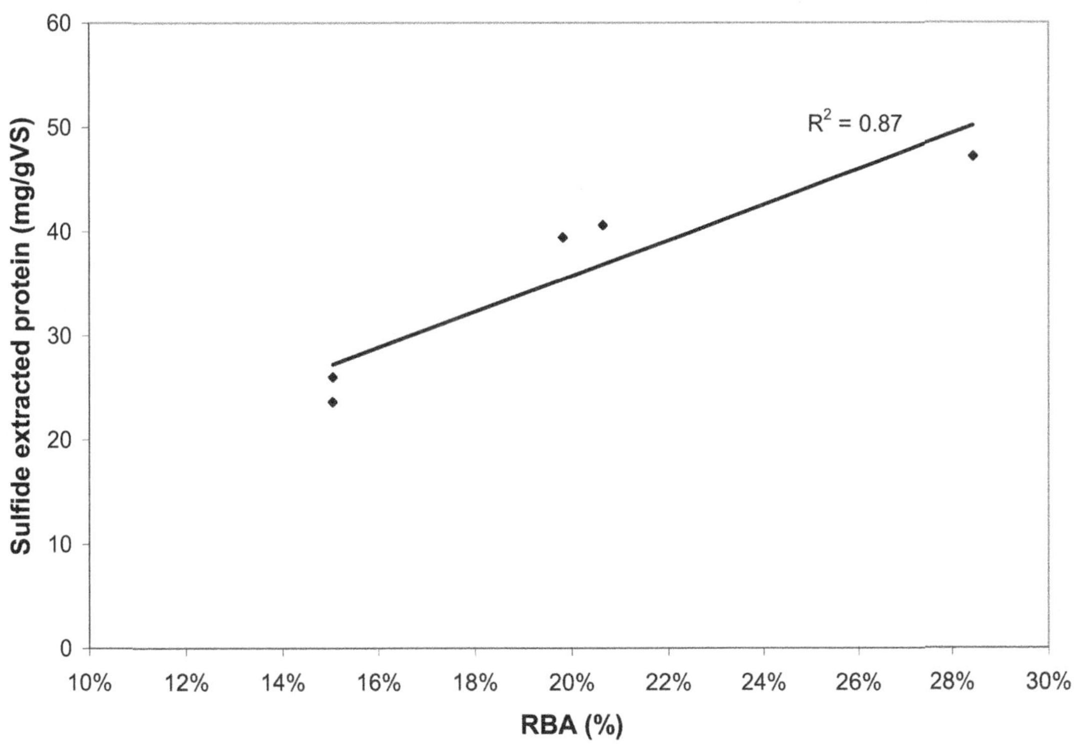

Figure 3-25. Sulfide Extracted Protein Versus RBA. (Data for JWPCP Sample Set 4).

3.2.2.6 Digestion SRT Study Summary and Conclusions

This aspect of the Phase 3 work was conducted to determine the effect of anaerobic digester SRT on the odor generation potential of dewatered biosolids cakes. A blend of 75% primary sludge and 25% WAS – the same mixture as in the LACSD Joint Water Pollution Control Plant – was digested in lab scale anaerobic digesters operated at SRT ranging from 10 days to 40 days. Residual biological activity (RBA) was measured by digesting biosolids samples for additional 40 days in each laboratory digester.

Digested biosolids were tested for odor generation using the centrifuge simulation method developed at Virginia Tech. In addition, Virginia Tech conducted EPS extraction studies whose goal was to improve understanding of what fractions of EPS are related to odor production in dewatered biosolids cakes.

The main conclusions that have been drawn from the Phase 3 research laboratory SRT studies are as follows:

1. Higher volatile solids reduction is achieved by operating anaerobic digesters at higher SRT. However, for the JWPCP sludge samples tested, the majority of the VS reduction occurred in the first 10 days of digestion (approximately 50% of the total 58% reduction achieved in a 40-day digestion period).
2. Dewatered cake headspace TVOSC concentration decreased with increasing laboratory anaerobic digester SRT for all four JWPCP sample sets studied. However, high TVOSC concentrations were observed even for biosolids cake from the digesters operated at 40 days detention time (as high as 1,000 ppmv). The digester SRT and cake TVOSC reductions also correlated with increases in VS reduction during the incubation period in the digesters.
3. RBA testing results indicated a reasonable correlation with VS reduction and also correlated well with the initial headspace TVOSC concentration measured following the initial (10 to 40 days) of digestion.
4. Lower cake headspace TVOSC concentrations were consistently observed from samples that had a lower cake solids concentration. This suggests that solids concentration influences the generation of TVOSC.
5. Amending samples with BESA produced better repeatability of results, suggesting that methanogenic activity has a key role in reducing sulfur odors.
6. Proteins extracted by different chemical methods behaved in a considerably different manner during anaerobic digestion. Headspace TVOSC generation correlated with the EPS bound to iron. As the iron-bound protein increased, the TVOSC concentration in the headspace increased. This finding was corroborated with an observation that lower solution protein concentrations were found in samples from digesters operated at higher SRT.

Further research needs to be conducted to identify the processes involved in sulfur gas generation and degradation from dewatered biosolids cake. The extraction data from the Phase 3 research suggests that a portion of iron-bound material (primarily proteins) are degraded during anaerobic digestion, leaving the remaining amount of iron-bound material bioavailable in the presence of shear (which may be imparted in the dewatering process). Once bioavailable, this material degrades to generate VOSC and produce odors.

3.2.3 CEBA Test Results

CEBA addition laboratory testing was performed at the Virginia Tech and Bucknell laboratories using the protocols discussed in Chapter 2.0 of this report. The following sections summarize the testing results for each type of CEBA product, categorized by pre-digestion applied CEBAs (testing conducted at Virginia Tech) and post-digestion or post-dewatering applied CEBAs (testing conducted at Bucknell University).

3.2.3.1 Pre-Digestion CEBA Application Testing Results

Four CEBAs (denoted as CEBA numbers 1, 4, 5, and 6) were added to sludge prior to laboratory anaerobic digestion at Virginia Tech. A summary of the Specific CEBA Test Protocol (SCTP) for each CEBA is contained in Appendix A. The Pepper's Ferry WWTP in Pepper's Ferry, VA provided primary and WAS samples for use in this CEBA testing, as described in Chapter 2. The detailed results for each pre-digestion CEBA are provided in the following sections:

CEBA No .1: CEBA No. 1 was added to the digester feed sludge according to the SCTP, which was developed by the WERF research team and approved by the vendor. The staged batch digester was operated according to the SCTP and the General CEBA Testing Protocol (GCTP). The total digestion period was 35 days, including two feed cycles.

Testing Protocol Synopsis

A summary of the testing protocol for CEBA No. 1 is as follows:

1. Initially, the digester contents were 2 liters of anaerobicly digested biosolids (to serve as seed sludge) plus eight 8 liters of a 50/50 ratio of primary sludge and WAS by dry weight.
2. The feed sludge received 100 mL of the CEBA No. 1 additive. The 100 mL was mixed with one 1 liter of sludge and then this liter of dosed sludge was mixed with the other nine 9 liters of sludge.
3. After the first 7 days of digestion, half the contents of the digester were removed and replaced with a 50/50 mix of primary sludge and WAS. Then 100 mL of CEBA No. 1 was added to the sludge in the same manner as before.
4. After another 7 days (on Day 14) the 50% feed sludge removal and replacement cycle was repeated, and 325 mL of CEBA No. 1 was added to the feed sludge.
5. The digester was operated for an additional 21 days (making a 35-day total testing period), and the digested sludge was then sheared, dewatered, stored, and analyzed for headspace TVOSC and other parameters per the GCTP.
6. The digester performance and TVOSC generation of the stored biosolids cake were compared to a set of control digesters that were operated in the same manner, but did not contain the CEBA additive.

Digester Performance with CEBA No. 1

The results of the digester performance with the addition of CEBA No. 1 compared to the controls (no CEBA addition) are shown in Figures 3-26 through Figure 3-31. Solids reduction, ammonia production, gas production, pH, and volatile fatty acid (VFA) production (as measured by acetic acid production) are all indicators of digester performance.

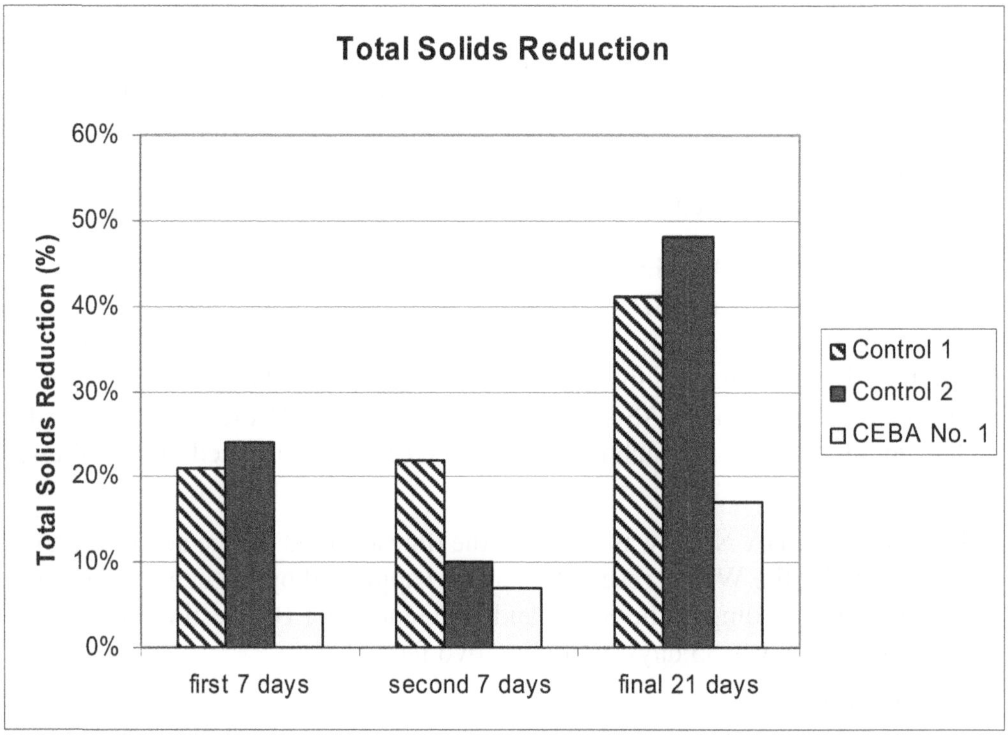

Figure 3-26. Total Solids Reduction in Laboratory Anaerobic Digesters (CEBA No. 1).

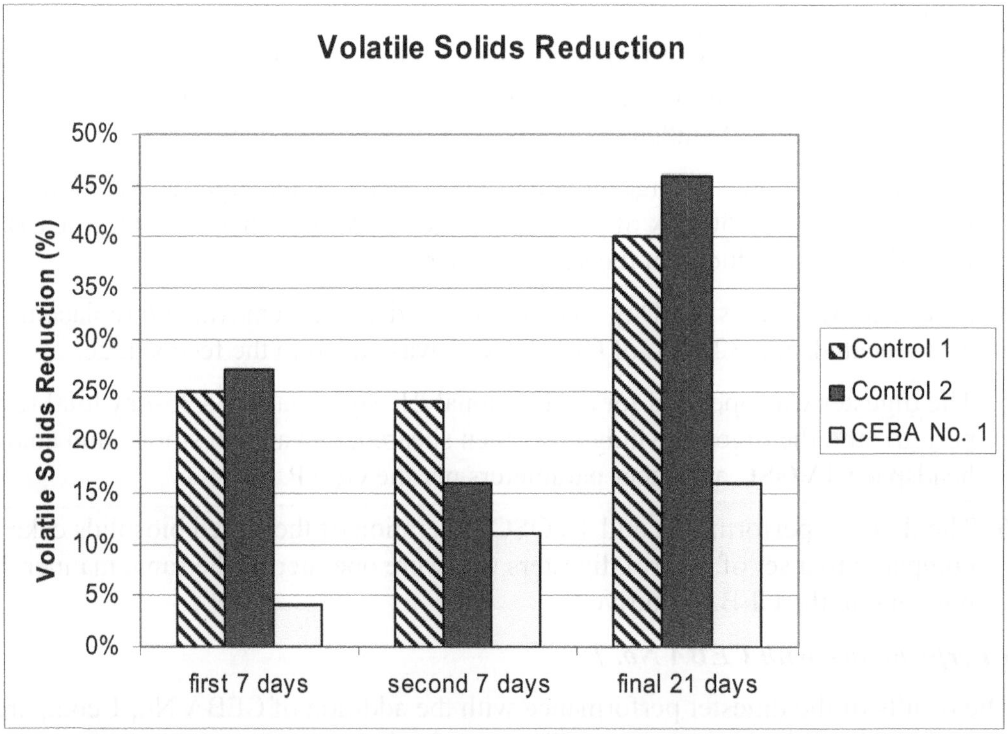

Figure 3-27. Volatile Solids Reduction in Laboratory Anaerobic Digesters (CEBA No. 1).

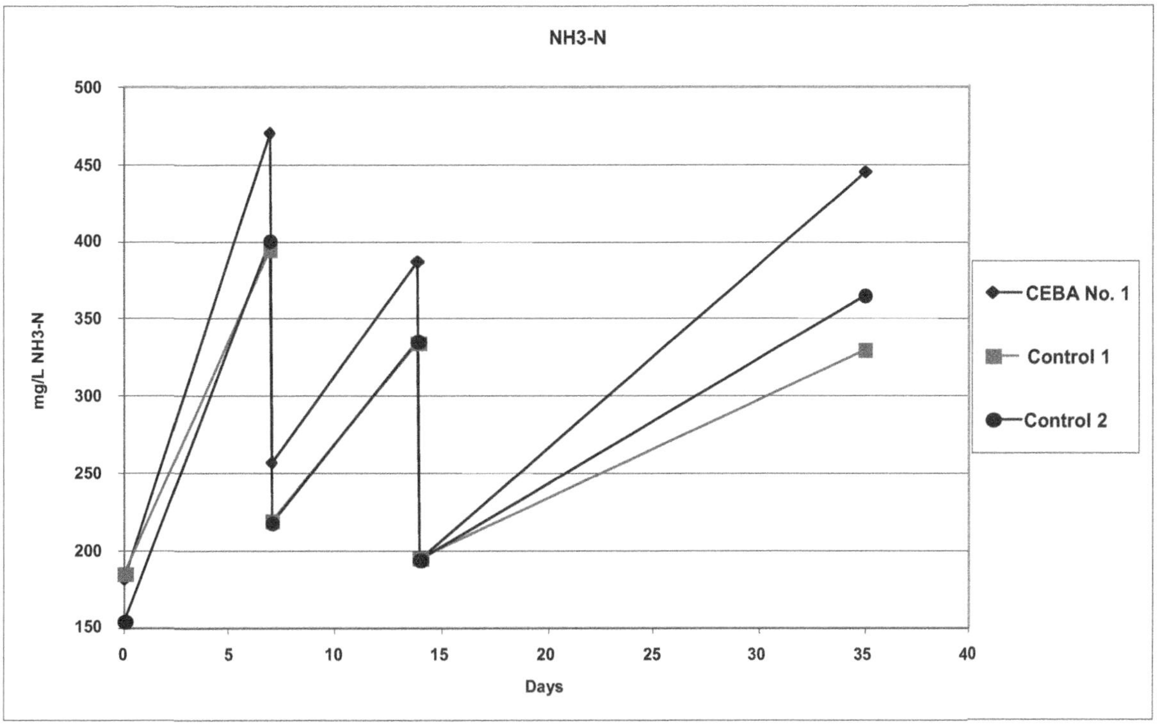

Figure 3-28. Ammonia-Nitrogen Production During Digestion (CEBA No. 1 Testing).

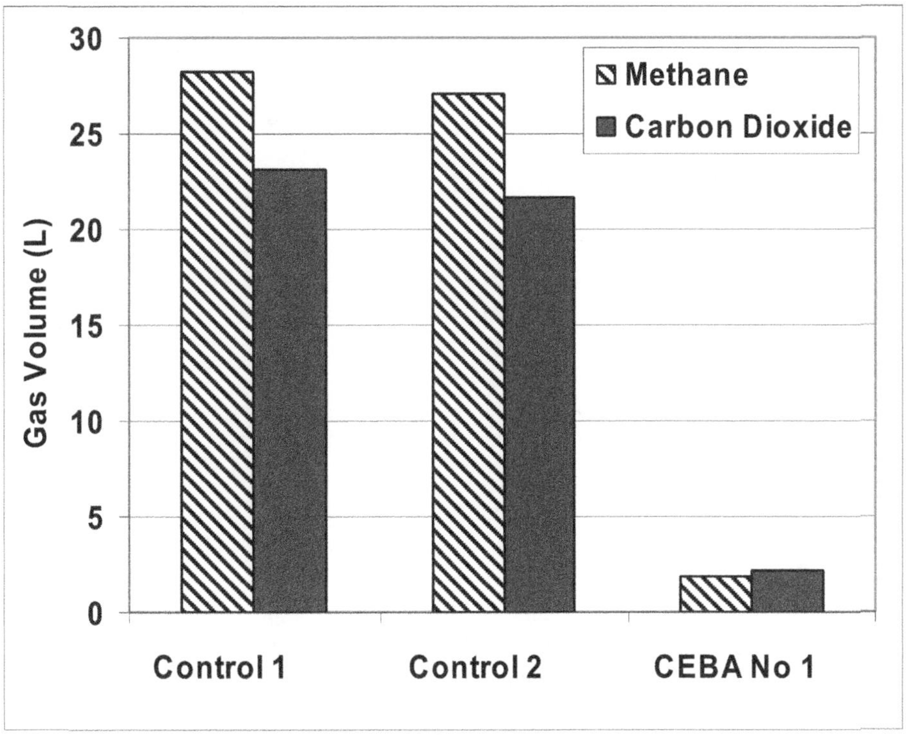

Figure 3-29. Gas Production from the Laboratory Anaerobic Digesters (CEBA No. 1 Testing).

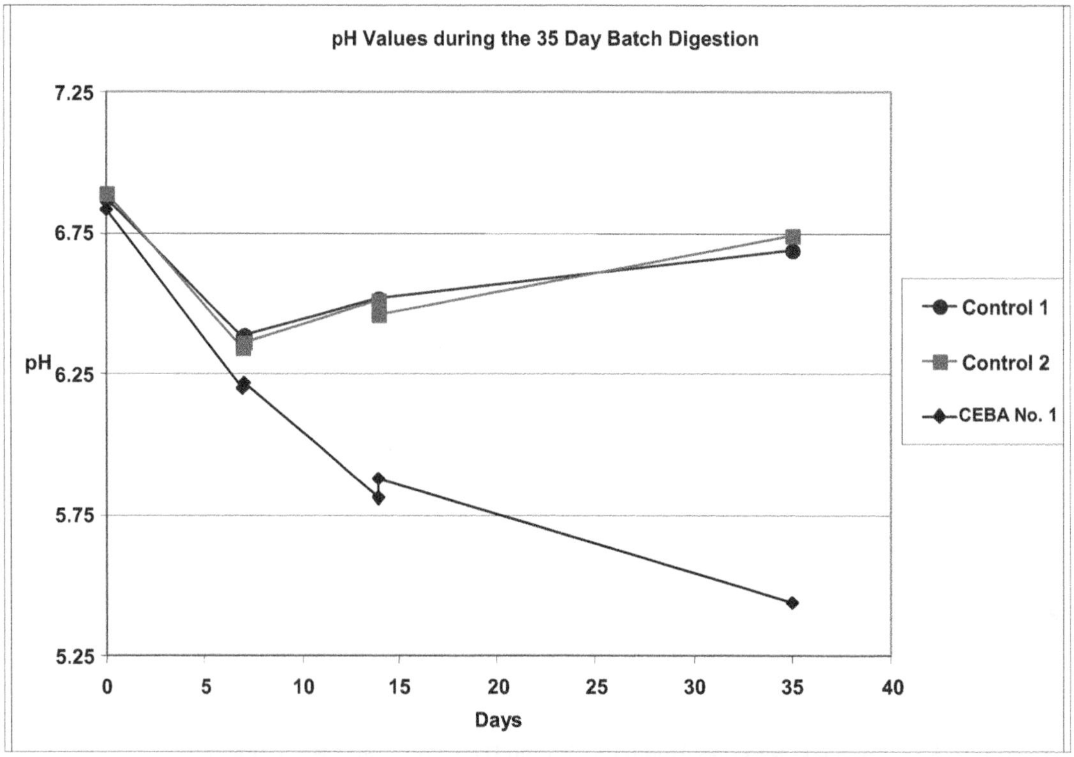

Figure 3-30. Digester pH Variation During Digestion (CEBA No. 1 Testing).

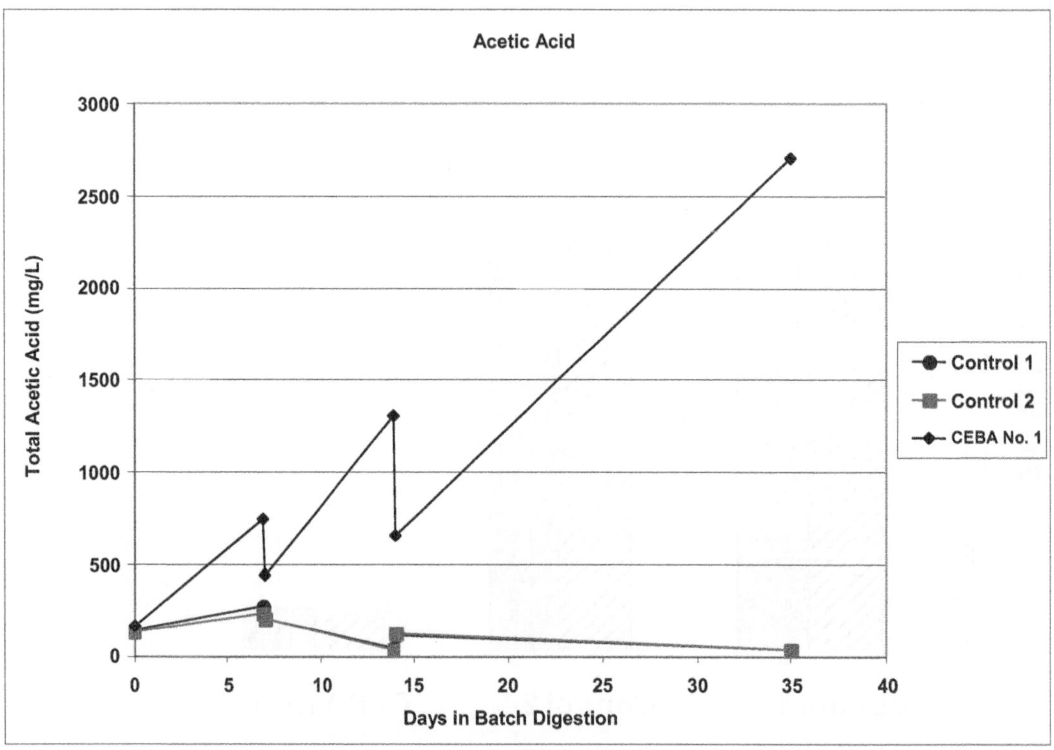

Figure 3-31. Effect of CEBA No. 1 on Acetic Acid Production.

It can be seen from the solids destruction data (Figures 3-26 and 3-27) that addition of CEBA No.1 did not improve digestion as evaluated by either total solids (TS) or volatile solids

(VS) reduction. In fact, both TS and VS reduction were dramatically *reduced* in the digesters containing CEBA No. 1 as compared to the two control digesters. The CEBA-treated sludge 35-day VS reduction was 18%, well below the minimum 38% required for compliance with the Vector Attraction Reduction requirement of the 40 CFR Part 503 rule.

There was a slight increase in ammonia nitrogen (Figure 3-28) for the CEBA No.1 digester, as compared to the control digesters, indicating some improved digestion. However, gas production (Figure 3-29) was significantly decreased in the CEBA No. 1 digester. The pH values in the CEBA digester (Figure 3-30) were significantly lower than both control digesters, with pH decreasing consistently throughout the 35-day digestion period. On the final day of digestion (Day 35) for the CEBA-amended sludge, the pH was 5.4, while the control digester sludges had a pH of approximately 6.7. Acetic acid production was also significantly different (Figure 3-31) in the CEBA No.1-dosed digester versus the control digesters, with a final (Day 35) acetic acid concentration of the CEBA-amended digested sludge of approximately 2,700 mg/L versus essentially zero acetic acid measured in the control digesters. This is an indication of VFA accumulation in the CEBA-amended digester.

CEBA No. 1 Impact on Headspace TVOSC

The generation of TVOSC in the headspace was evaluated using the shearing and dewatering method as described in Section 2.2 General Protocols. Results for the sample headspace TVOSC analyses are presented in Figures 3-32 and 3-33. The data for the BESA-dosed samples (Figure 3-36) shows the effect of inhibition of methanogenic organisms on TVOSC production, which reflects the maximum odor potential of the sludge. Results show that there was little TVOSC generated at any time during digestion in the digester amended with CEBA No. 1 as compared to the controls, which generated peaks of 70-90 ppmv.

CEBA No. 1 Overall Evaluation

Figures 3-32 and 3-33 indicate that CEBA No. 1 was highly effective at reducing odors (as represented by TVOSC) in the digested biosolids dewatered cake to essentially zero. However, the reason that there is so little TVOSC in the dewatered cake is that it appears that the organisms that generate all gases (methane, CO_2, and all VOSCs) were inhibited as a result of CEBA No. 1 addition. CEBA No. 1 was also characterized by slightly elevated ammonia levels in the sludge, significantly low pH in the digester and very high acetic acid production. During digestion, the breakdown of proteins into amino acids and removal of ammonia from the amino acids appears to have been increased by the addition of CEBA No. 1, which explains the elevated ammonia levels depicted in Figure 3-28. The degradation of proteins beyond this point appears to be inhibited by CEBA No. 1.

The increase in ammonia compared to the controls suggests that CEBA No. 1 might possibly be effective under aerobic conditions or with pH buffering under anaerobic conditions, and possibly if added at a different concentration. However, in this anaerobic digestion application, it appears that the high dosage (5 ppmv per volume of digester, or 40,000 ppmv) of CEBA No. 1 resulted in a low pH, which inhibits those organisms that are responsible for gas production. This resulted in low VS destruction, low gas production, and accumulation of VFAs, which are indications of a "stuck" (biologically inactive) digester.

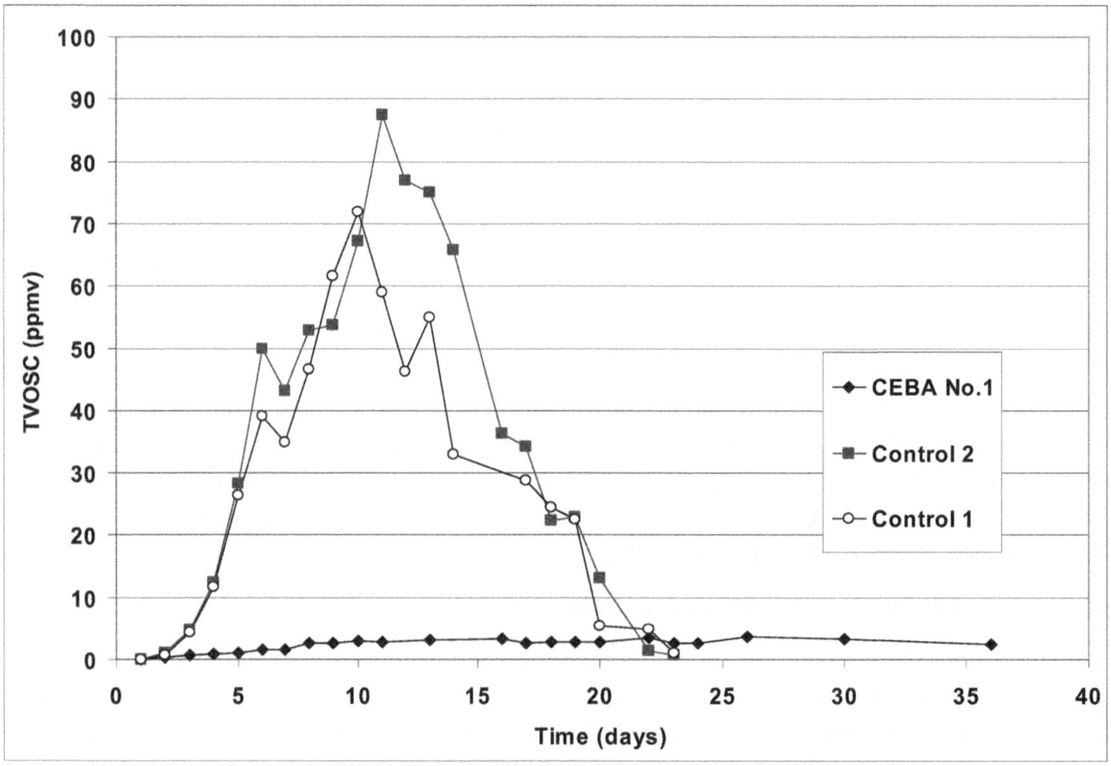

Figure 3-32. CEBA No. 1 Testing: Headspace TVOSC Concentrations (No BESA Added).

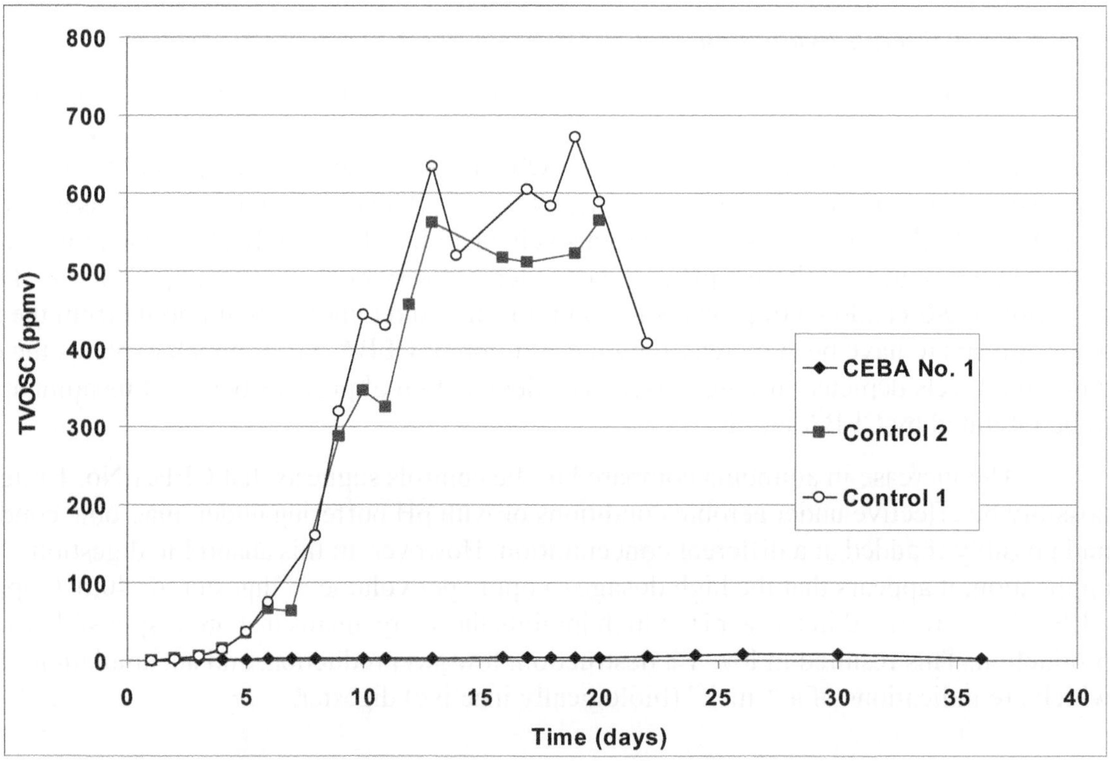

Figure 3-33. CEBA No. 1 Testing: Headspace TVOSC Concentrations (BESA Added).

CEBA No. 4: CEBA No. 4 was added to the digester feed sludge according to the SCTP, which was developed by the WERF research team and approved by the vendor. The staged batch digester was operated according to the approved SCTP and the GCTP. The total digestion period was 35 days with two feed cycles.

Testing Protocol Synopsis

A summary of the testing protocol for CEBA No. 4 is as follows:

1. Initially, the digester contents were 2 liters of anaerobicly digested biosolids (to serve as seed sludge) and 8 liters of digester feed sludge, which was a 50/50 ratio of primary sludge and WAS by dry weight.
2. The CEBA No. 4 product was added to the digester at a daily dosage rate of 5 ppmv (per digester volume). The CEBA No. 4 product was diluted by a ratio of 1:100 with deionized water prior to its addition to the feed sludge. Therefore, for a 10-liter digester, the daily dosage of CEBA No. 4 product was 5mL of diluted CEBA No. 4 per day.
3. After 7 days, 50% of the contents were removed and replaced with a 50/50 mixture of primary sludge and WAS. This cycle was repeated after another 7 days (on Day 14 of the digestion period).
4. The digester was operated for an additional 21 days (making a 35-day total testing period), and the digested sludge was then sheared, dewatered, stored, and analyzed for headspace TVOSC and other parameters per the GCTP. After each sludge exchange, the contents of the digester were purged with nitrogen to insure the maintenance of anaerobic conditions.
5. The treated digester performance and odor generation from the dewatered sludge cakes were compared to a pair of control digesters that were operated in the same manner but did not contain CEBA No. 4.

CEBA No. 4 Digester Performance

The results of the digester performance compared to the controls are shown in Figure 3-34 to Figure 3-39. Figures 3-34 and 3-35 show that CEBA No. 4 overall did not improve digestion as measured by either TS or VS reduction, as both controls had higher reduction after 35 days of digestion. The sludge amended with CEBA No. 4 also did not meet the 38% requirement for compliance with the Part 503 rule (achieving 35% reduction after 35 days).

There was a slight increase in gas production for the CEBA No. 4 dosed digester (Figure 3-37), indicating some improved digestion. For all other digestion indicators plotted in the figures below, there was little difference between the CEBA No. 4-dosed and the control digesters.

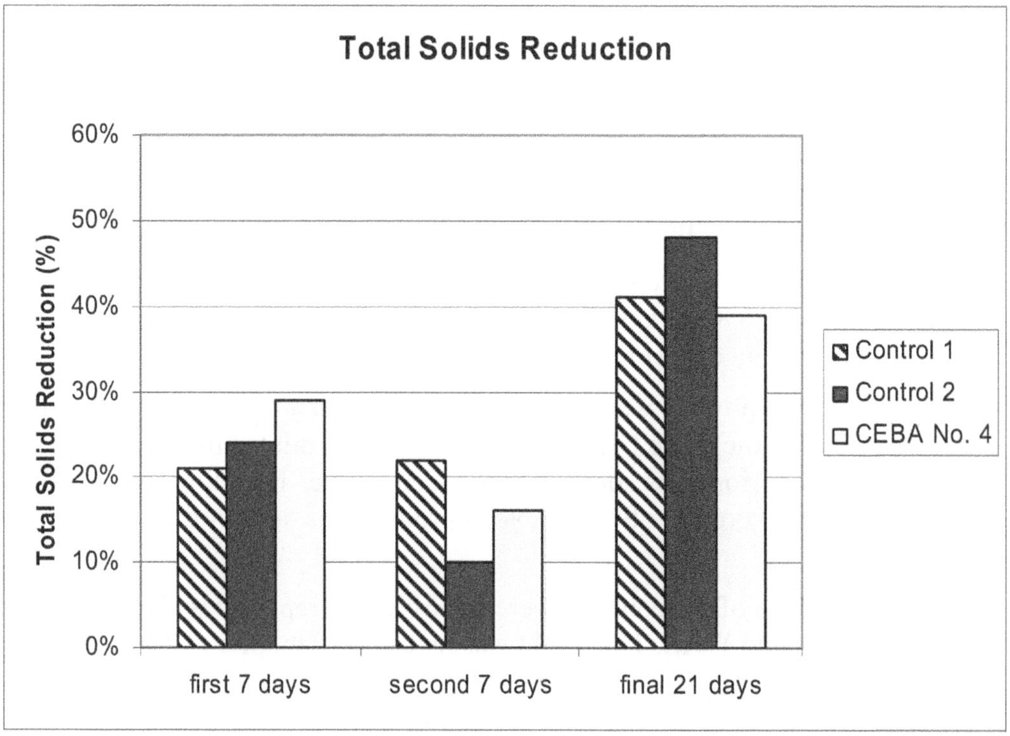

Figure 3-34. Total Solids Reduction in Laboratory Anaerobic Digesters (CEBA No 4).

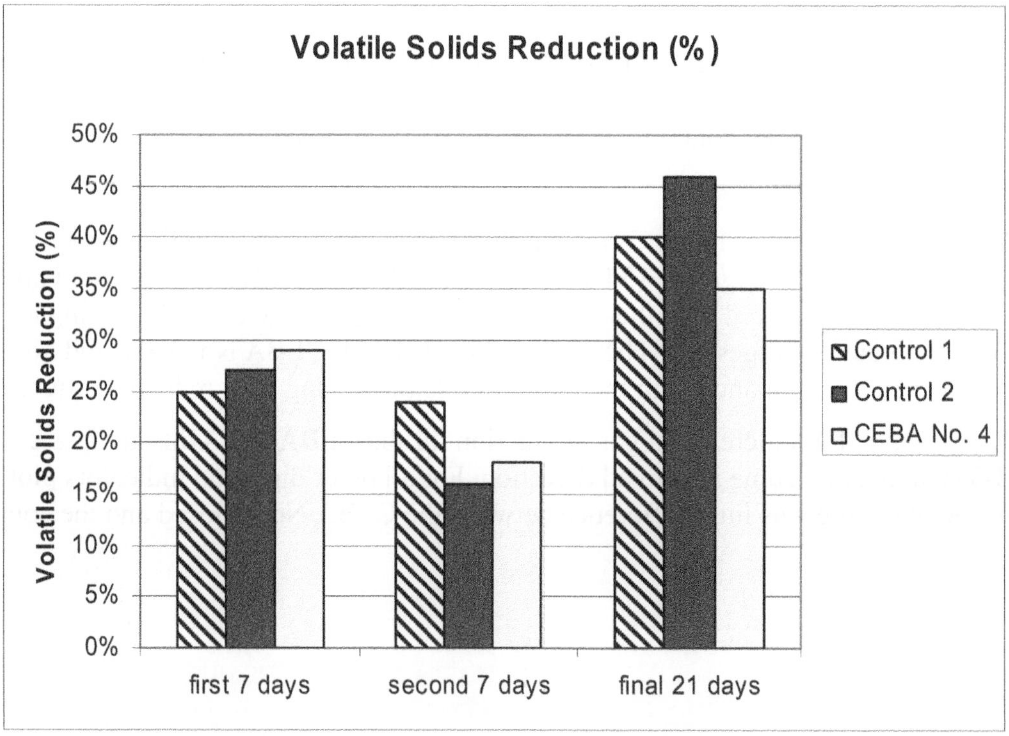

Figure 3-35. Volatile Solids Reduction in Laboratory Anaerobic Digesters (CEBA No 4).

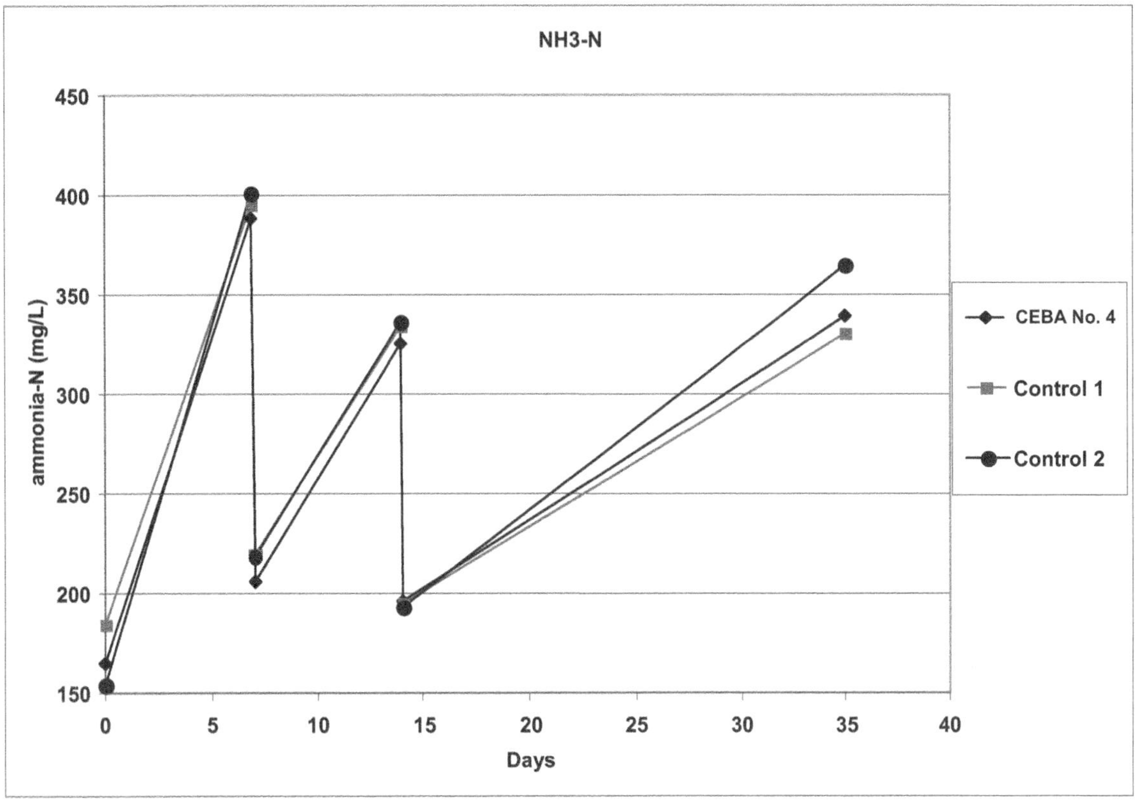

Figure 3-36. Ammonia-Nitrogen Production During Digestion (CEBA No. 4 Testing).

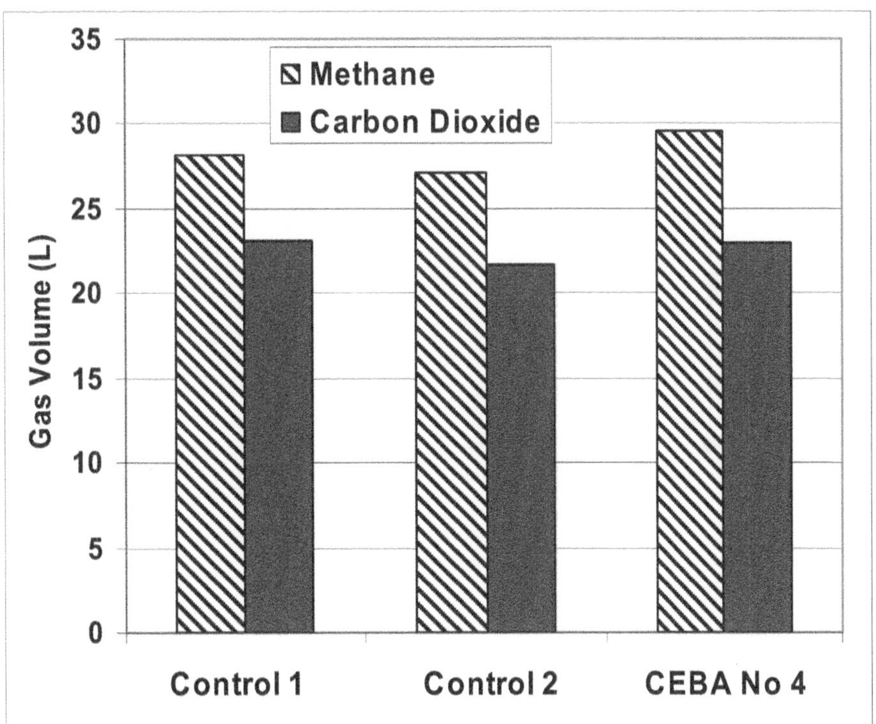

Figure 3-37. Gas Production from the Laboratory Anaerobic Digesters (CEBA No. 4 Testing).

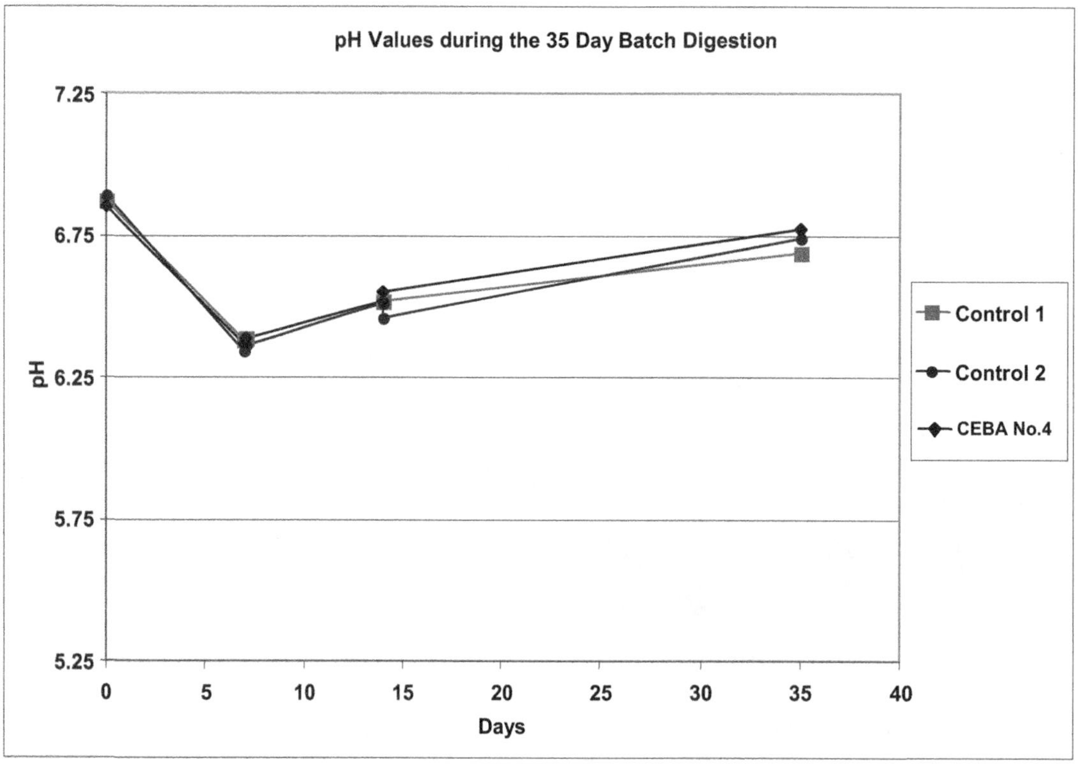

Figure 3-38. Digester pH Variation During Digestion (CEBA No. 4 Testing).

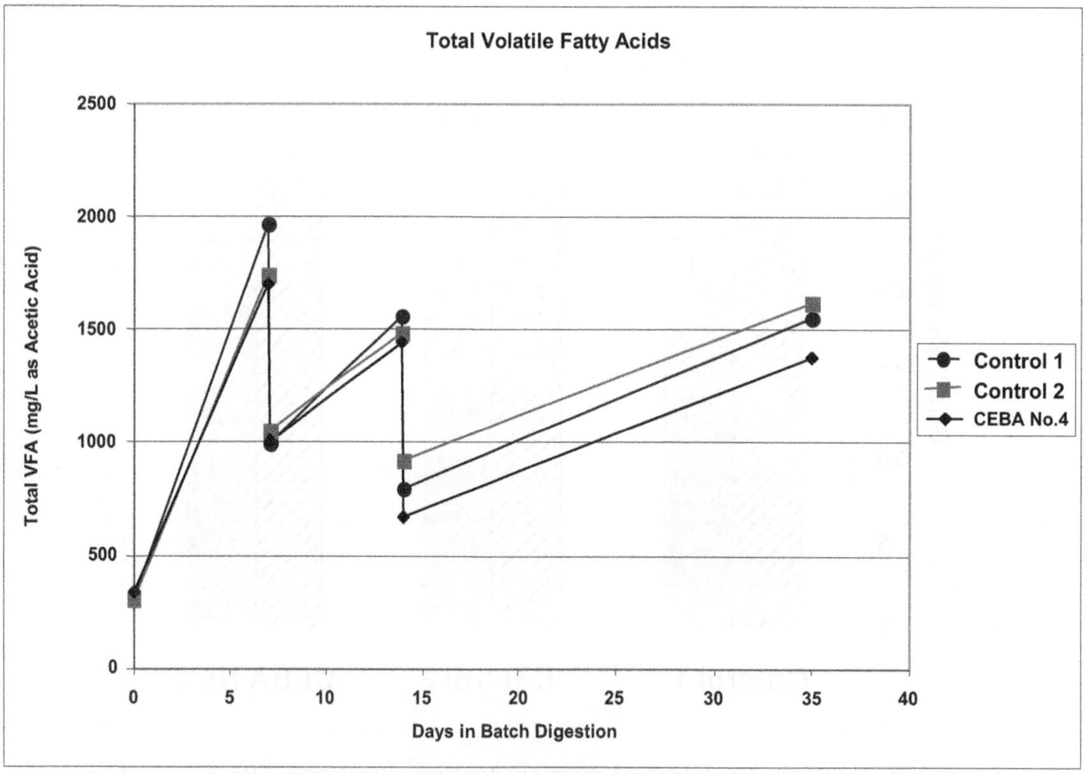

Figure 3-39. Effect of CEBA No. 4 on Acetic Acid Production.

CEBA No. 4 Impact on Headspace TVOSC

The generation of TVOSC in the headspace was evaluated using the shearing and dewatering method as described in Section 2.2 General Protocols. The results for the headspace TVOSC analyses are presented in Figure 3-40 and Figure 3-41. The figures show that there was little impact on the TVOSC odor potential generation for the CEBA No. 4-dosed digester as compared to the control digesters. In addition to the data shown, hydrogen sulfide headspace measurements were also collected, which also showed no difference between the control digesters and the CEBA No. 4–dosed digester.

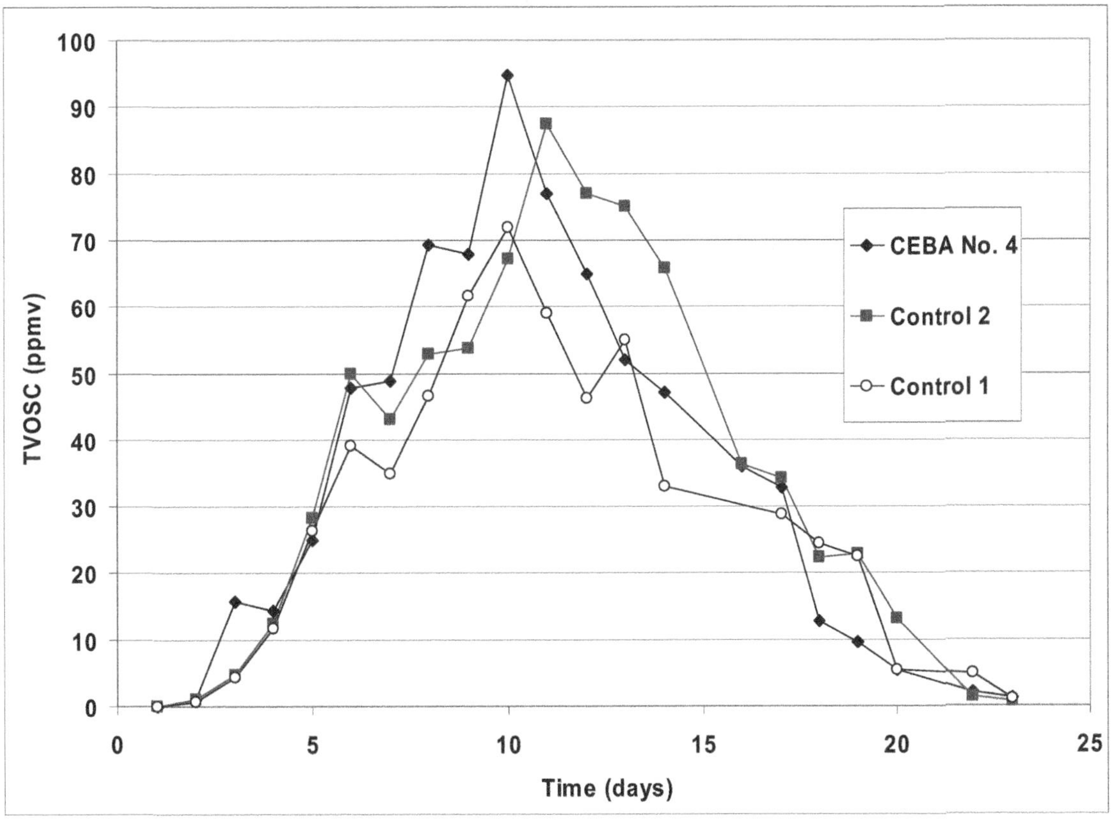

Figure 3-40. CEBA No. 4 Testing: Headspace TVOSC Concentrations (No BESA Added).

CEBA No. 4 Overall Evaluation

The results of only two tests performed in the analysis of CEBA No. 4 showed a difference between the digester that received CEBA No. 4 and the control digesters: solids reduction and gas production. Both TS and VS reduction were decreased in the CEBA-amended digester, with VS reduction below the 38% Part 503 requirement. Gas production (methane and carbon dioxide) was increased marginally over controls.

Because of the detrimental effect on TS and VS reduction and the fact that CEBA No. 4 showed no measurable improvement in TVOSC reduction over controls, CEBA No. 4 is not recommended for odor control in anaerobic digestion. The WERF team notes that it is possible that CEBA No. 4 could provide benefits if it were added at a different concentration, but modifications in concentration were not made in the Phase 3 research (the concentration used was the level prescribed by the vendor).

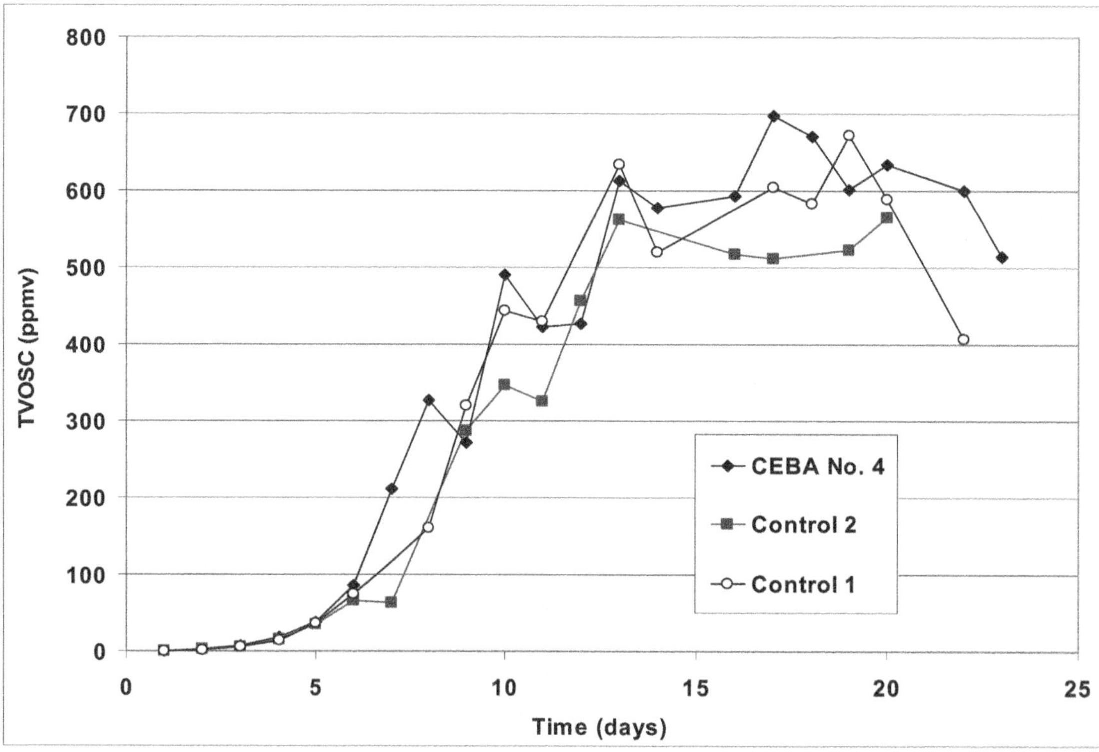

Figure 3-41. CEBA No. 4 Testing: Headspace TVOSC Concentrations (BESA Added).

CEBA No. 5: CEBA No. 5 product added to the feed sludge according to the SCTP, which was developed by the WERF team and approved by the vendor. The staged batch digester was operated according to the SCTP and the GCTP. The total digestion period was 35 days with two feed cycles.

Testing Protocol Synopsis

A summary of the testing protocol for CEBA No. 5 is as follows:

1. Initially, the digester liquid contents were 2 liters of anaerobicly digested biosolids (to serve as seed sludge) and 8 liters of a 50/50 ratio of primary and waste-activated sludges by dry weight.
2. When adding the sludge to each reactor, CEBA No. 5 was combined with the sludge using the SCTP specified dosage and dilution. The CEBA No. 5 product was added at 5 grams per liter of feed sludge or 5000 ppmv and mixed thoroughly by hand.
3. After 7 days, half of the digester contents were removed and replaced with a 50/50 mixture of primary and secondary sludge, again containing the agent at 5 grams per liter of sludge. This cycle was repeated after another 7 days (on Day 14).
4. Following the second sludge exchange, the digester was operated for an additional 21 days (making a 35-day total testing period), and the digested sludge was then sheared, dewatered, stored, and analyzed for headspace TVOSC and other parameters per the GCTP.

5. The treated digester performance and odor generation from the dewatered sludge cakes were compared to a pair of control digesters that were operated in the same manner but did not contain CEBA No. 5.

CEBA No. 5 Digester Performance

The results of the digester performance compared to the controls are shown in Figure 3-42 through Figure 3-47. It can be seen from the solids destruction data in Figures 3-42 and 3-43 that CEBA No. 5 significantly improved digestion as indicated by average increases (over the controls) in solids reductions of about 31% for total solids (TS) reduction and about 35% for volatile solids (VS) reduction. This was accompanied by a slight increase in ammonia-nitrogen production (Figure 3-44) and an average 18% increase in methane production (Figure 3-45), as compared to the two control digesters. These results indicate substantially improved digestion for the CEBA-amended digester.

In the CEBA-amended digester, the pH was higher than the controls (Figure 3-46), which was more favorable for digestion. Figure 3-47 indicates that the acetic acid concentration was higher in the feed sludge after adding CEBA No. 5 (before digestion commenced). Additionally, acetic acid production was significantly higher in the measurement taken after 14 days of digestion. This data suggests that CEBA No. 5 improves hydrolysis of complex organic matter and the generation of volatile fatty acids, which may account for the improved digestion. It should also be noted that, although the CEBA additive increased the volatile fatty acid concentration, it did not reduce the pH.

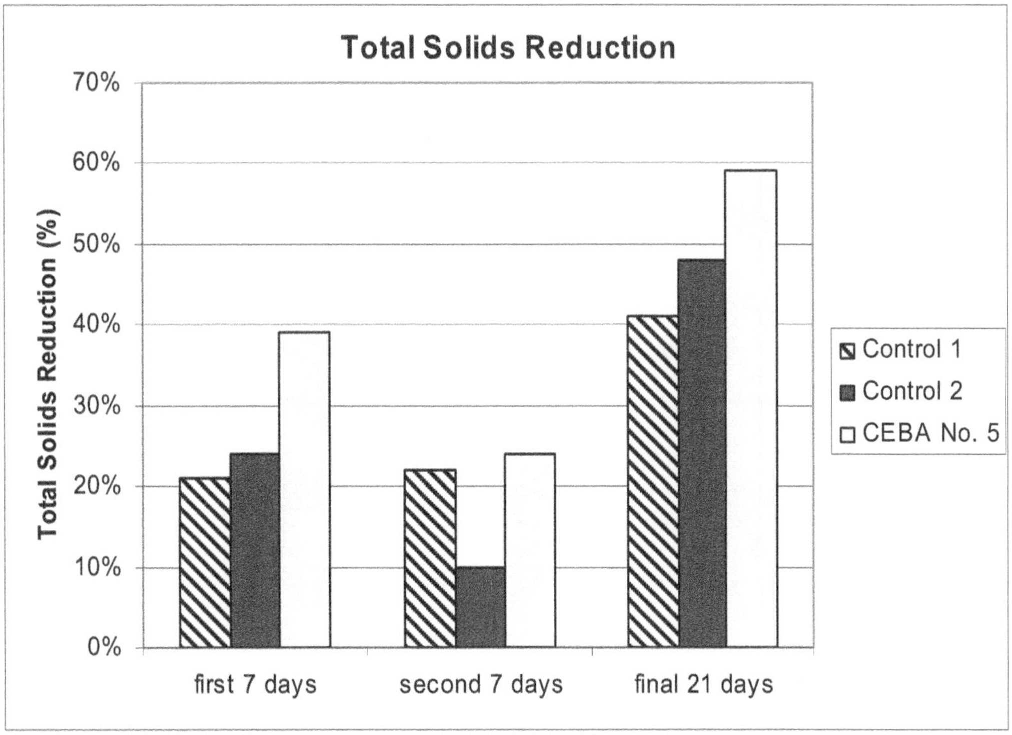

Figure 3-42. Total Solids Reduction in Laboratory Anaerobic Digesters (CEBA No. 5).

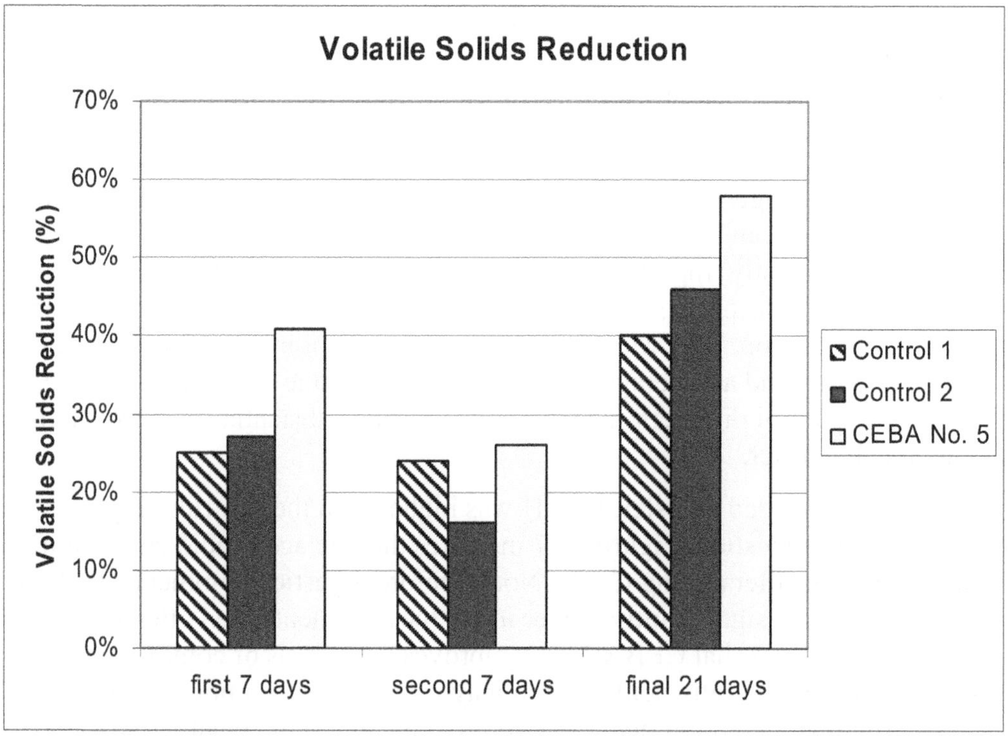

Figure 3-43. Volatile Solids Reduction in Laboratory Anaerobic Digesters (CEBA No. 5).

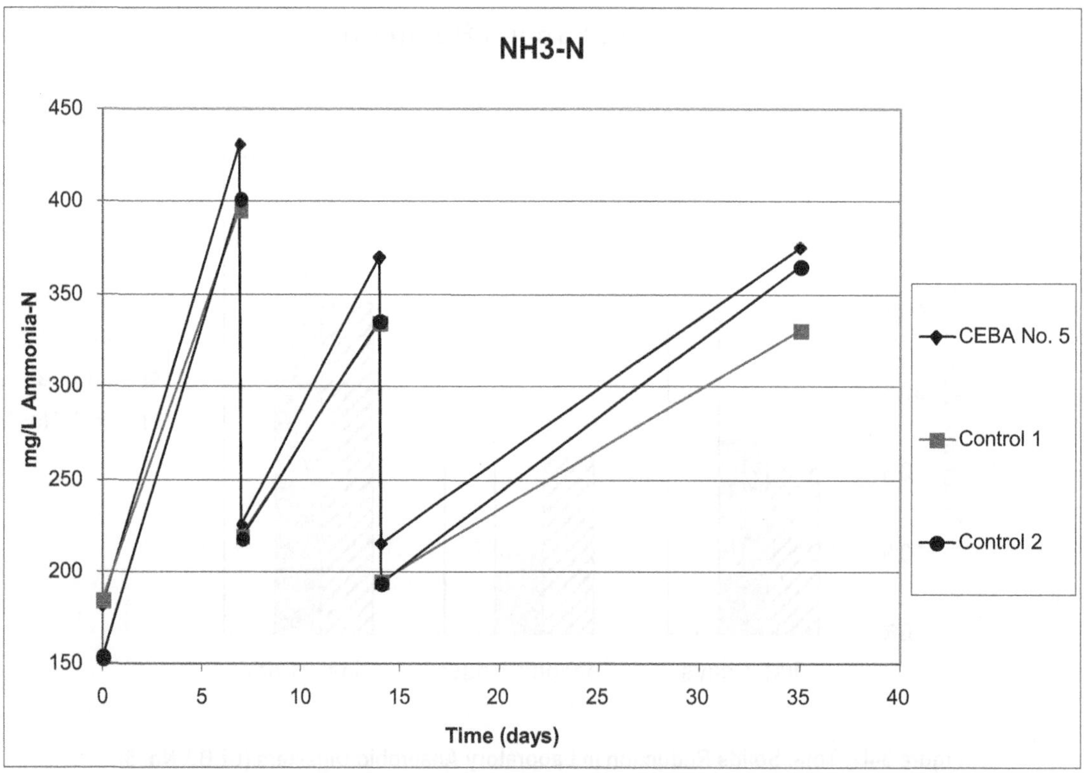

Figure 3-44. Ammonia-Nitrogen Production During Digestion (CEBA No. 5 Testing).

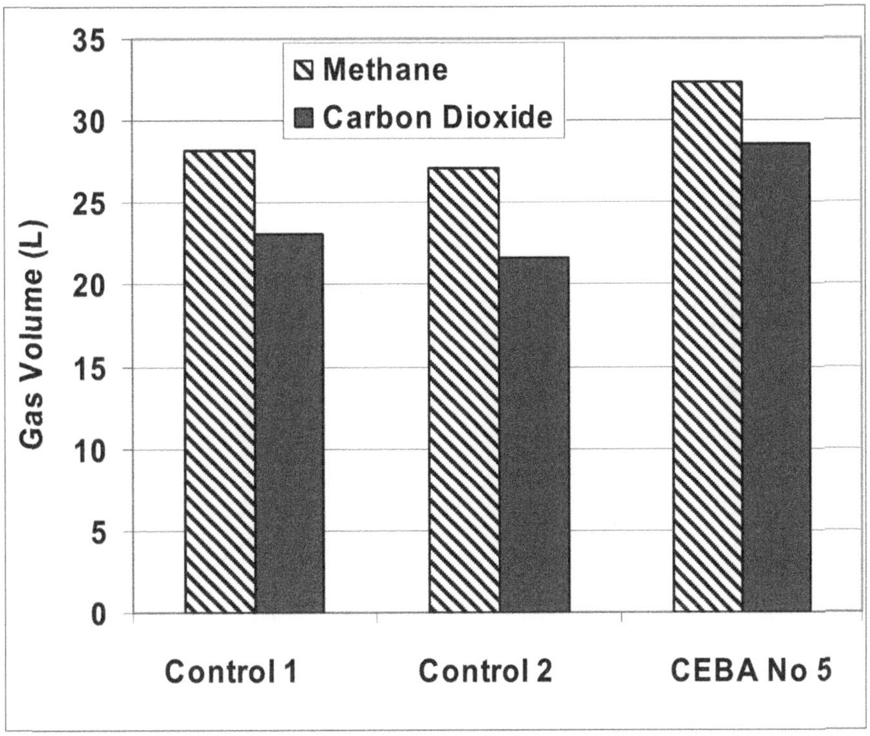

Figure 3-45. Gas Production from the Laboratory Anaerobic Digesters (CEBA No. 5 Testing).

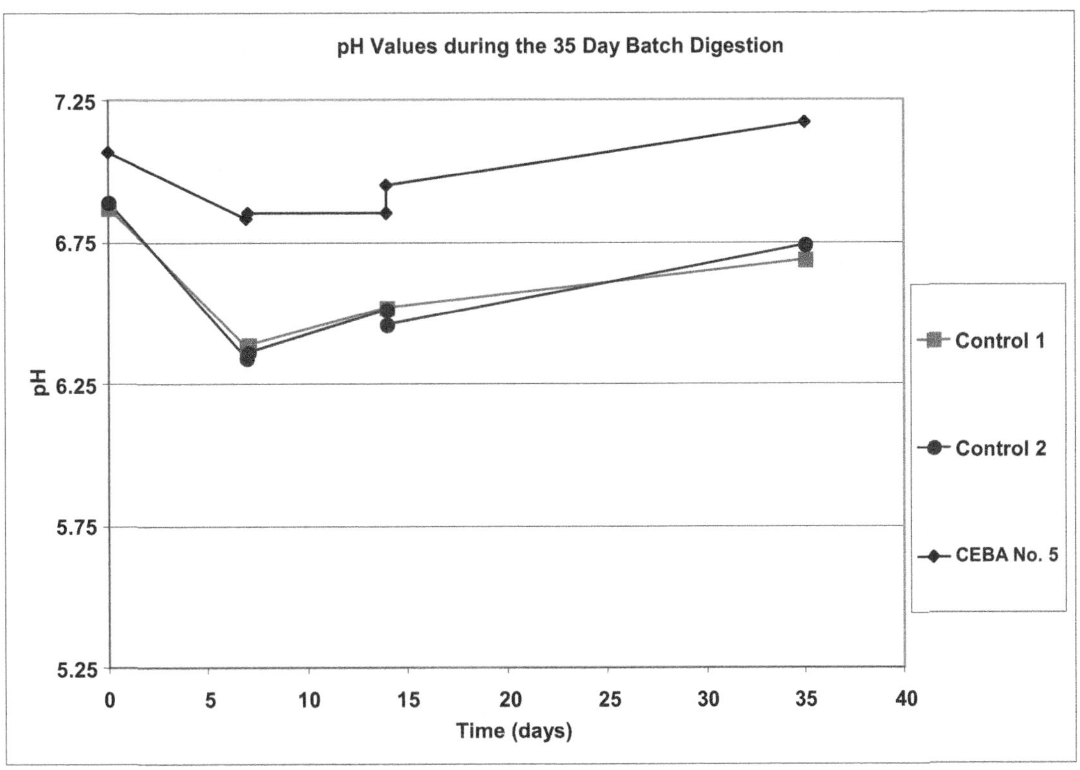

Figure 3-46. Digester pH Variation During Digestion (CEBA No. 5 Testing).

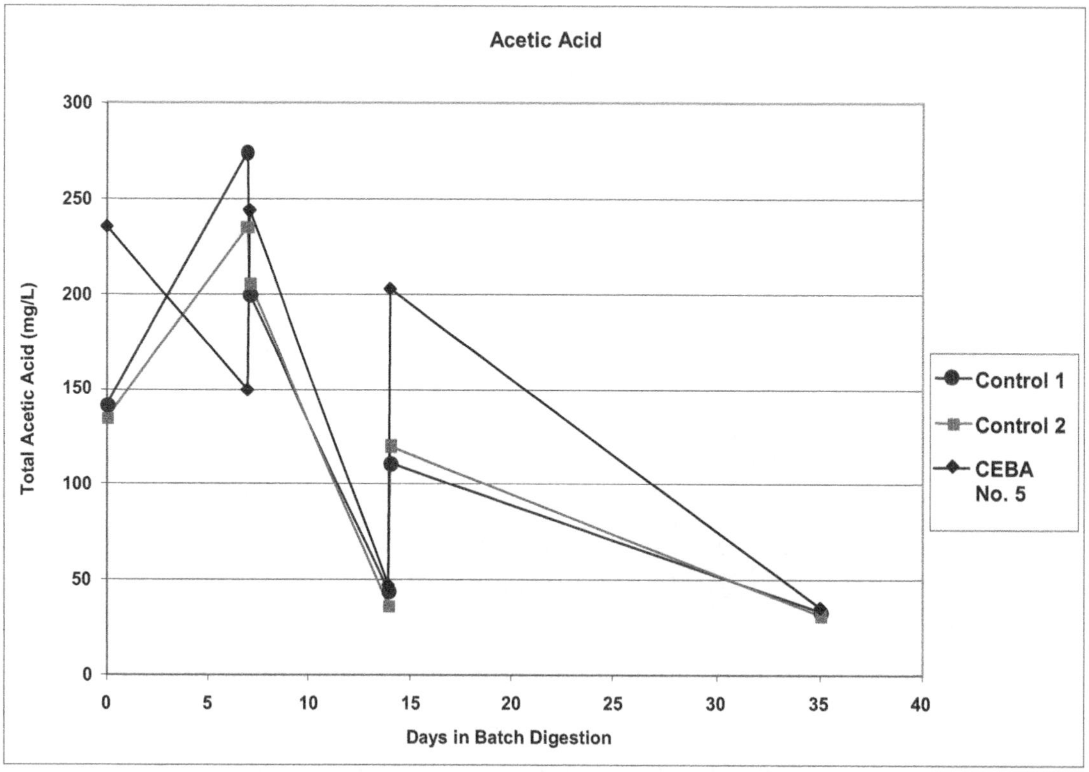

Figure 3-47. Effect of CEBA No. 5 on Acetic Acid Production.

CEBA No. 5 Impact on Headspace TVOSC

The generation of TVOSC in the headspace was evaluated using the shearing and dewatering method as described in Section 2.2 General Protocols. The results for the headspace TVOSC analyses are presented in Figure 3-48 and Figure 3-49. It can be seen from Figure 3-48 (when no BESA was added) that there was a significant increase in TVOSC in the dewatered cakes from the digesters that were dosed with the CEBA No. 5.

The CEBA additive increased the production of volatile fatty acids and likely also increased the rate of VOSC generation in the digester, since both the fatty acids and organic sulfur compounds are produced from complex organics. Therefore, VOSCs were generated faster than they could be converted to non-volatile sulfides, so the TVOSC peaked higher than did the controls. The WERF team considers this peak in TVOSC to have a relatively minor impact on overall cake odors since the VOSC compounds were rapidly dissipated within a few days after peaking.

Because this additive reduced the volatile solids content of the sludge, one would expect that the TVOSC generation would also be less. However, if one considers the ammonia-nitrogen data (Figure 3-44), the TVOSC data is not unreasonable. The ammonia nitrogen percentage increase was much less than the VS destruction increase (Figure 3-43), indicating that the additional material that was digested was probably not substantially proteinaceous (with its amino groups). Since proteins are precursors of TVOSC in sludges, the small increase in ammonia suggests a smaller amount of additional protein destruction.

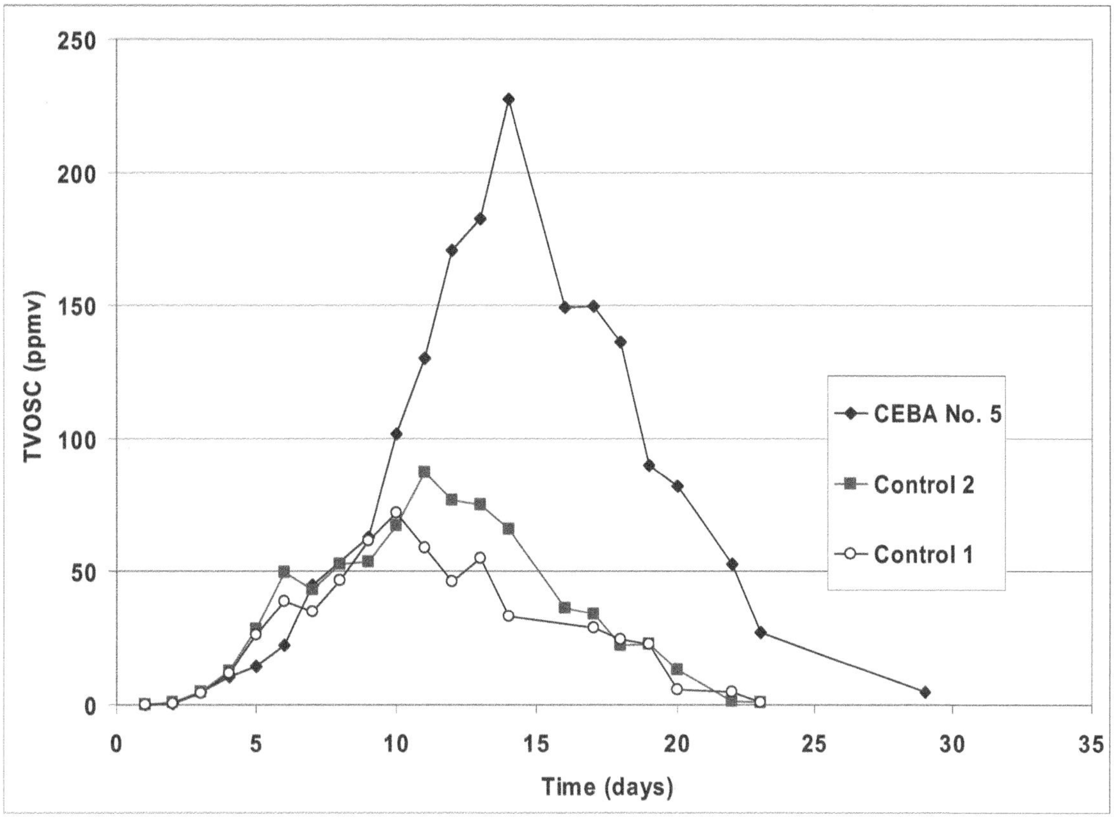

Figure 3-48. TVOSC CEBA No. 5 Testing: Headspace TVOSC Concentrations (No BESA Added).

Figure 3-49 depicts the variation in TVOSC through digestion for the CEBA-amended digester and the control digesters with BESA added. BESA inhibits the methanogenic organisms that convert organic sulfur to sulfide, therefore the TVOSC with BESA added can be considered to be the maximum odor potential. The figure shows that there was little difference between the controls and the CEBA No. 5-dosed sludges with regard to TVOSC odor potential when BESA was added. The similarity between the controls and the CEBA No. 5-dosed samples with BESA added supports the idea that the added VS reduction was not due to protein degradation, but was instead due to the degradation of some other organic component of sludge, possibly polysaccharides.

CEBA No. 5 Overall Evaluation

The addition of CEBA No. 5 resulted in much improved solids destruction (31% for TS and 35% for VS) and a beneficial (for digestion) elevated pH. The reductions in VS and TS were accompanied by increased carbon dioxide and methane production. Overall, based on the digestion indicators, it appears that the CEBA No. 5 agent is beneficial to digestion at the dose specified by the vendor for use in this study.

However, addition of CEBA No. 5 did not reduce VOSC emissions from the dewatered sludge cake. Conditions without the addition of BESA resulted in a large spike in TVOSC generation that was dissipated within a few days, but overall was reflective of increased TVOSC throughout the entire digestion period. A plot of TVOSC with BESA added showed that CEBA No. 5 was essentially the same as the controls with regard to TVOSC emissions from the dewatered cake.

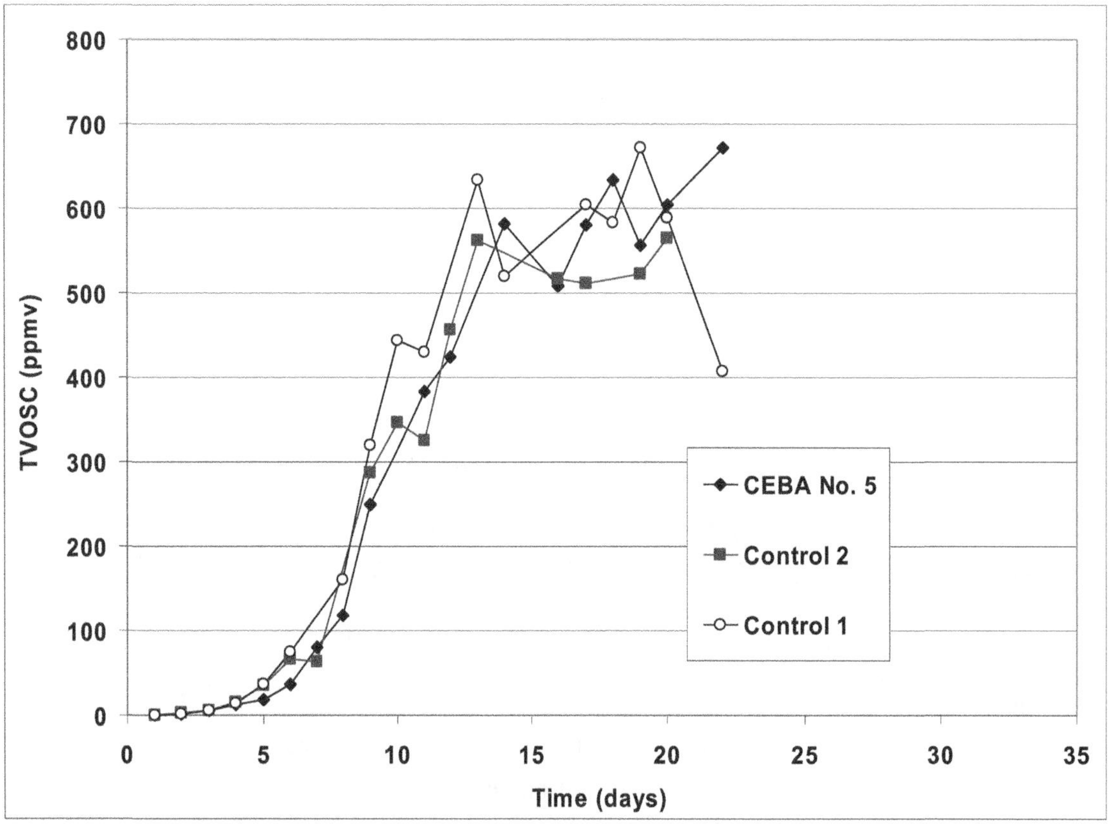

Figure 3-49. CEBA No. 5 Testing: Headspace TVOSC Concentrations (BESA added).

CEBA No. 6: CEBA No. 6 was added to the feed sludge according to the SCTP, which was developed by the WERF team and approved by the vendor. The staged batch digester was operated according to the vendor-approved SCTP and the GCTP. The total digestion period was 35 days with two feed cycles.

Testing Protocol Synopsis

A summary of the testing protocol for CEBA No. 6 is as follows:

1. Initially, the digester contents were two (2) liters of anaerobicly digested biosolids (to serve as seed sludge) plus eight (8) liters of a 50/50 ratio mixture of primary and waste-activated sludges by dry weight.
2. When adding the sludge to the reactor, CEBA No. 6 was also combined with the sludge using the specified dosage and dilution. CEBA No. 6 product was diluted 1:10,000 using the two-step procedure indicated by the vendor in their application instructions, which was also shown in the vendor-approved SCTP. To achieve this mix, 10mL was diluted into 1 liter of distilled water, and then 10 mL of the initial mix was added to another liter of distilled water. The diluted CEBA No. 6 product was then added to the sludge at 50mL to 10 liters of sludge and thoroughly mixed by hand.
3. After 7 days, 50% of the contents were removed and replaced with a 50/50 mixture of primary and secondary sludge containing the additive dosed at 25mL in the replacement 5 liters of sludge. This cycle was repeated after another 7 days (Day14).

4. The digester was then operated for an additional 21 days (making a 35-day total testing period), and the digested sludge was then sheared, dewatered, stored, and analyzed for headspace TVOSC and other parameters per the GCTP.
5. The treated digester performance and odor generation from the dewatered sludge cakes were compared to a pair of control digesters that were operated in the same manner but did not contain CEBA No. 6.

CEBA No. 6 Digester Performance

The results of the digester performance compared to the controls are shown in Figure 3-50 through Figure 3-55. The solids destruction data indicates that CEBA No. 6 did not significantly improve digestion as evaluated by either TS or VS reduction (Figures 3-50 and 3-51, respectively), with the final VS destruction being less than the 38% requirement put forth the 503 rule. There was a slight increase in ammonia-nitrogen, carbon dioxide, and methane in the CEBA No. 6 digester (Figures 3-52 and 3-53), indicating some improved digestion. However, based on all digestion indicators tested, it appears that overall there was little difference in digestion performance parameters between the CEBA No. 6-dosed and control digesters.

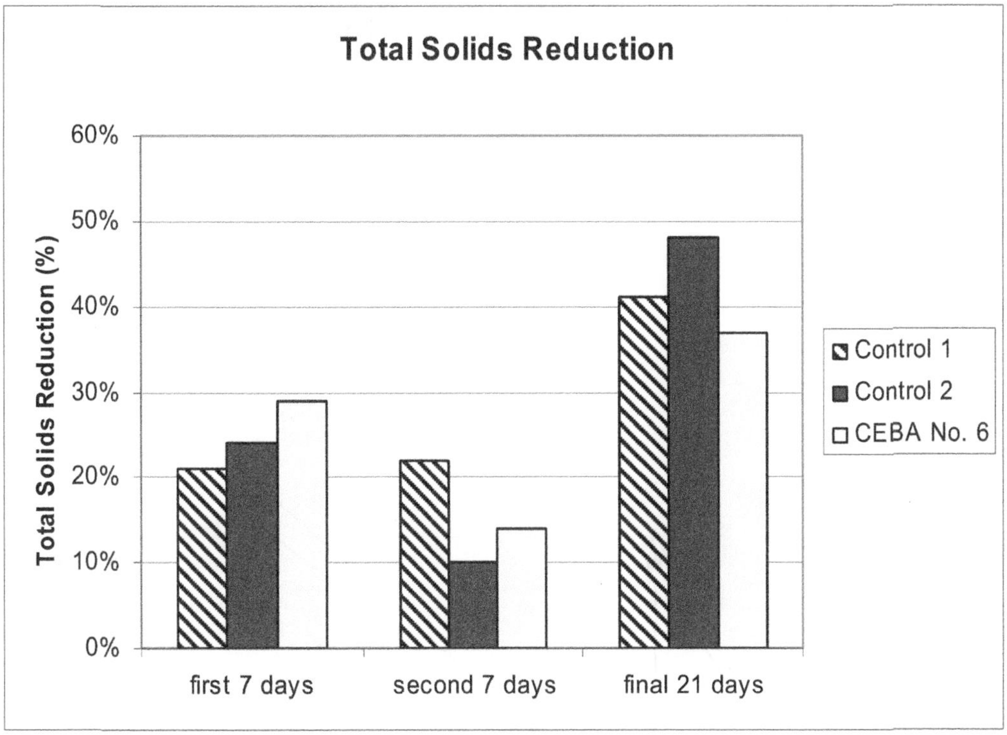

Figure 3-50. Total Solids Reduction in Laboratory Anaerobic Digesters (CEBA No. 6).

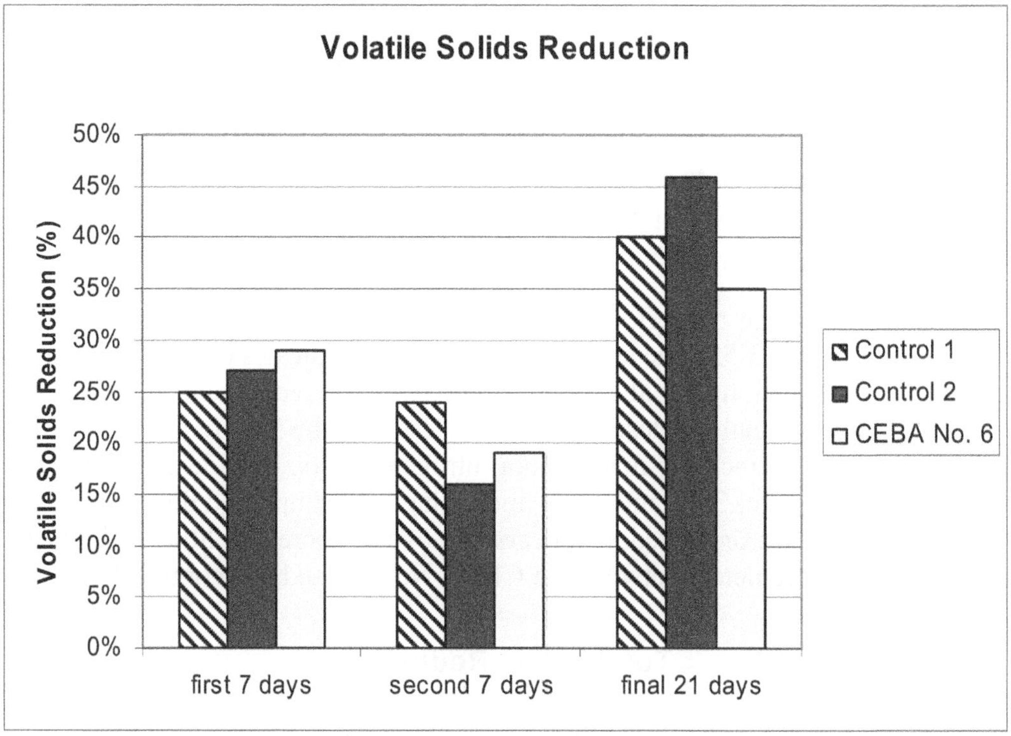

Figure 3-51. Volatile Solids Reduction in Laboratory Anaerobic Digesters (CEBA No. 6).

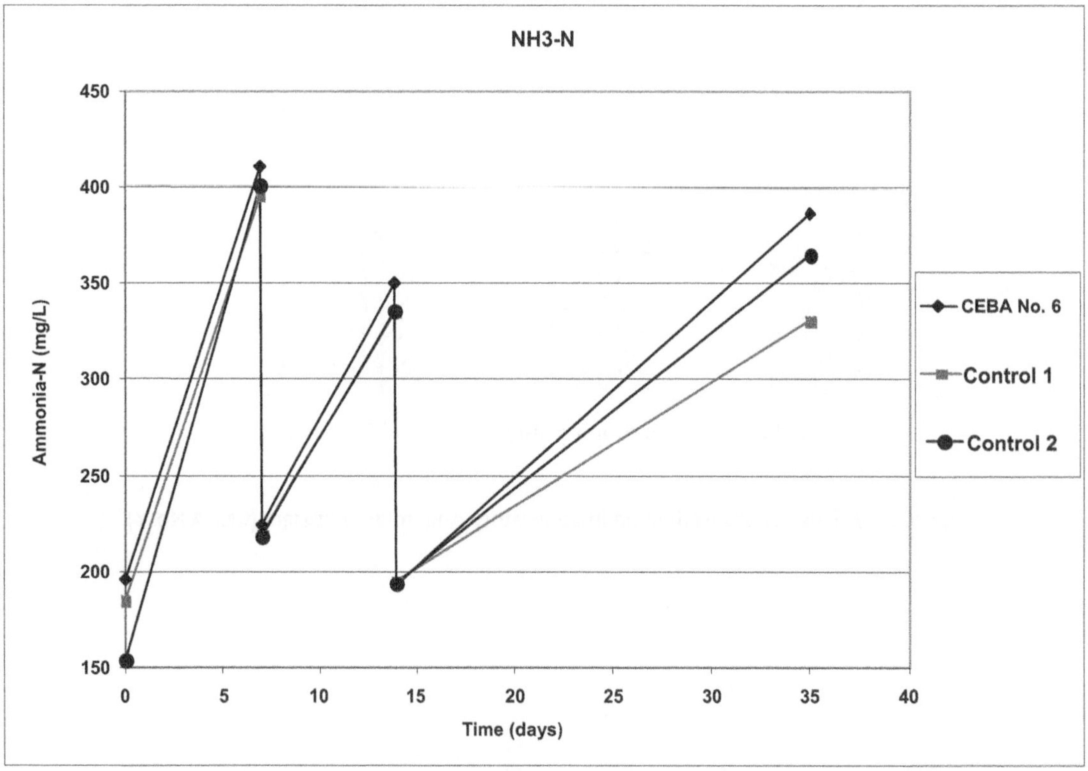

Figure 3-52. Ammonia-Nitrogen Production During Digestion (CEBA No. 6 Testing).

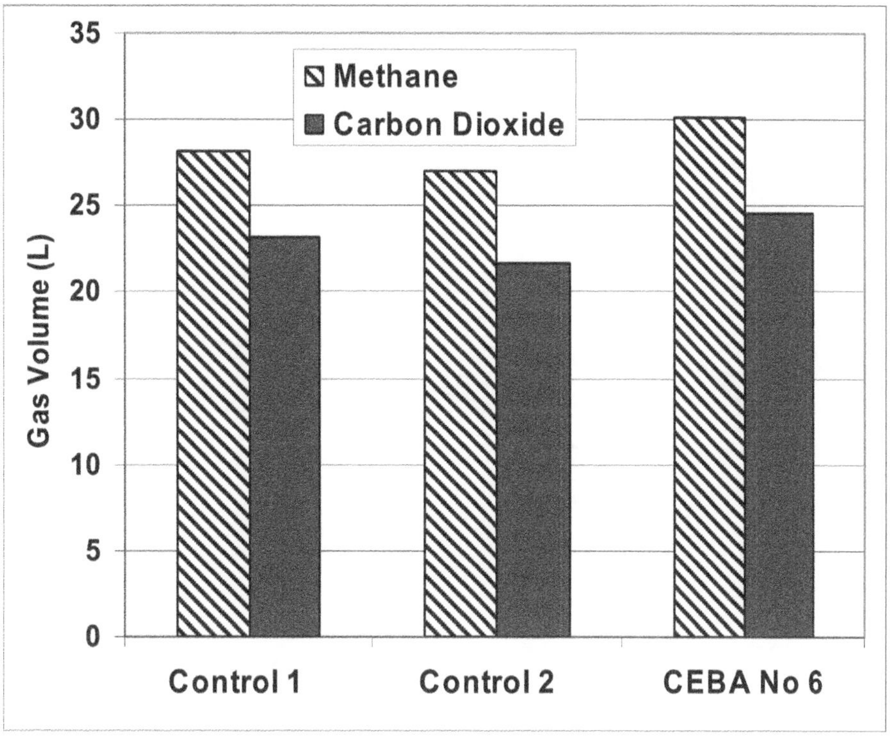

Figure 3-53. Gas Production from the Laboratory Anaerobic Digesters (CEBA No. 6 Testing).

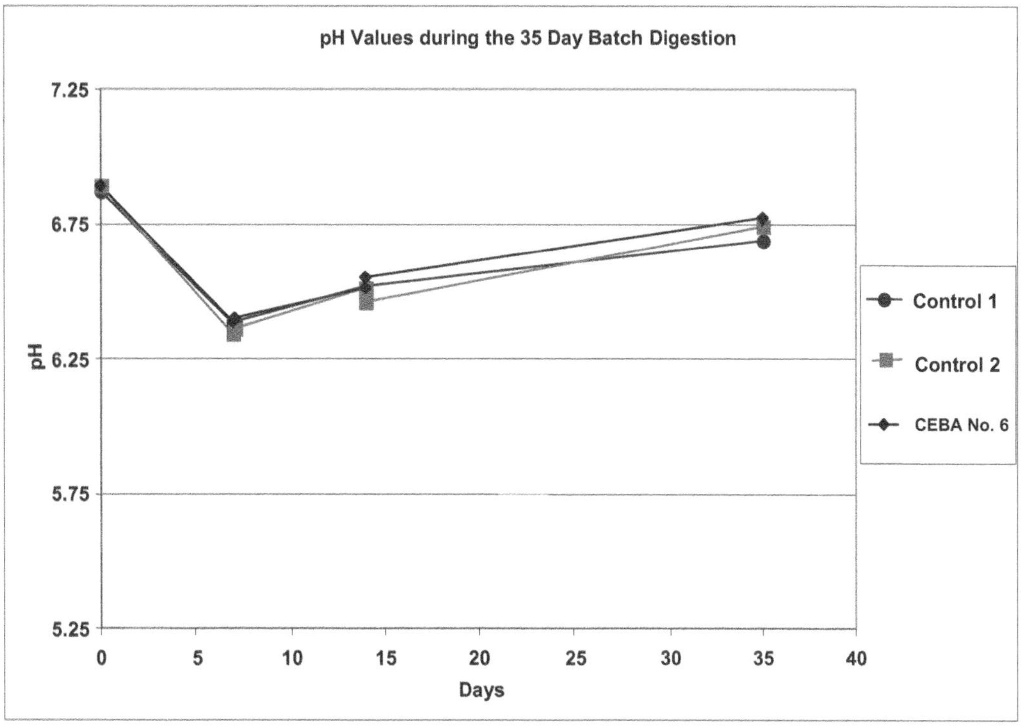

Figure 3-54. Digester pH Variation During Digestion (CEBA No. 6 Testing).

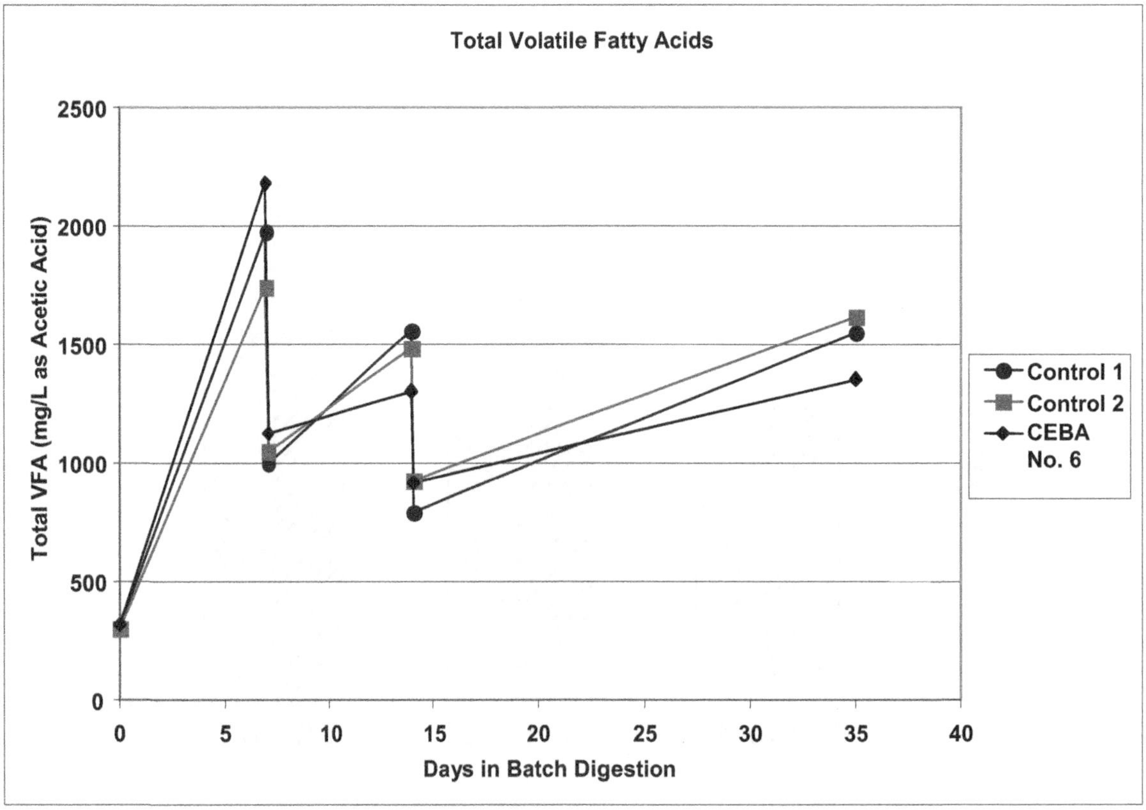

Figure 3-55. Effect of CEBA No. 6 on Acetic Acid Production.

CEBA No. 6 Impact on Headspace TVOSC

The generation of TVOSC in the headspace was evaluated using the shearing and dewatering method as described in Section 2.2 General Protocols. The results for the headspace TVOSC analyses are presented in Figure 3-56 and Figure 3-57. The figures show that the addition of CEBA No. 6 had little impact on dewatered cake TVOSC generation in the sample bottle headspace. The data for the BESA–dosed samples (Figure 3-57) shows the effect when the methanogenic organisms are inhibited and the volatile organic sulfur compounds are allowed to accumulate. This reflects the maximum odor potential and thus indicates that no additional odor generating organic matter has been degraded in the digester as a result of CEBA No. 6 addition.

The data in Figures 3-56 and 3-57 shows that the CEBA No. 6 had little impact on the generation of TVOSC at the prescribed dose. Based on the data collected, it appears that, for the recommended chemical dose of CEBA No. 6 applied, there was no benefit for reduction of the TVOSC odor potential. In addition to the data shown, hydrogen sulfide was collected that also showed no difference between the controls and the CEBA No. 6-dosed digester.

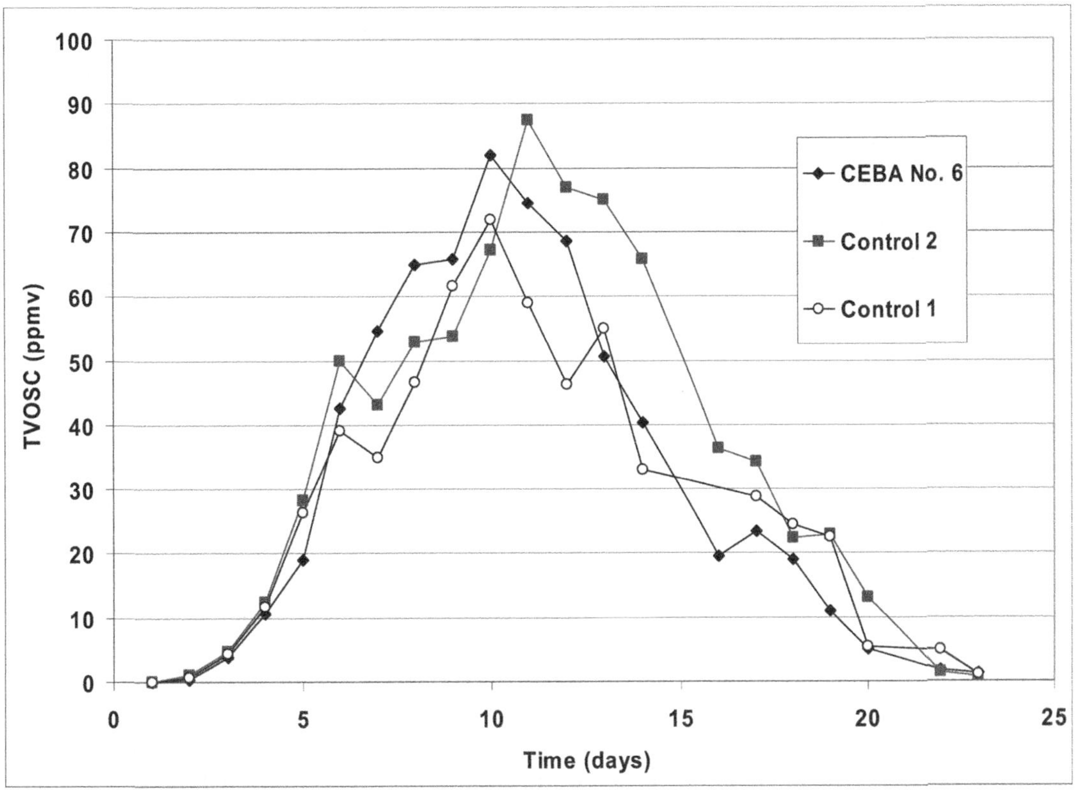

Figure 3-56. TVOSC CEBA No. 6 Testing: Headspace TVOSC Concentrations (No BESA Added).

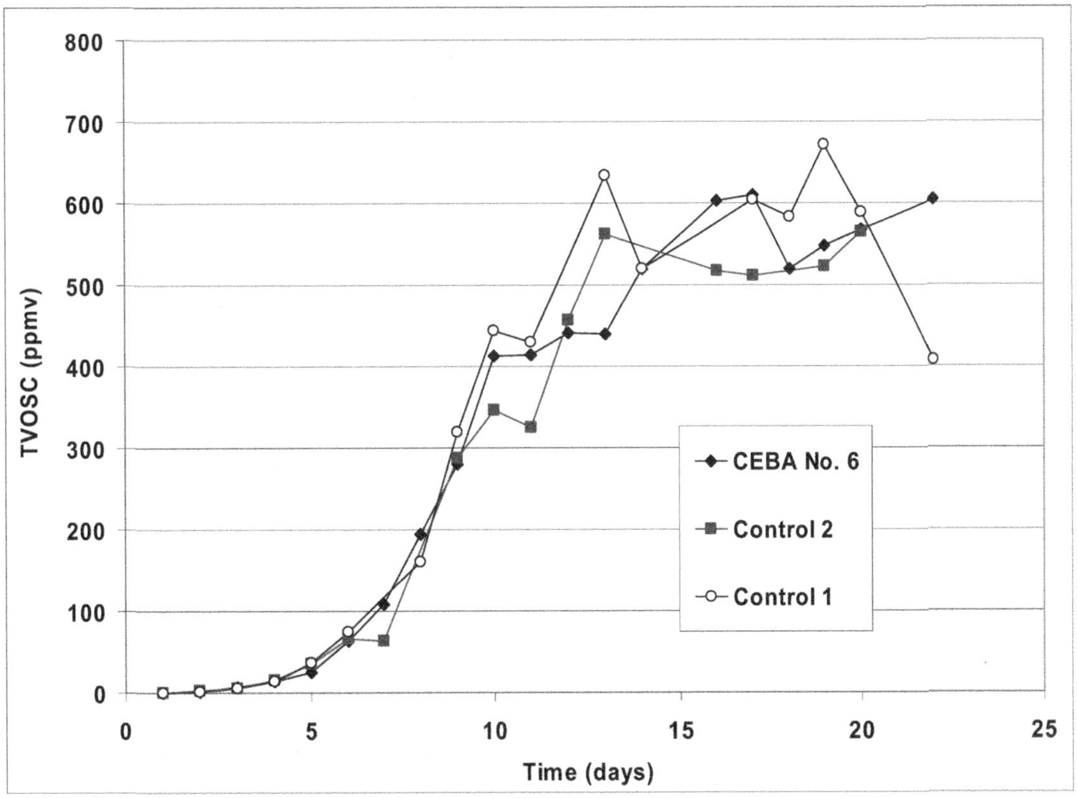

Figure 3-57. TVOSC CEBA No. 6 Testing: Headspace TVOSC Concentrations (BESA Added).

CEBA No. 6 Overall Evaluation

There was a marginal increase in carbon dioxide and methane production in the digester in which CEBA No. 6 was added, but other than those slight benefits to digestion, other measured parameters were either similar to the control digesters or were indicative of a detrimental change (TS and VS reduction was less in the CEBA-amended digester). CEBA No. 6 showed no odor control benefit for the final dewatered biosolids cake. It is possible that CEBA No. 6 could show improvement if added at a higher concentration than the 0.5 ppmv per volume of digester feed specified for this test by the vendor.

3.2.3.2 Post-Digestion CEBA Application Test Results

Two CEBAs (Nos. 7 and 8) were tested as post-digestion agents in the Phase 3 research at the Bucknell University laboratory. The overall test protocol summary for the post-digestion CEBAs is provided in Appendix A. The detailed GCTP and the individual SCTPs are located in Appendix A. Laboratory bench-scale tests were performed to determine the effect of CEBA addition on the production of odorous compounds from anaerobicly digested biosolids that are dewatered using a process that simulates a high (percentage) solids centrifuge. Anaerobicly digested biosolids samples were obtained from the Philadelphia Water Department SW Wastewater Treatment Plant for use in this testing.

CEBA No. 7: CEBA No. 7 was added to the digested sludge according to the SCTP, which was developed by the WERF team and approved by the vendor. The biosolids were conditioned and dewatered according to the vendor-approved SCTP for CEBA No. 7, a CEBA provided for testing by the vendor.

Testing Protocol Synopsis

A summary of the testing protocol for CEBA No. 7 is as follows:

1. CEBA No. 7 was added to the liquid sludge, and allowed to incubate at room temperature for the prescribed time period of six hours.
2. After incubation, the biosolids were conditioned and dewatered using a high-solids centrifuge simulation process described in the GCTP. After conditioning and dewatering, a cake was produced with 27% dry solids content. This is within the typical range for cake solids from high-solids centrifuges.
3. After the dewatering process, the cake that was generated was placed in sample bottles and incubated at 25°C for a series of headspace analyses. Samples of the headspace were taken at regular intervals and analyzed for VOSCs using gas chromatography according to the methods described in Section 2.2 General Protocols. A control sample was prepared that was treated the same as the CEBA No. 7 sample except no agent was added. All samples were prepared in duplicate.

CEBA No. 7 Impact on Headspace TVOSC

The results of the headspace analyses varied with sludge storage time are depicted in Figure 3-58 through Figure 3-61. Bucknell University analyzed for individual TVOSC components methane thiol (MT) and dimethyl sulfide (DMS) in addition to TVOSC and methane in the biosolids headspace samples with and without the addition of CEBA No. 7.

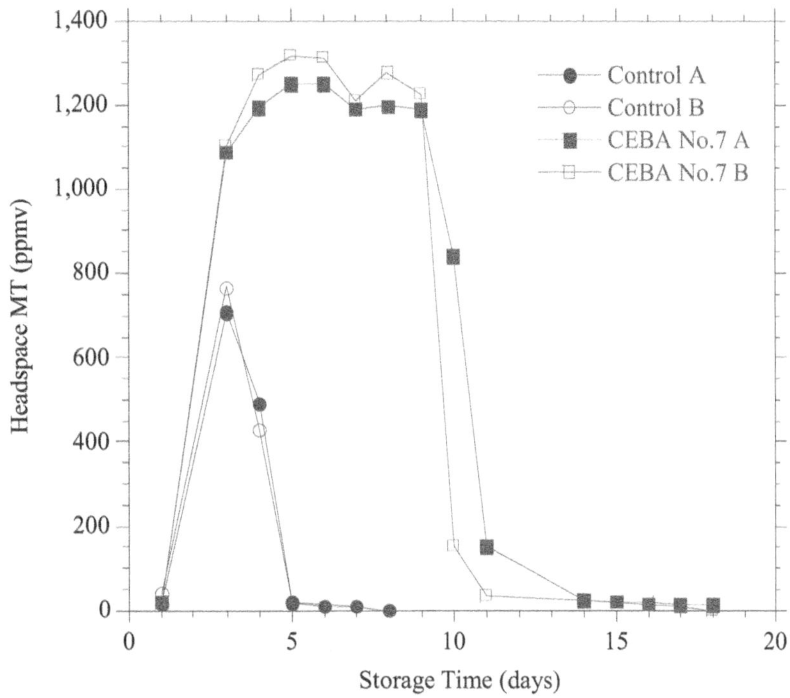

Figure 3-58. Headspace Methanethiol during Storage of Control and CEBA No. 7– Amended Sludge.

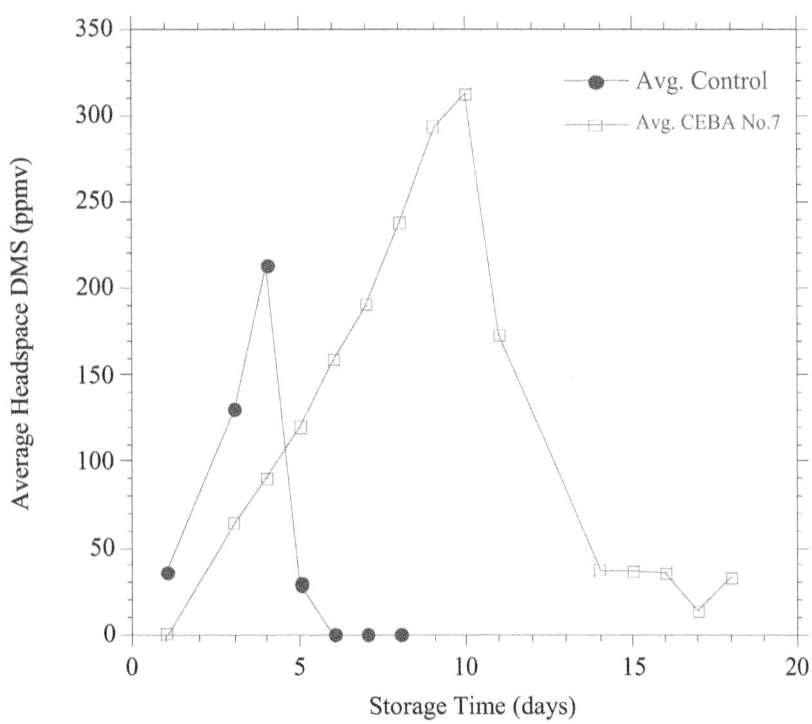

Figure 3-59. Headspace Dimethyl Sulfide (DMS) During Storage of Control and CEBA No. 7- Amended Sludge.

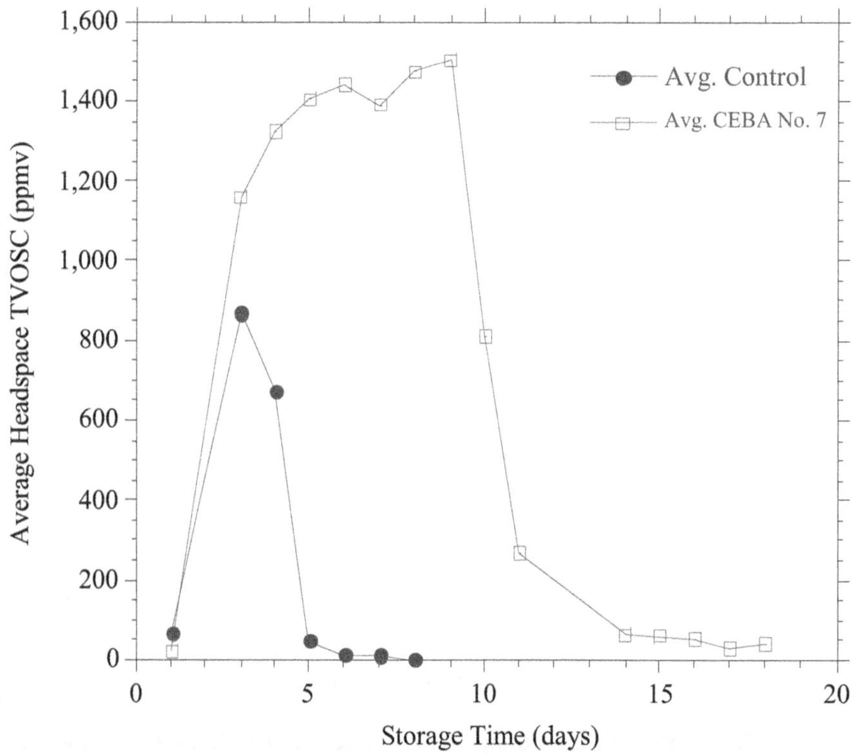

Figure 3-60. Headspace TVOSC During Storage of Control and CEBA No. 7- Amended Sludge.

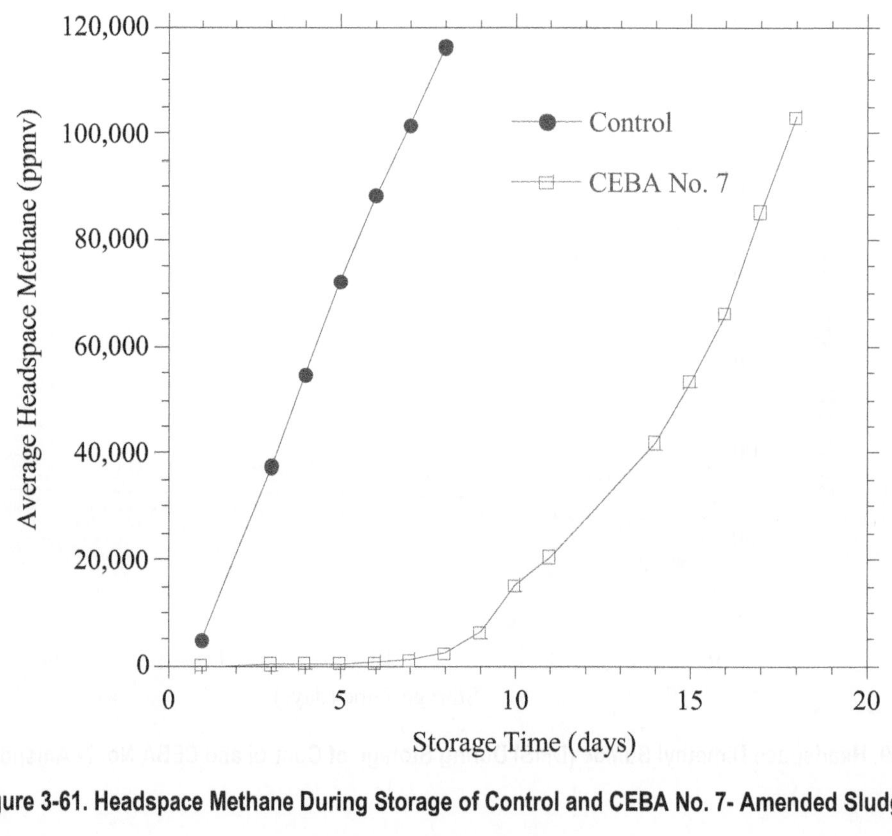

Figure 3-61. Headspace Methane During Storage of Control and CEBA No. 7- Amended Sludge.

The variations in headspace MT concentration for the CEBA No. 7-added sample and the control sample is shown in Figure 3-58. For this comparison, the results from the duplicate samples (A and B) are shown to provide some data on the good reproducibility of the results. As shown in Figure 3-58, the production of headspace MT was substantially increased by the addition of CEBA No. 7, and the peak MT concentration of the CEBA No. 7-amended sample was 74% greater than the corresponding control sample.

Similarly, the average headspace DMS concentration measured during storage is shown in Figure 3-59, and the results showed greater production of DMS during storage of the CEBA-amended sample in addition to a time-to-peak shift. The CEBA No. 7-amended sample had a peak DMS concentration approximately 56% greater than the control.

The TVOSC concentration in the headspace during storage is shown in Figure 3-60. Similar to the MT results (which is understandable as MT is the dominant component of TVOSC in this sample), the peak concentrations of TVOSC in the headspace was about 73% greater than the control sample. The addition of CEBA No. 7 resulted in a substantial suppression in methane production concurrent with the peaking of TVOSC, as indicated in Figures 3-63 and 3-61. The reduction in methane production may help to explain the increased TVOSC concentrations produced by the CEBA No. 7–amended cake samples.

In summary, the addition of CEBA No. 7 resulted in greater production of odorous organic sulfur compounds compared to control samples. The results seem to indicate that the inhibition of methane production may have contributed to the increased production of the organic sulfur compounds. Additional testing of different dosages and conditions may be warranted to more fully investigate the effect to this agent on reducing odorous compounds, since this bench-scale experiment tested only one vendor-approved dosage rate.

CEBA No. 8: CEBA No. 8 was added to the digested sludge according to the SCTP, which was developed by the WERF team and approved by the vendor. The biosolids were conditioned and dewatered according to the vendor-approved SCTP for CEBA No. 8, a CEBA provided for testing by the vendor.

Testing Protocol Synopsis

A summary of the testing protocol for CEBA No. 8 is as follows:

1. CEBA No. 8 was added to the liquid sludge, just prior to conditioning, at a dosage of 350 ppmv.
2. After addition, the sample was conditioned and dewatered using the high solids centrifuge simulation process described in the GCTP. After conditioning and dewatering, a cake was produced with 27% dry solids content. This is within the typical range for cake solids from high solids centrifuges.
3. After the dewatering process, the cake that was generated was placed in sample bottles and incubated at 25°C for headspace analysis. Samples of the headspace were taken at regular intervals and analyzed for VOSCs using gas chromatography according to the methods described in Section 2.2 General Protocols. A control sample was prepared that was treated the same as the CEBA No. 8 sample, except no agent was added. All samples were prepared in duplicate.

CEBA No. 8 Impact on Headspace TVOSC

The headspace analysis results are shown graphically in Figure 3-62 through Figure 3-65. The headspace concentration of MT for the CEBA No. 8-amended and the associated control sample is shown in Figure 3-62 with duplicate sample results shown to indicate the level of reproducibility of the results. As shown in the plot, the production of MT was greater in the sample with CEBA No. 8 addition than was measured in the control, similar to what was observed for the CEBA No. 7 testing. The peak MT concentration of the CEBA No. 8-amended sample was about 38% greater than the corresponding control sample.

The average headspace DMS concentration measured during storage (Figure 3-63) indicates that the peak DMS concentration was slightly less for the CEBA No. 8 amended sample compared to the control sample. The CEBA No. 8-dosed sample had a peak DMS concentration approximately 18% less than the control.

The TVOSC (which includes both MT and DMS) concentration in the headspace during storage is shown in Figure 3-64. Similar to the MT results, the peak concentrations of TVOSC in the headspace was about 33% greater than the control sample. The addition of CEBA No. 8 resulted in a slight decrease in the methane production, as depicted in Figure 3-65. The reduction in methane production may help to explain the increased TVOSC concentrations produced by the CEBA No. 8-amended cake samples.

In summary, the addition of CEBA No. 8 to the digested sludge samples resulted in greater production of MT but less DMS production. Overall, the TVOSC concentrations were greater in the CEBA No. 8 amended cake samples. The results seem to indicate that the inhibition of methane production may have contributed to the increased production of the organic sulfur compounds; however, the decrease in methane production was relatively small. Additional testing of different dosages and conditions may be warranted to more fully investigate the effect of this agent on reducing odorous compounds, since this experimental testing relied only on one dosage rate, provided by the vendor.

3.2.3.3 Post-Dewatering CEBA Application Test Results

Two CEBAs, Nos. 2 and 3, were tested as post-dewatering agents in the Phase 3 research at the Bucknell University laboratory. The overall test protocol summary for the post-dewatering CEBAs is provided in Section 2.3.3.2. The detailed GCTP and the individual SCTPs are in Appendix A. Laboratory bench-scale tests were performed to determine the effect of CEBA addition on the production of odorous compounds from anaerobicly digested biosolids that are dewatered using a process that simulates a high (percentage) solids centrifuge. Anaerobicly digested biosolids samples were obtained from the Philadelphia Water Department SW Wastewater Treatment Plant for use in this testing.

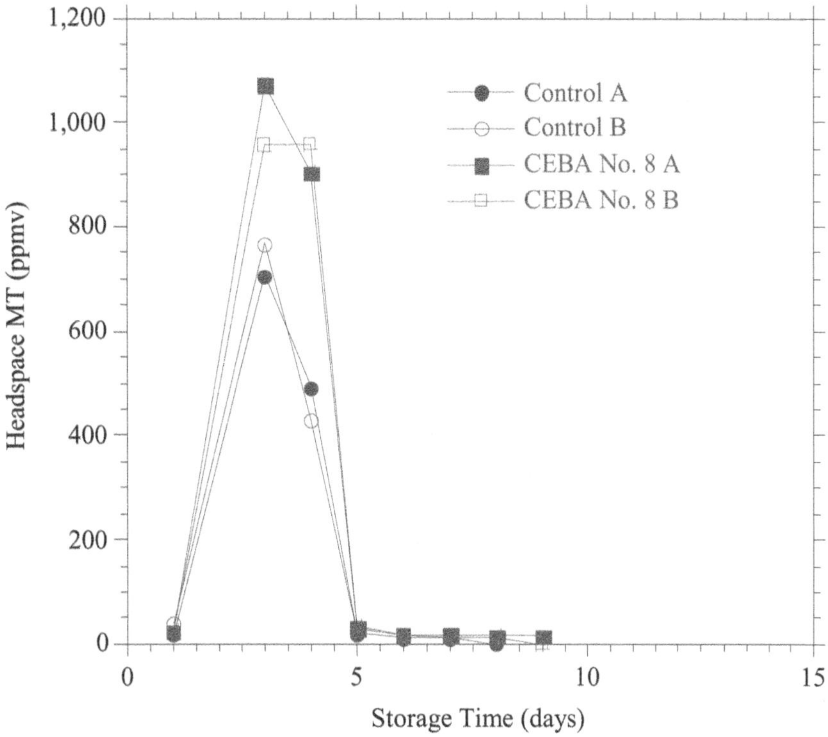

Figure 3-62. Headspace Methanethiol During Storage of Control and CEBA No. 8- Amended Cake.

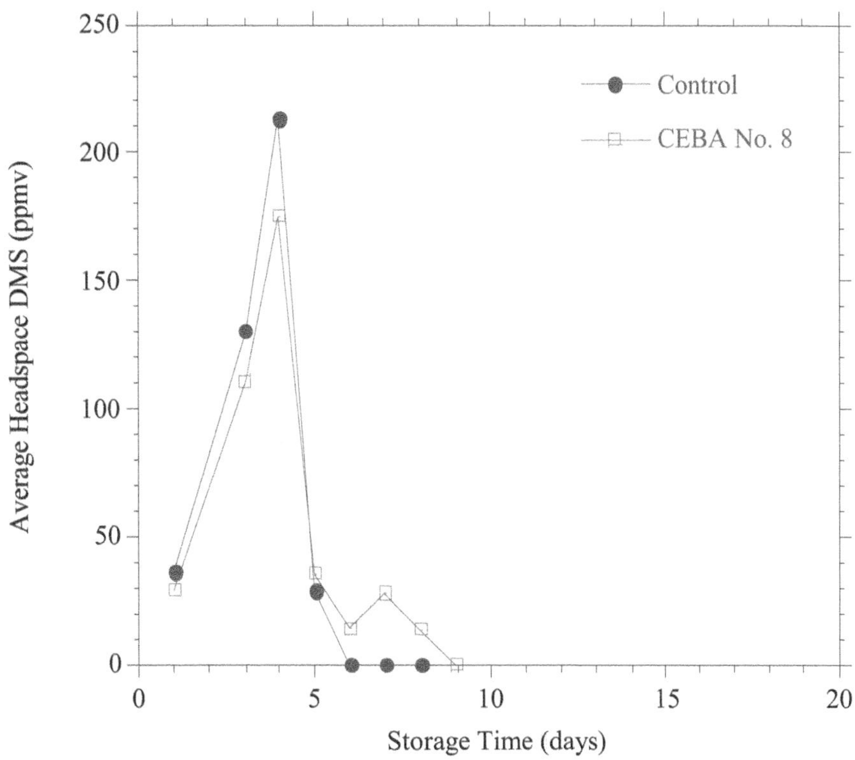

Figure 3-63. Headspace DMS During Storage of Control and CEBA No. 8- Amended Cake.

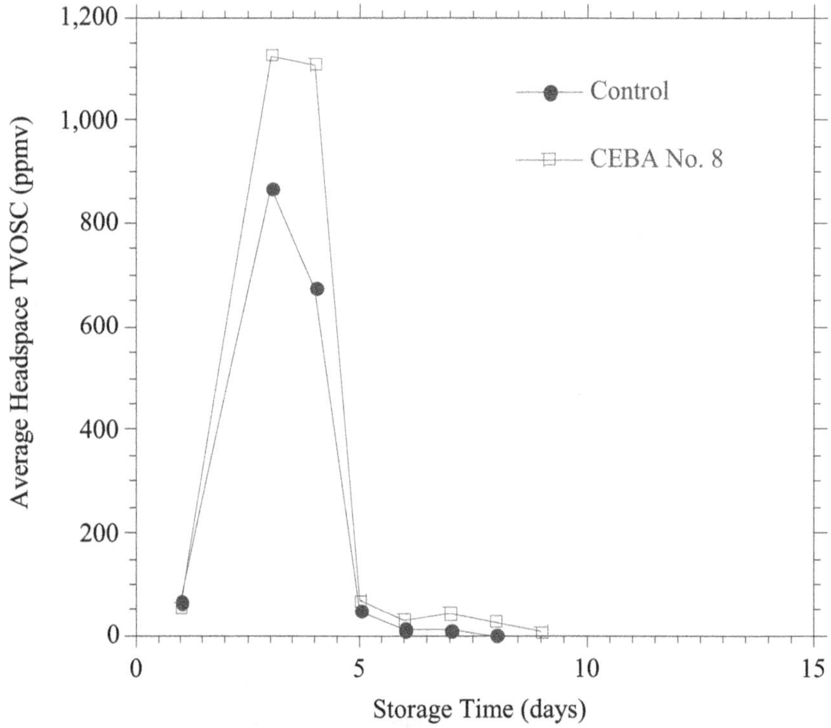

Figure 3-64. Headspace TVOSC During Storage of Control and CEBA No. 8- Amended Cake.

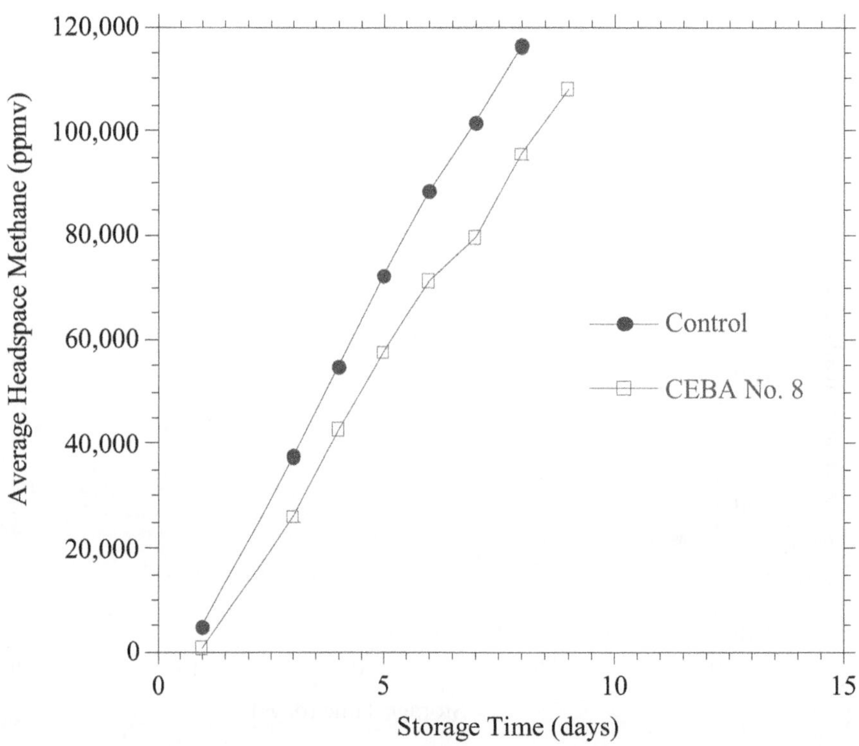

Figure 3-65. Headspace Methane during Storage of Control and CEBA No. 8-Amended Cake.

CEBA No. 2: CEBA No. 2 was added to the dewatered sludge according to the SCTP, which was developed by the WERF team and approved by the vendor. The biosolids were conditioned and dewatered according to the SCTP for CEBA No. 2, a CEBA provided by the vendor.

Testing Protocol Synopsis

A summary of the testing protocol for CEBA No. 2 is as follows:

1. The biosolids were conditioned and dewatered using the Bucknell high-solids centrifuge simulation process. After conditioning and dewatering, a cake was produced with 28% dry solids content.
2. The cake that was generated was sprayed with the prepared CEBA No. 2 product, using a spray bottle. The dosage was 10 mL of CEBA No. 2 extract per kg of wet cake. The CEBA No. 2 extract was prepared in accordance with the method described in the vendor-approved SCTP, which was based on the vendor's instructions.
3. After the agent was added, the cake sample was mixed by hand, placed in sample bottles, and incubated at 25 °C for a series of headspace analyses. Samples of the headspace were taken at regular intervals and analyzed for VOSCs using gas chromatography according to the methods described in Section 2.2 General Protocols. A control sample was prepared that was processed the same as the CEBA No. 2 sample, except that no agent was added. All samples were prepared in duplicate.

CEBA No. 2 I mpact on Headspace TVOSC

The results of the CEBA No. 2 testing are shown graphically in Figure 3-66 through Figure 3-69. The headspace concentrations of MT for CEBA No. 2 and the associated control sample are shown in Figure 3-66. For this graph, the results from the duplicate samples are shown to provide some data on the good reproducibility of the results. As shown in the plot, the production of MT was not impacted by the addition of CEBA No. 2 in any way, as results were nearly identical.

The average headspace DMS concentration of the CEBA-amended cake and control (Figure 3-67) were also similar, with the control sample having a slightly lower peak DMS concentration than the CEBA No. 2-amended sample. The TVOSC concentrations of the CEBA-amended cake and the controls during storage are depicted in Figure 3-68. Similar to the MT results, the concentrations of TVOSC in the headspace were similar for both the CEBA No. 2-dosed sample and the control sample. The addition of CEBA No. 2 did have a slight positive impact on the methane production as shown in Figure 3-69, although this improvement did not translate into a reduction in the headspace TVOSC concentrations of the CEBA-amended cake.

In summary, CEBA No. 2 did not reduce the production of VOSCs from anaerobicly digested biosolids. Additional testing of different dosages and conditions may be warranted to more fully investigate the effect to this agent on reducing odorous compounds, since this experimental testing relied only on one vendor-approved dosage rate.

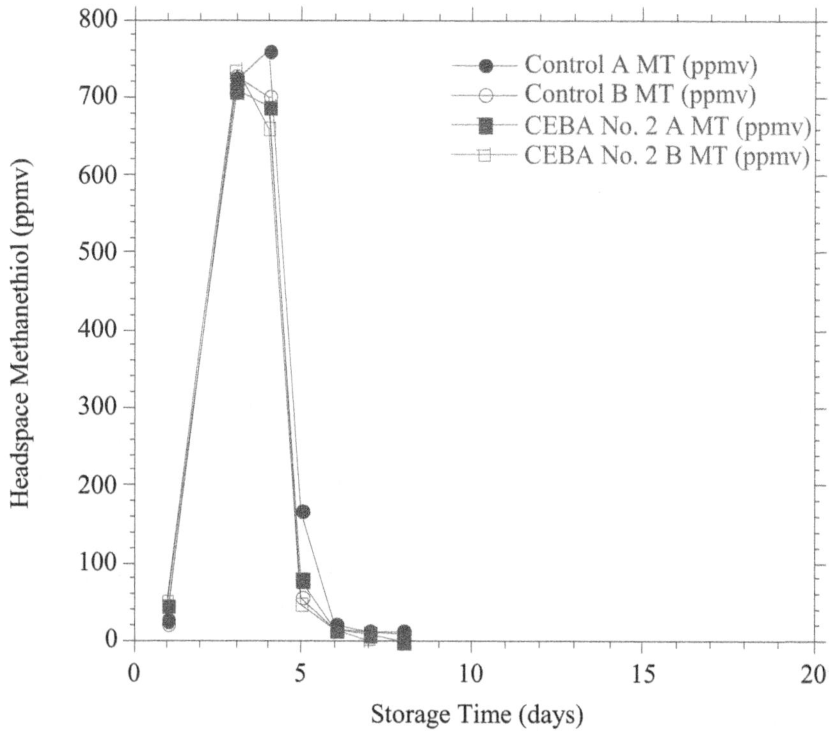

Figure 3-66. Headspace Methanethiol During Storage of Control and CEBA No. 2-Amended Cake.

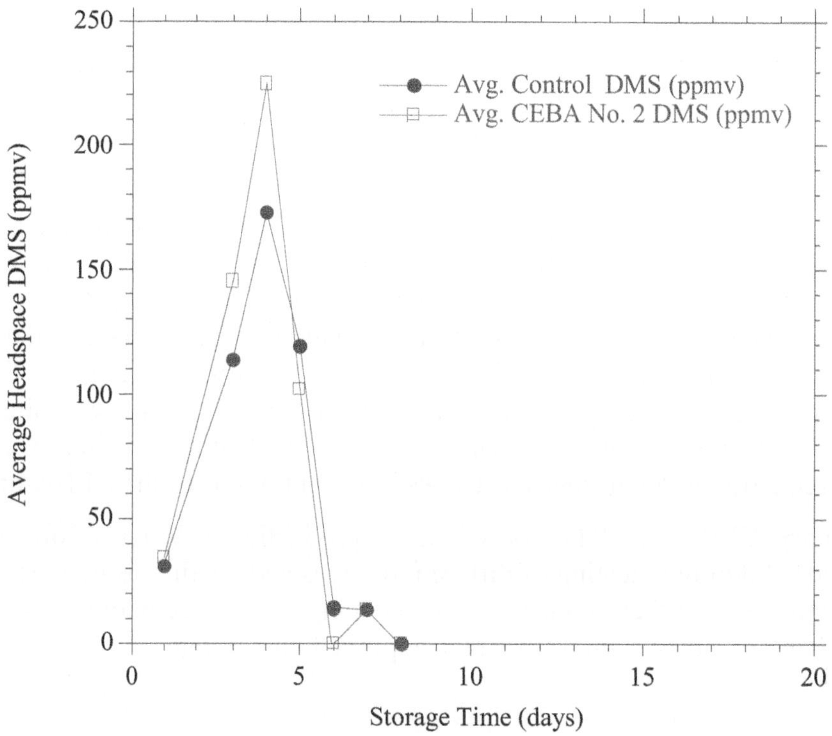

Figure 3-67. Headspace Dimethyl Sulfide During Storage of Control and CEBA No. 2–Amended Cake.

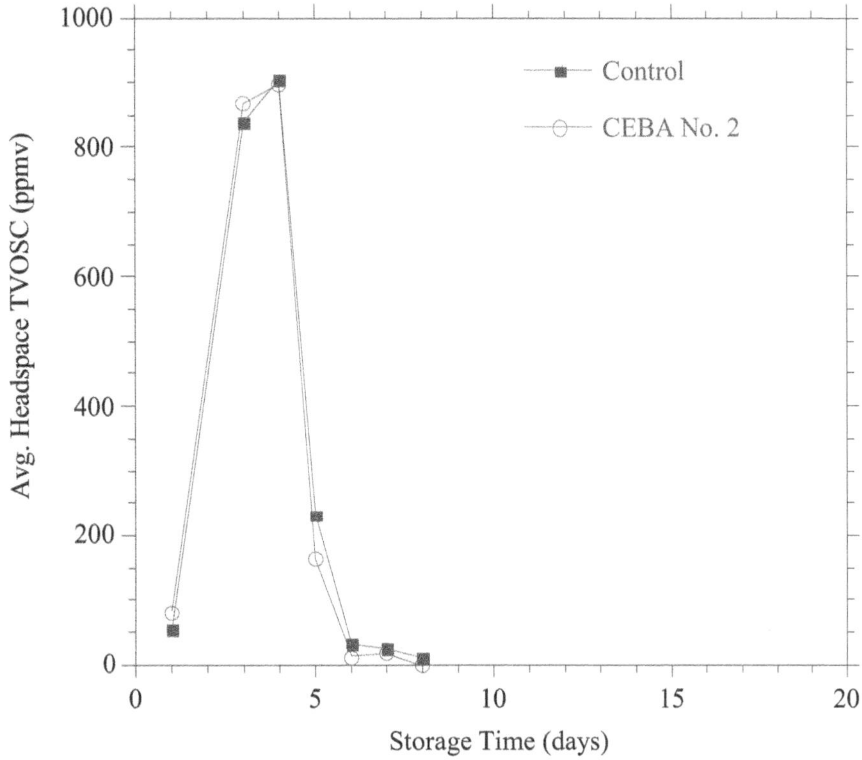

Figure 3-68. Headspace TVOSC During Storage of Control and CEBA No. 2-Amended Cakes.

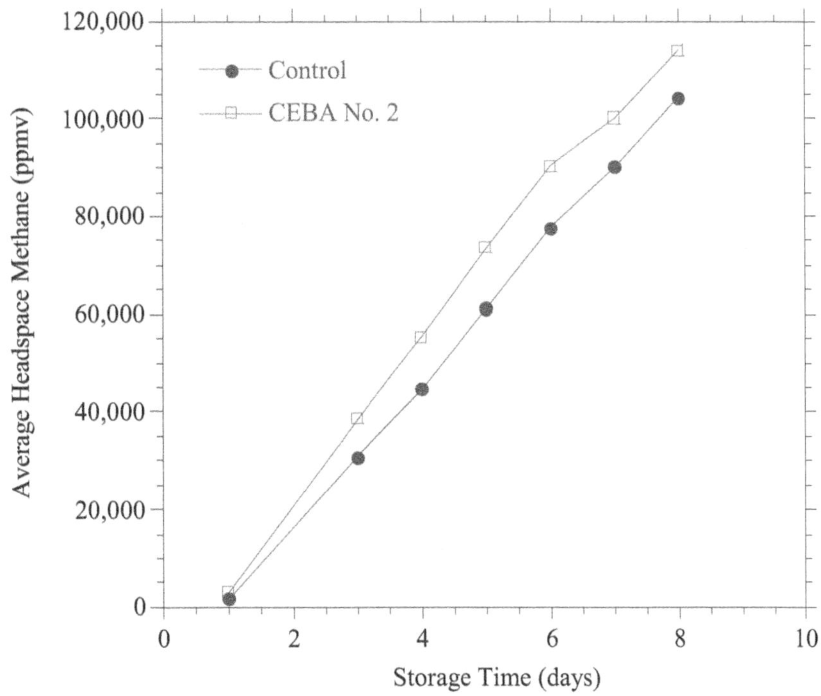

Figure 3-69. Headspace Methane During Storage of Control and CEBA No. 2-Amended Cakes.

CEBA No 3: CEBA No. 3 was added to the dewatered sludge according to the SCTP, which was developed by the WERF team and approved by the vendor. The biosolids were conditioned and dewatered according to the SCTP for CEBA No. 3, a CEBA provided by the vendor.

Testing Protocol Synopsis

A summary of the testing protocol for CEBA No. 3 is as follows:

1. The biosolids were conditioned and dewatered using the Bucknell high-solids centrifuge simulation process. After conditioning and dewatering, a cake was produced with 27% dry solids content. This is within the typical range for cake solids from high solids centrifuges.
2. The cake that was generated was sprayed with the prepared CEBA No. 3 product, using a spray bottle. The dosage was 368 ppmv of CEBA No. 3 product using a wet weight basis for the biosolids.
3. After the agent was added, the cake sample was processed through the mixer one time, and was placed in sample bottles and incubated at 25°C for a series of headspace analyses. Samples of the headspace were taken at regular intervals and analyzed for VOSCs using gas chromatography according to the methods described in Section 2.2 General Protocols. A control sample was prepared that was treated the same as the CEBA No. 3 sample, except no that agent was added. All samples were prepared in duplicate.

CEBA No. 3 Impact on Headspace TVOSC

The results are shown graphically on Figure 3-70 through Figure 3-73. The MT headspace concentration for CEBA No. 3 and the associated control sample (Figure 3-70) were essentially the same, with the CEBA measurements being slightly higher. In the comparative plot of DMS concentrations (Figure 3-71), the control sample had a slightly greater peak DMS concentration than the CEBA No. 3 amended sample, resulting in a 17% reduction in DMS due to CEBA No. 3 addition. The TVOSC concentration in the headspace during storage is shown in Figure 3-72.

Similar to the MT results, the concentrations of TVOSC in the headspace were similar for both the CEBA No. 3 amended sample and the control sample. The addition of CEBA No. 3 resulted in a small decrease in the methane production as shown in Figure 3-73, similar to CEBA No. 2.

In summary, the addition of CEBA No. 3 did not substantially reduce the production of odorous VOSCs from anaerobicly digested and dewatered biosolids. Additional testing of different dosages and conditions may be warranted to more fully investigate the effect to this agent on reducing odorous compounds, since this experimental testing relied only on one vendor-approved dosage.

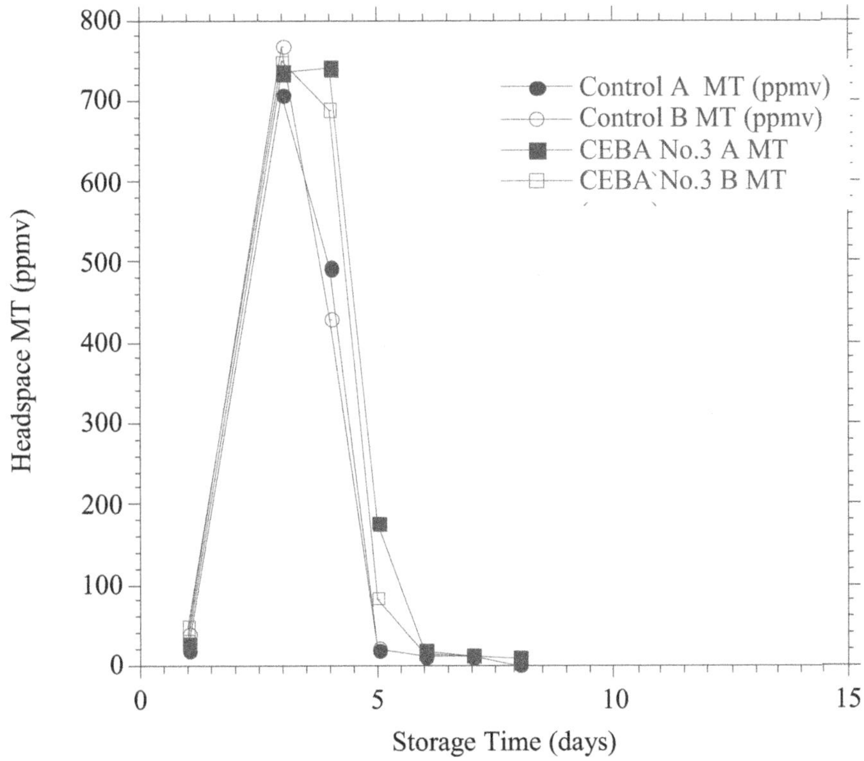

Figure 3-70. Headspace Methanethiol During Storage of Control and CEBA No. 3-Amended Cake.

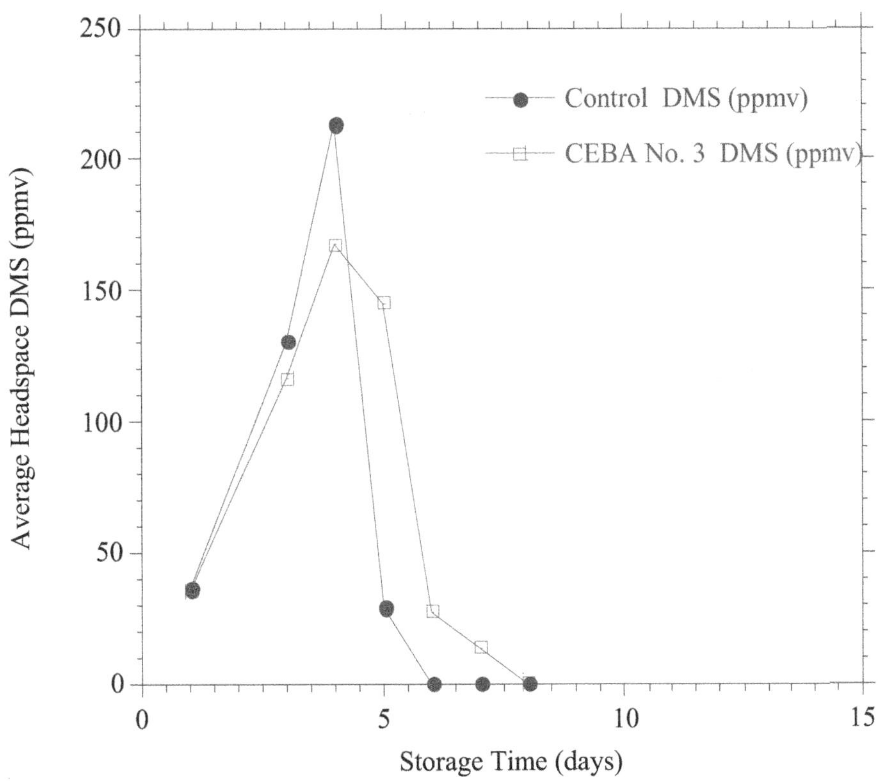

Figure 3-71. Headspace Dimethyl Sulfide During Storage of Control and CEBA No. 3-Amended Cake.

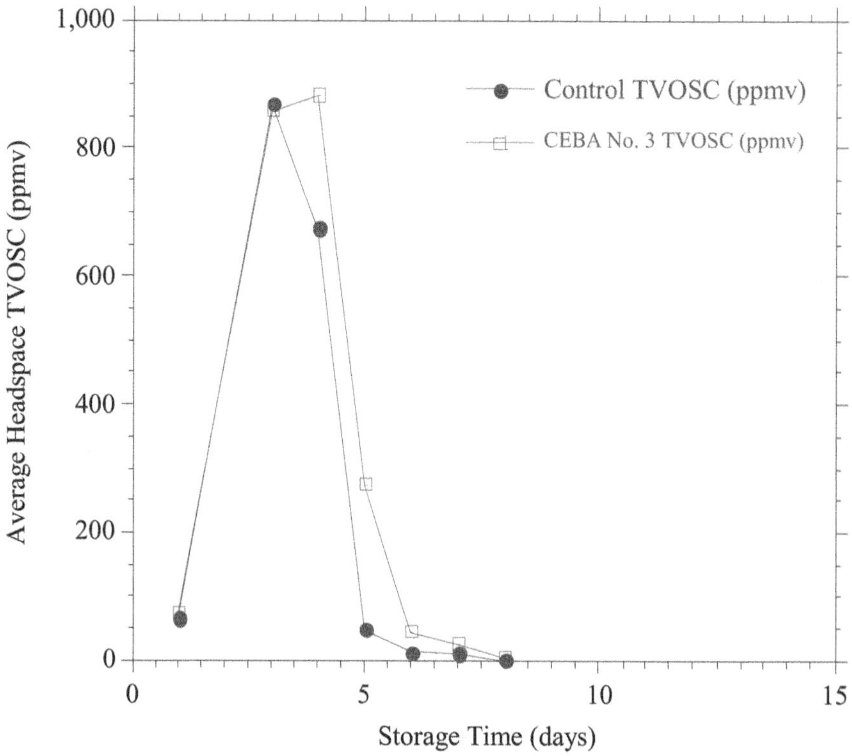

Figure 3-72. Headspace TVOSC During Storage of Control and CEBA No. 3-Amended Cake.

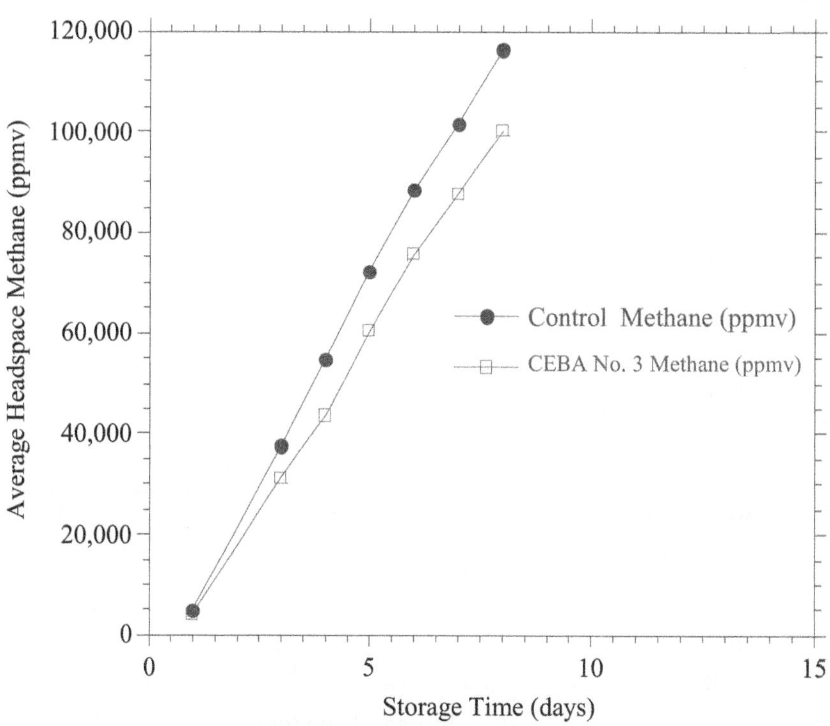

Figure 3-73. Headspace Methane During Storage of Control and CEBA No. 3-Amended Cake.

3.2.3.4 CEBA Application Test Results Summary

The following conclusions have been made regarding the 8 CEBAs tested in the Phase 3 research:

- Of the four pre-digestion CEBAs tested, the headspace TVOSC results of three (CEBAs Nos. 4, 5, and 6) were essentially no different than controls at the doses tested. Addition of CEBA No. 5 to the digester sludge feed resulted in an improvement in volatile solids reduction (indicating improved digestion) but the VS reduction improvement was not accompanied by a reduction in TVOSC.
- The application of pre-digestion CEBA No. 1 produced a sludge that was inhibitory to methanogens due to reduced pH. As a result, VS reduction was retarded and little gas was produced. The TVOSC generation was very low following addition of CEBA No. 1, but the level of digestion accomplished was not acceptable. The TVOSC reduction was attributed to inhibition of overall microbial activity in the digester.
- Headspace analysis of cake produced with sludge dosed with post-digestion applied CEBAs No. 7 and 8 showed that there was an increase in TVOSC upon treating with the CEBAs (about a 70% increase for CEBA No. 7 and a 33% increase in CEBA No. 8 over controls).
- Both CEBA No. 2 and No. 3-dosed samples showed little change in peak TVOSC readings as compared to control cake samples.

The results summarized above indicate that none of the eight CEBAs tested had an overall positive effect on the sludges that were used for testing. For each of the CEBAs, additional testing of different dosages and conditions may be warranted to more fully investigate any impact the agents may have on reducing odorous compounds or on improvement of digestion.

3.2.4 Aluminum Addition Laboratory Test Results

In the Phase 3A research, Bucknell University performed several laboratory-scale trials with alum addition to the dewatering process feed sludge. These trials used the high solids centrifuge simulation procedure described in Section 2.2 General Protocols to produce cake that had similar solids content as cake samples from full-scale high solids centrifuges. In the laboratory testing, alum is added to digested sludge prior to polymer addition, and the samples are mixed and then dewatered using a laboratory centrifuge.

Bucknell University completed alum testing in four trials using sludge from three different WWTPs. Two trials were performed using the Philadelphia Water Department (PWD) Biosolids Recycling Center (BRC) sludge, one trial was performed using the San Francisco Southeast Wastewater Treatment Plant sludge, and one trial was performed using the LACSD Joint Water Pollution Control Plant sludge. The following sections summarize results from each of these trials.

3.2.4.1 Philadelphia Water Department Alum Testing Trial 1: Constant Polymer Dose

In the first alum addition trial using the BRC sludge, the polymer dose used for dewatering in the BRC centrifuges was kept constant. This decision was made because polymer dose itself has been shown to impact the production of odor-causing compounds (Higgins et al., 2005). Bucknell University conducted laboratory analysis of sludge samples. A testing protocol synopsis is provided in Section 2.3.4. The effect of alum dosage on the TVOSC production profile is shown in Figure 3-74. The figure indicates that addition of alum greatly reduced the production of TVOSCs compared to the control with no alum added. In fact, the lowest dosage examined (0.5% as Al) reduced the peak TVOSC concentration by about 83%.

In the Philadelphia BRC testing, alum addition did not appreciably affect methanogenic activity, as indicated in Figure 3-75. Therefore, alum addition could have reduced the bioavailability of protein, and thereby decreased the biodegradation of protein that leads to the production of odorous compounds. The effect of the alum dosage on the dewatered cake solids content is shown in Figure 3-76. This figure shows that alum addition did reduce the solids content in the dewatered cake from about 18% to about 16.5%.

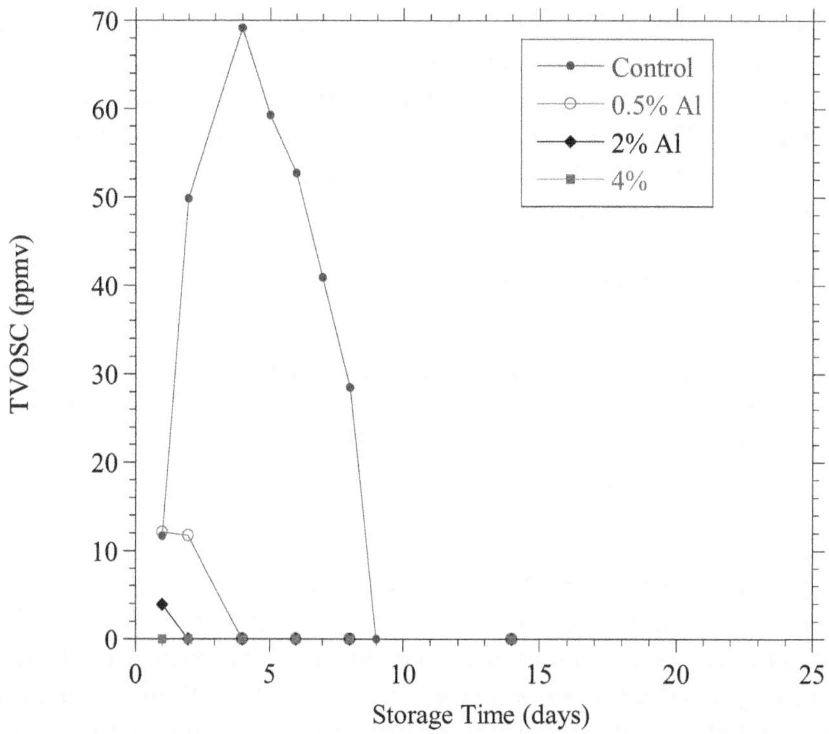

Figure 3-74. Production Profile for TVOSCs for Different Alum Dosages Used During Conditioning.

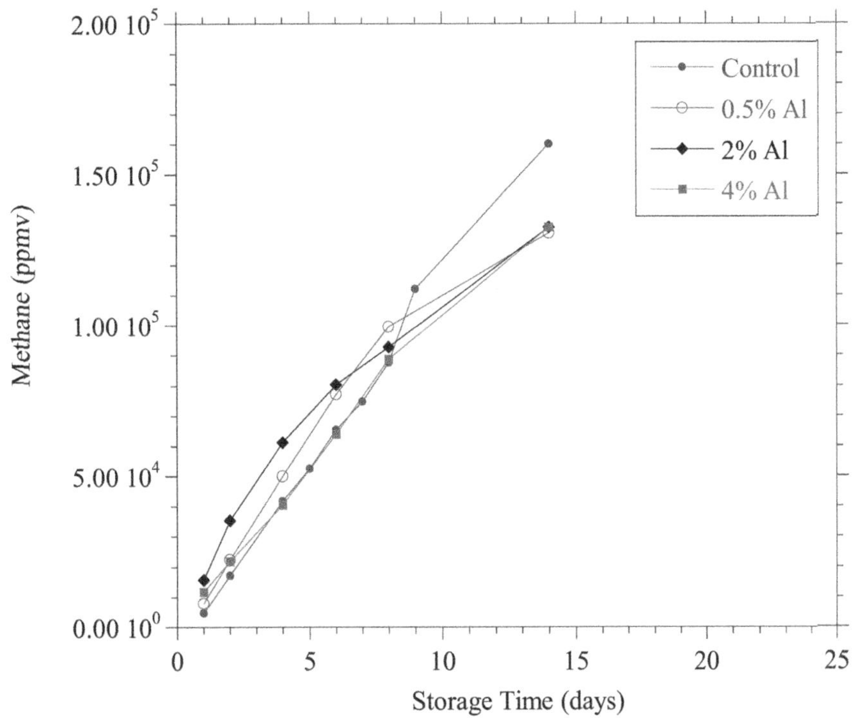

Figure 3-75. Methane Production by Cakes with Different Alum Dosages Used During Conditioning.

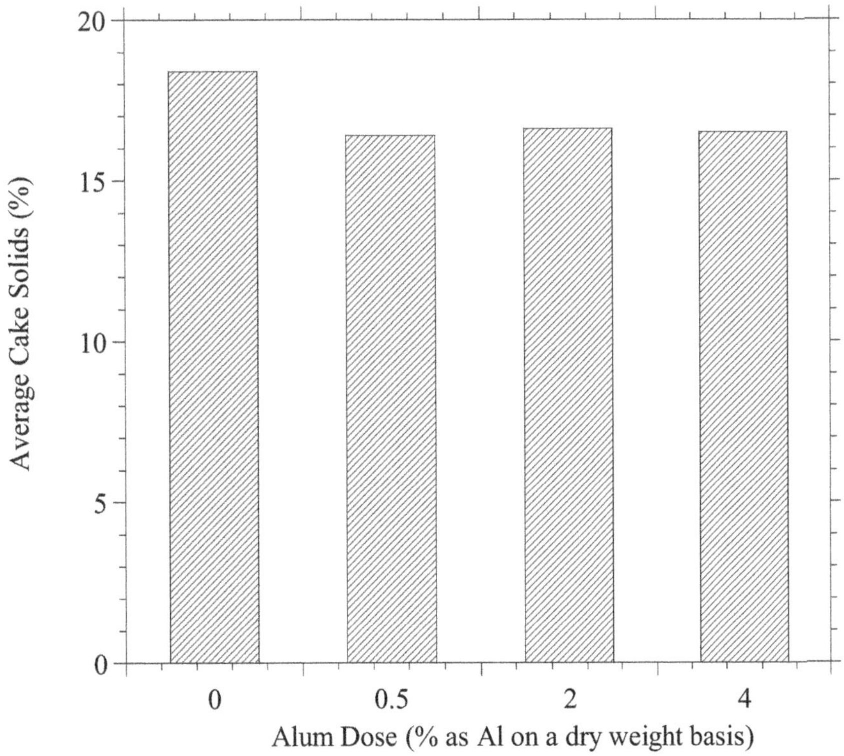

Figure 3-76. Average Cake Solids for Different Alum Dosages Used During Conditioning.

3.2.4.2 Philadelphia Water Department Alum Testing Trial 2: Optimum Polymer Dose

In the second trial at the BRC, the optimum polymer dose was determined for each alum dose. This trial was added to the Phase 3A research to simulate conditions that would occur in a full-scale application, where a POTW that chose to incorporate alum addition at a specified percentage as Al to the dewatering feed would also want to determine the optimum polymer dose for that alum dose. The WERF team has theorized that one potential benefit of alum addition is that it may be able to lower the polymer dose needed for conditioning in addition to any benefit that may be realized for odor control.

The OPD for each alum dose is shown in Figure 3-77. The increased alum dose resulted in a substantial decrease in OPD, which could represent an important reduction in costs for POTWs. The effect of the alum addition on the TVOSC production profile is depicted in Figure 3-78, and the peak TVOSC measured from each sample is shown in Figure 3-79. Similar to Trial 1 at the BRC, alum addition resulted in a significant reduction in the peak TVOSC concentration. In these trials, methane production increased compared to the control for the first two dosages (0.5 and 2%), and decreased slightly at the highest dosage of 4% compared to the control. The results indicate that alum addition in the low dosage range used for these trials does not seem to negatively impact beneficial methanogenic activity (Figure 3-80). This is important since inhibiting methanogenic activity has been shown to increase VOSC concentrations, therefore, greater odor reduction could be achieved if inhibition of methanogenic activity is minimized.

Field and laboratory data demonstrated that polymer dosage could be reduced with low alum doses up to 1%. However, the studies also showed that differences in sludges from different facilities make it difficult to establish a "low" dosage point where polymer consumption reduction would be definitively observed along with odor reduction.

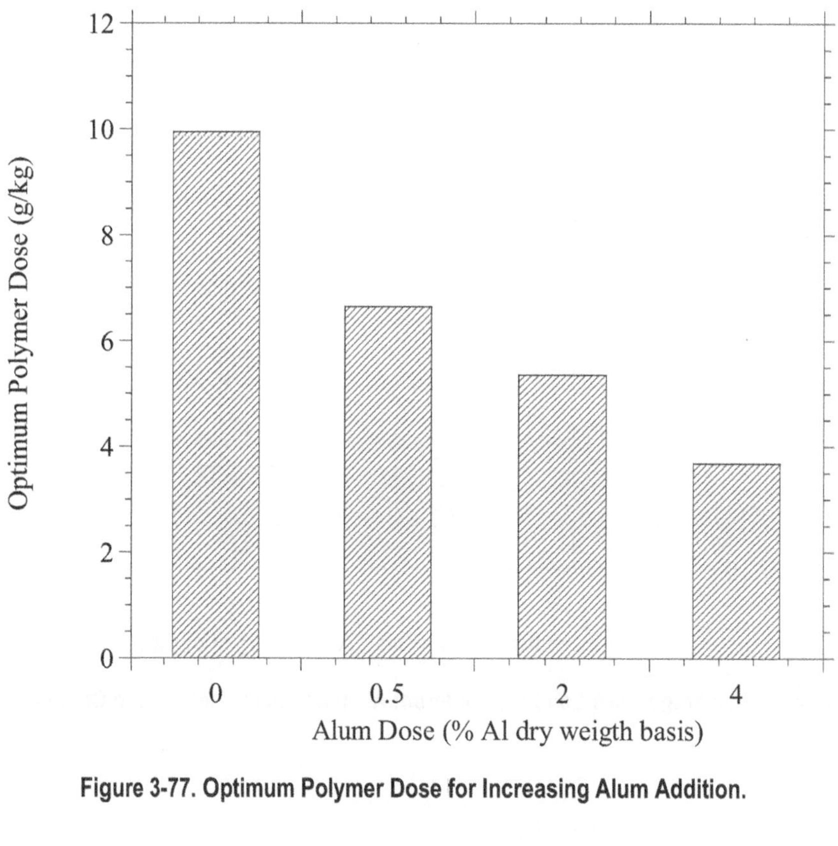

Figure 3-77. Optimum Polymer Dose for Increasing Alum Addition.

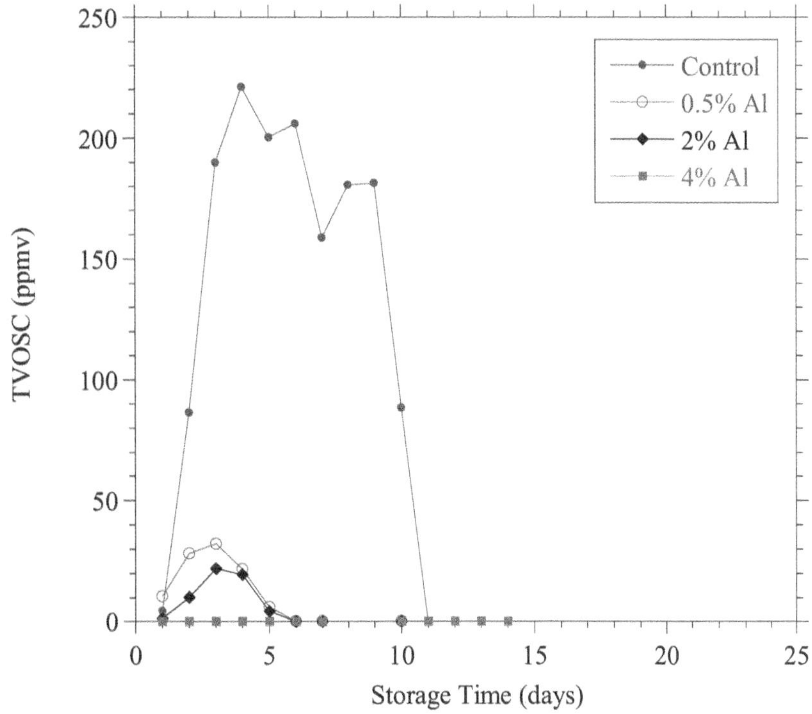

Figure 3-78. Production of TVOSC During Storage for Different Alum Dosages.

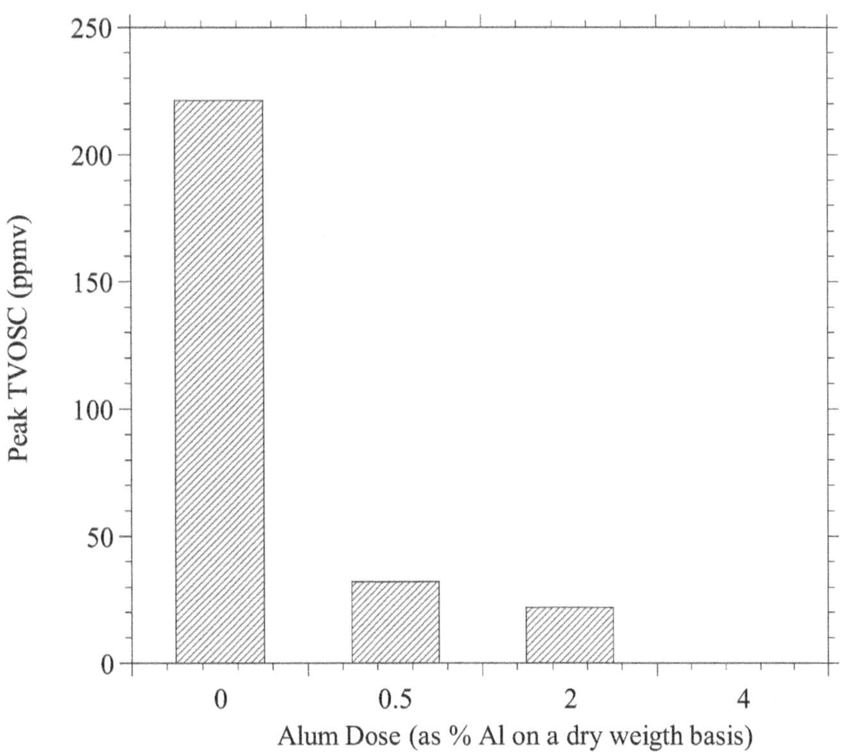

Figure 3-79. Peak TVOSC Concentrations from Cake Samples Produced with Different Alum Dosages.

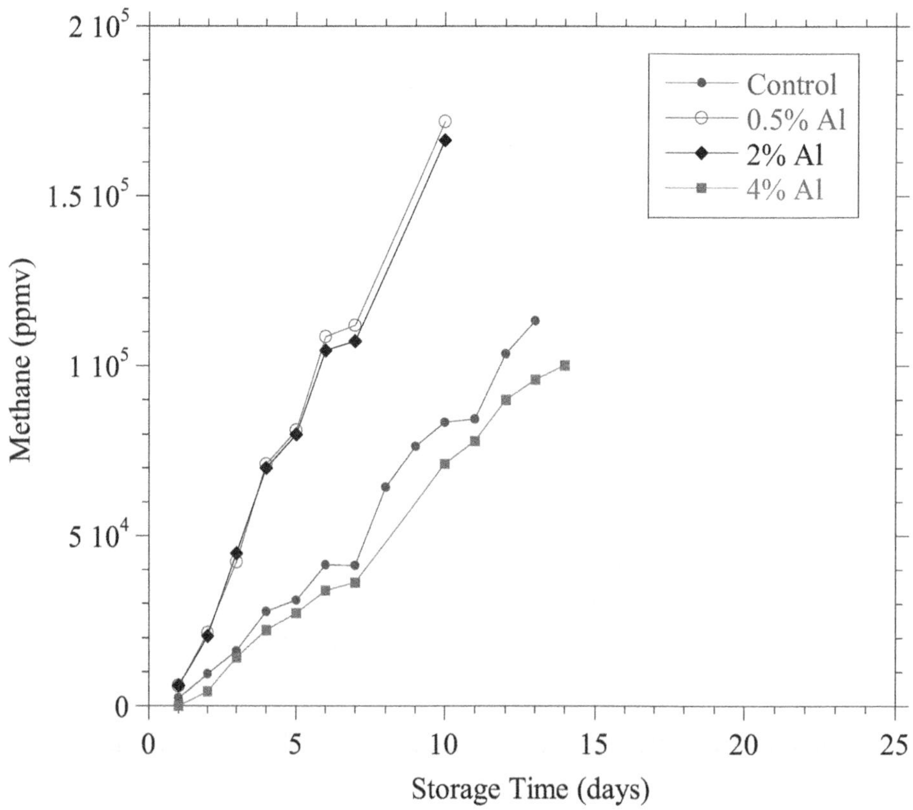

Figure 3-80. Methane Production from Cake Samples Produced with Different Alum Dosages.

3.2.4.3 Effect of Alum Addition – San Francisco

Biosolids from the San Francisco Southeast WWTP were sampled and tested in the same manner as described in the PWD trials to determine if the effects of alum addition would be repeated using a different sludge. Both liquid-digested biosolids and dewatered cake samples were shipped from the plant to Bucknell for testing. The liquid biosolids samples were subjected to alum doses ranging from 0-4% (in terms of aluminum mass to dry solids mass), as in the PWD trials.

The results in terms of peak TVOSC versus alum dose are shown in Figures 3-81 through Figure 3-83. Note that the dewatered cake sample sent to Bucknell from the WWTP ("Centrifuge Cake" data in Figure 3-81) had lower TVOSC concentrations than the laboratory-sheared and dewatered corresponding sample (0% Al), indicating that the laboratory shearing device likely exerted higher shear than the full-scale centrifuge. Increasing aluminum dosage rates were once again effective at reducing the peak TVOSC in the laboratory-dewatered cake, similar to the results of Philadelphia BRC Trials 1 and 2. In the San Francisco testing, the most effective aluminum dose appeared to be in the range of 2% aluminum to mass of dry biosolids, as methane production decreased upon application of the 4% Al dose (Figure 3-84).

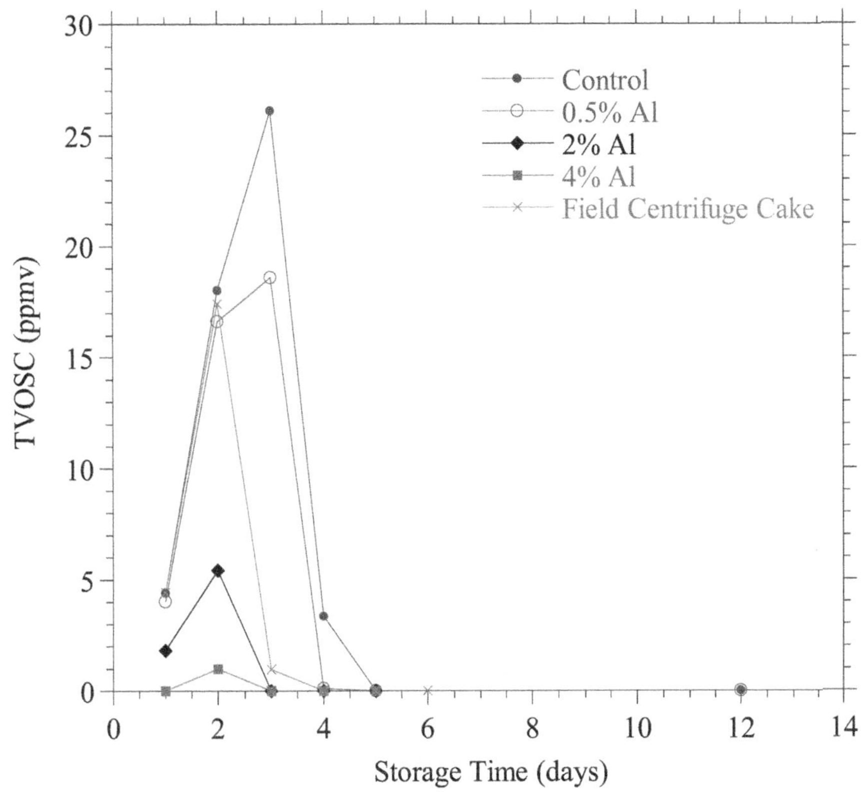

Figure 3-81. Headspace TVOSC Concentrations in San Francisco Alum Testing. (Note the "Field Centrifuge Cake" Data Represents the Cake Sample Sent Directly from the Plant).

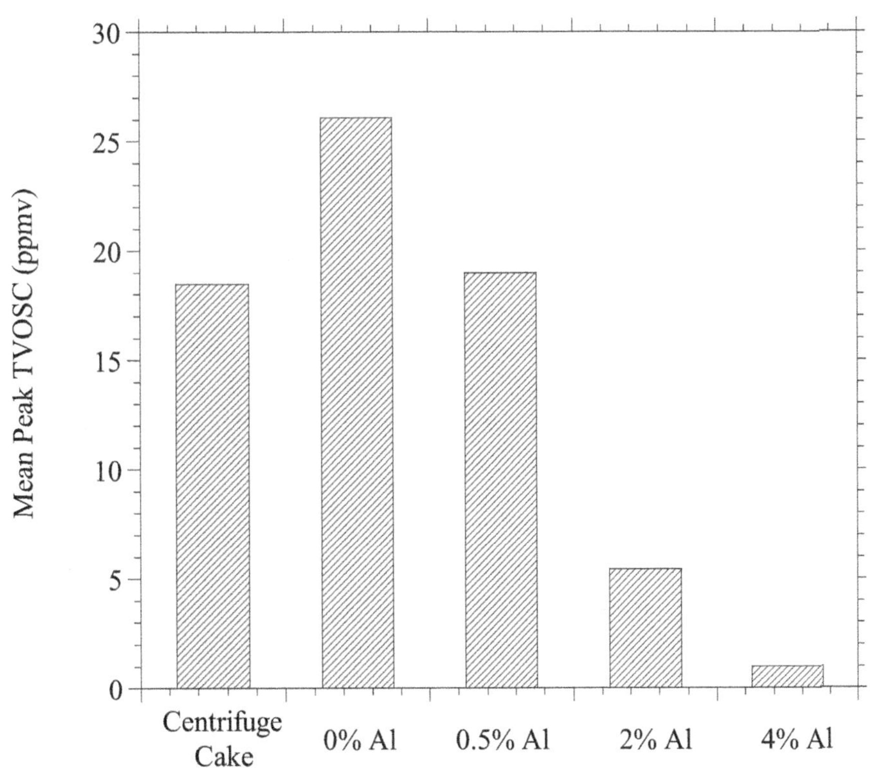

Figure 3-82. Mean Peak Headspace TVOSC for Each Sample in San Francisco Alum Testing.

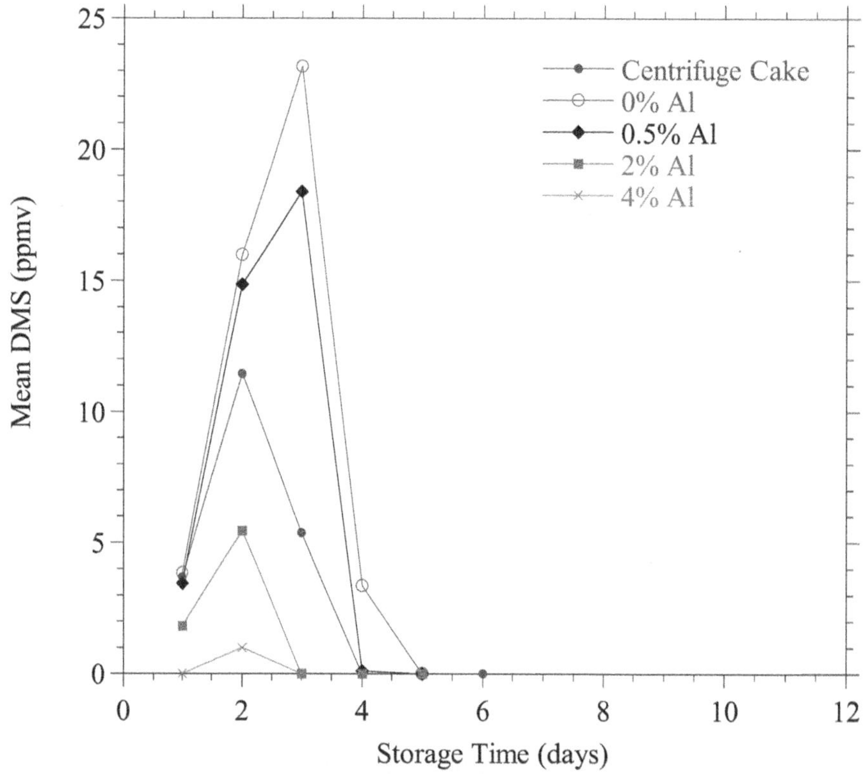

Figure 3-83. Headspace Dimethyl Sulfide Concentration for Different Samples in San Francisco Alum Testing.

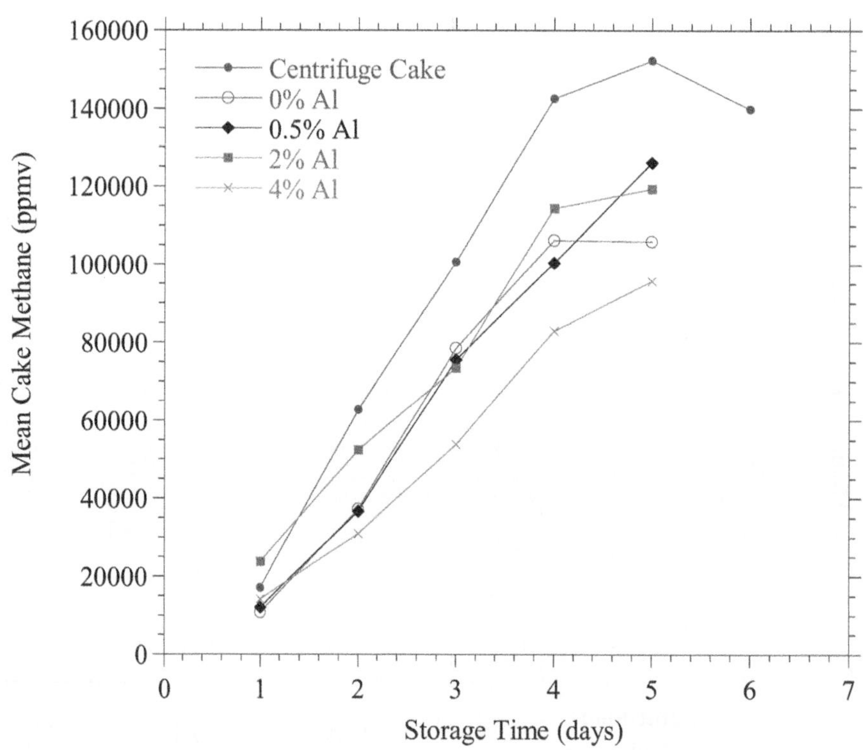

Figure 3-84. San Francisco Headspace Methane Concentrations for Different Alum Additions.

3.2.4.4 Effect of Alum Addition – LACSD

A fourth trial studying alum addition to dewatering feed sludge was conducted using sludge from the LACSD JWPCP. The same testing protocol and alum dosages were used as in the previous alum trials. The TVOSC production profiles for cakes with varying alum dosages made in the Bucknell laboratory is shown in Figure 3-85. The control TVOSC profile is indicated as 0% as Al alum addition. The figure indicates that after a slight increase in TVOSC during storage for a 0.5% as Al alum dose (though the peak TVOSC for the 0.5% as Al dose was slightly lower than the control), the TVOSC profile was depressed significantly with greater alum doses (2% and 4% as Al). Furthermore, the peak TVOSC was reduced from approximately 100 ppmv in the control to approximately 54 ppmv and 34 ppmv with 2% and 4% as Al doses, respectively.

However, the control sample (0% alum) laboratory cake TVOSC measurement in these tests was significantly lower than the TVOSC measured in the centrifuge cake dewatered at full-scale at the plant and shipped to Bucknell for analysis (a laboratory peak TVOSC of 100 ppmv versus 2,600 ppmv measured in the field cake). During the testing, the Bucknell University laboratory experienced difficulties in effectively dewatering the digested sludge samples using the polymer sample provided by LACSD, which may have contributed to the discrepancy between TVOSC profiles produced at bench-scale versus full-scale. Therefore, the results in Figure 3-85 are not comparable to full-scale dewatered cake at the JWPCP, but the relative results comparing alum doses at bench-scale are encouraging.

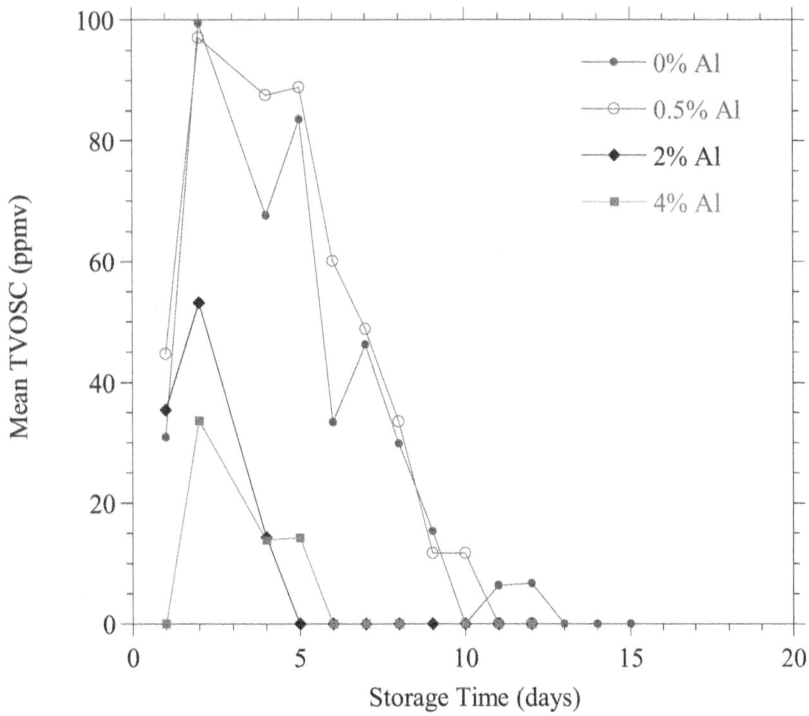

Figure 3-85. LACSD Sludge Mean TVOSC for Different Alum Doses. (Note the TVOSC Profile Produced by the JWPCP-Dewatered Cake Sample is Not Included to Provide an Adequate Scale).

3.3 Phase 3C Full-Scale Experiments Test Results

Full-scale dewatered biosolids odor tests that were conducted as part of Phase 3C included experiments on advanced digestion, dewatering equipment comparisons,

MicrosludgeTM evaluations, CEBA addition, and chemical addition (lime, ferric chloride, and alum). The following sections present the test results for these portions of the Phase 3 research and provide conclusions based on those results.

3.3.1 Advanced Digestion Studies

In these experiments wastewater treatment plants that had operational advanced digestion processes, either pilot or full-scale, were studied to determine how advanced digestion processes influenced odor potential from centrifugally dewatered sludges. Tests were conducted at three treatment plants that had operational advanced digestion systems: Inland Empire Utilities Agency, Baltimore Back River WWTP, and Western Lake Superior Sanitation District. Results from the Phase 3C tests are described in the following sections.

3.3.1.1 Inland Empire Utilities Agency, California, Phased Digestion

The WERF team conducted full-scale advanced digestion testing at Inland Empire Utilities Agency (IEUA) from three distinct treatment trains:

1. Acid-gas mesophilic-thermophilic (AGMT)
2. Acid-gas mesophilic-mesophilic (AGMM)
3. Acid-gas mesophilic-thermophilic-mesophilic (AGMTM)

The AGMM and AGMT systems had a total SRT of 19 days and the AGMTM had a total SRT of 26 days, as detailed in Section 2.4.1.1. Three anaerobic treatment schemes were compared for odor generation from dewatered biosolids cake. Figure 3-86 depicts the variation in TVOSC in the headspace of a dewatered high-solids cake from the three anaerobic digestion schemes when BESA was added, which reflects the odor potential of the cake sample. Figure 3-87 depicts the TVOSC concentrations when BESA was not added.

The dewatered sludge cake from the BESA-amended AGMT system was approximately 15% lower in headspace TVOSC compared to the AGMM system. This suggests that a combination of mesophilic and thermophilic digestion is more effective in reducing odors from dewatered biosolids cake. However, the amount of the reduction is not likely to be sufficient enough to eliminate odor complaints, as TVOSC still peaks at approximately 1,300 ppmv.

The three phase digestion (AGMTM) resulted in the lowest TVOSC of the three digestion trains sampled, as the headspace peak TVOSC of the AGMTM system was reduced by 50% compared to the AGMT dewatered cake. Therefore, results indicate that the three stage acid-gas system with a combination of mesophilic-thermophilic-mesophilic temperatures is an effective process scheme for reducing odor potential from dewatered biosolids. This could be due to the longer SRT coupled with an additional digestion step or from the unique combination of the MTM configuration, or both. This reduction is considered by the WERF team to be significant and suggests that multi-stage digestion as a means of odor control of dewatered biosolids should be investigated further.

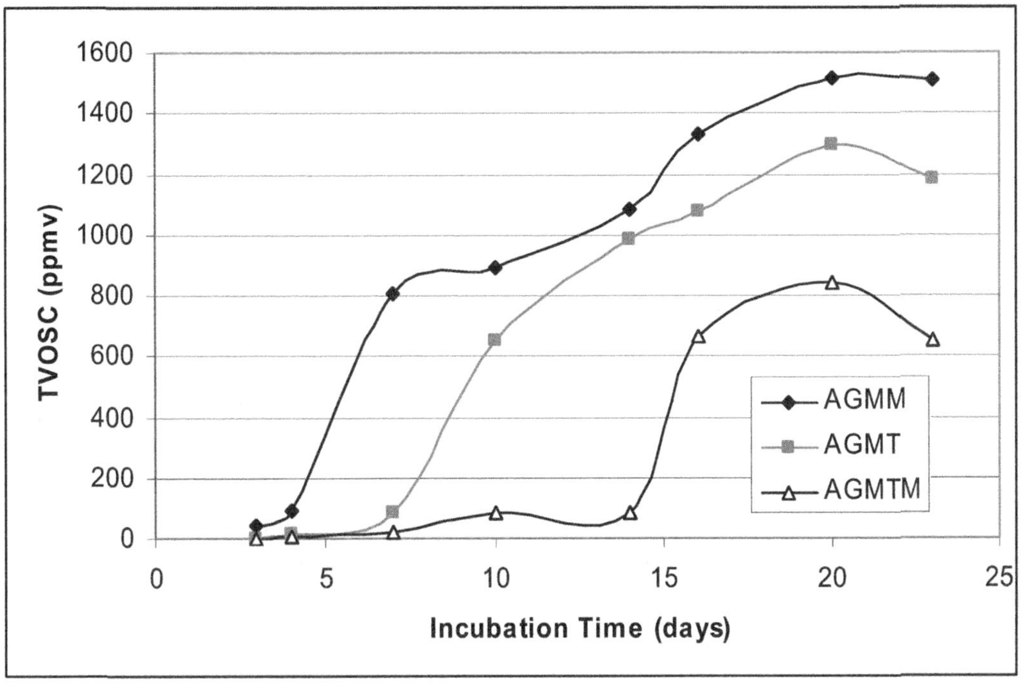

Figure 3-86. IEUA Headspace TVOSC Concentrations for Three Digestion Trains (Amended with BESA).

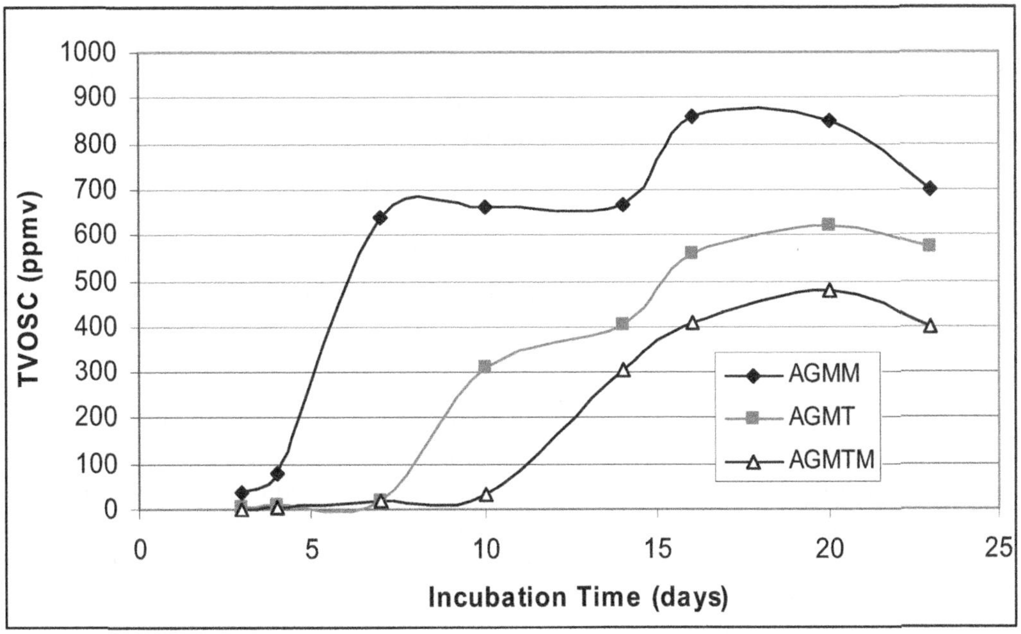

Figure 3-87. IEUA Headspace TVOSC Concentrations for Three Digestion Trains (Not Amended with BESA).

3.3.1.2 Baltimore Back River WWTP Phased Digestion

The WERF team conducted full-scale advanced digestion testing at the Baltimore Back River WWTP from three distinct treatment trains:

1. Conventional mesophilic digestion (pancake-shaped digesters)
2. Conventional mesophilic digestion (egg-shaped digesters)
3. Acid-gas mesophilic-mesophilic (AGMM) digestion

The Baltimore Back River plant was tested on two occasions. The first set of tests were on sludge from the conventional mesophilic egg-shaped digesters, the acid phase of the AGMM digestion train, and the gas phase of the AGMM digestion train (which were pancake-shaped digesters). Figures 3-88 and 3-89 show the variation in TVOSC from the dewatered cake for the conventional mesophilic digestion (egg-shaped) samples versus the AGMM digestion train samples with and without the addition of BESA. The plots indicate that the AGMM system did not have much benefit with regard to reduction in TVOSC generation compared to the conventional mesophilic egg-shaped digester. The acid phase sludge was high in TVOSC, which was expected because of the limited digestion time.

A second set of experiments was conducted to compare headspace TVOSC results for sludge digested by conventional mesophilic digestion using a pancake digester versus sludge digested by the two-stage AGMM system. These samples were comprised of sludge from the gas phase of the acid-gas system (where the gas phase digester was pancake shaped), conventional mesophilic digestion (pancake-shaped digester) and conventional mesophilic digestion (egg-shaped digester).

Figure 3-90 depicts the variation in headspace TVOSC concentration for the dewatered sludge cakes as a function of incubation time for the three sludges with BESA and Figure 3-91 depicts the headspace TVOSC concentration with BESA added. As indicated in the plots, the sludge from conventional mesophilic, pancake-shaped digester generated the highest amount of TVOSC in the headspace while the conventional egg-shaped digester had the lowest amount of TVOSC. These results suggest that an acid/gas advanced digestion system produces sludge with lower TVOSC compared to conventional mesophilic digestion, but *higher than conventional mesophilic egg-shaped digesters*.

The acid-gas system reduced the TVOSC odor potential by 40-50% compared to the conventional pancake digester. However, the egg-shaped digester in a conventional mode reduced TVOSC by about 30% compared to the acid-gas system. A comparison of TVOSC concentrations suggest that the conventional mesophilic egg-shaped digester would reduce TVOSC by approximately 70% compared to the conventional mesophilic pancake-shaped digester. The results of this testing appears to offer a strategy to reduce odors through selection of the shape of the digester in addition to additional evidence that advanced digestion produces a less odorous dewatered biosolids cake.

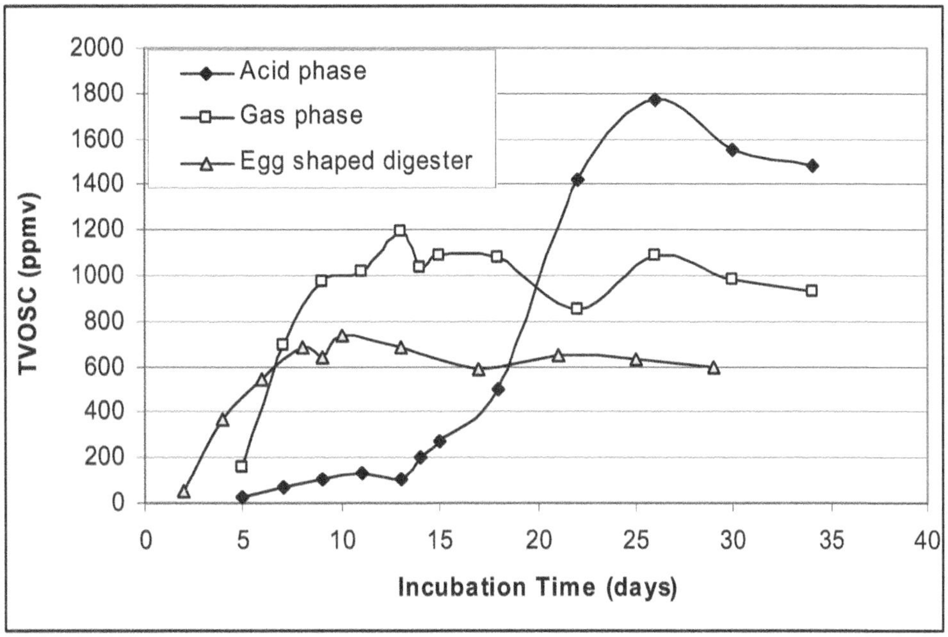

Figure 3-88. Baltimore Back River WWTP Digestion Train Variations in Headspace TVOSC for Sample Set 1 (BESA Added).

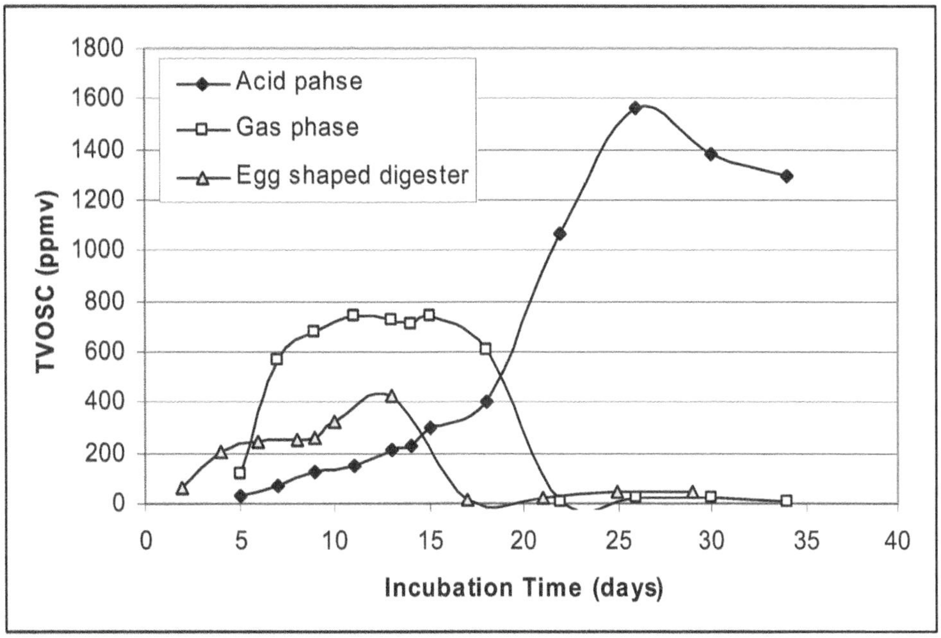

Figure 3-89. Baltimore Back River WWTP Digestion Train Variations in Headspace TVOSC for Sample Set 1 (No BESA Added).

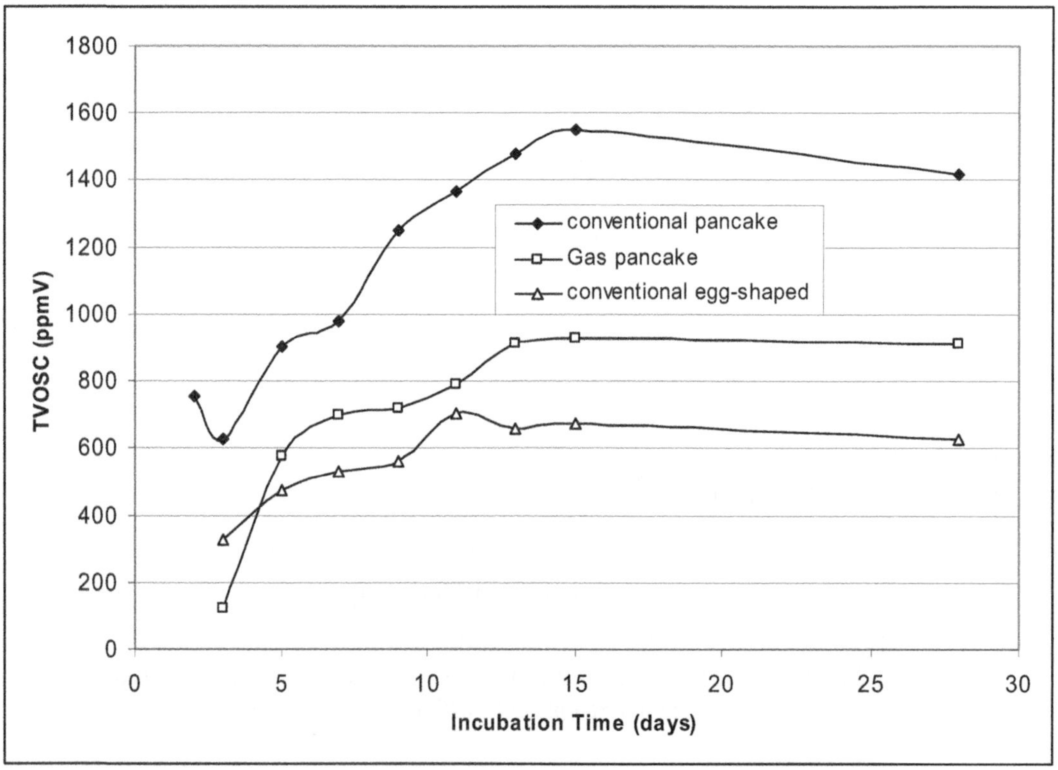

Figure 3-90. Baltimore Back River WWTP Digestion Train Variations in Headspace TVOSC for Sample Set 2 (BESA Added).

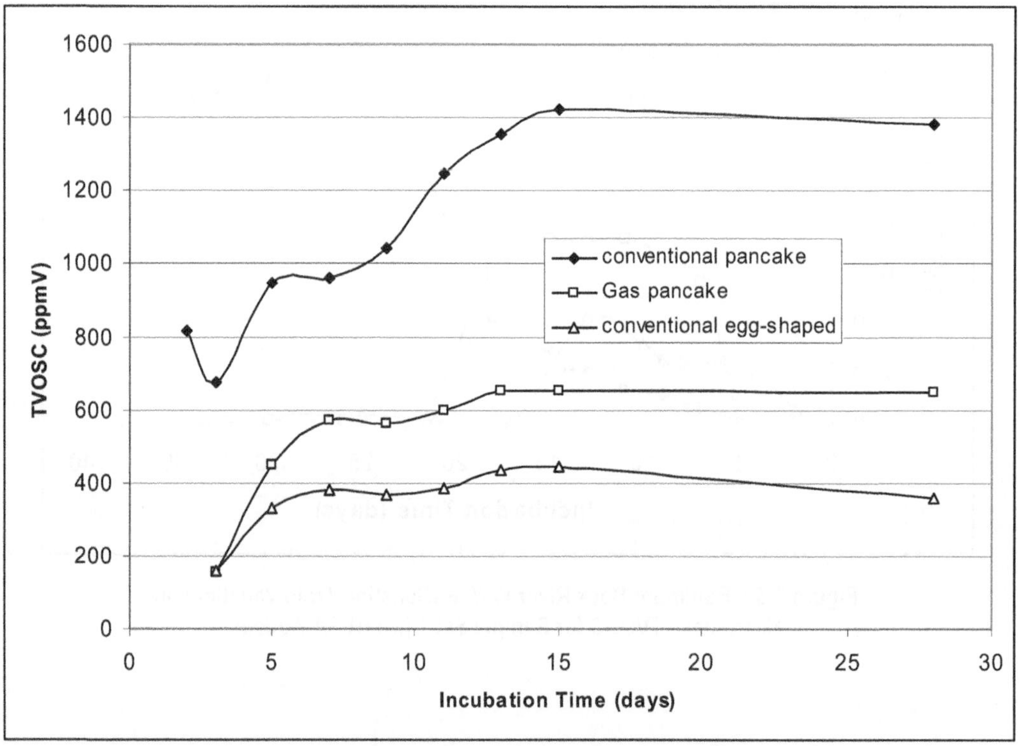

Figure 3-91. Baltimore Back River WWTP Digestion Train Variations in Headspace TVOSC for Sample Set 2 (No BESA Added).

3.3.1.3 Western Lake Superior Sanitation District Duluth WWTP Temperature-Phased Digestion

The final advanced digestion study was conducted using sludge samples from a temperature-phased anaerobic digestion (TPAD) system at the Western Lake Superior Sanitation District (WLSSD) in Duluth, MN. During the full-scale Phase 3C testing, WLSSD was employing a second stage mesophilic digester that was higher in temperature (42-48°C) than is "typical" for a TPAD system.

The study at WLSSD was undertaken to determine if operating the second mesophilic digestion stage at a "normal" mesophilic temperature could reduce TVOSC generation in the dewatered biosolids cake. To accomplish this, bench-scale testing was conducted at Virginia Tech on a sample from the first digestion stage that was digested in the laboratory for an additional 30 days at 37°C and compared to the dewatered cake from the second stage of the full-scale process (which was operating at a temperature of 42°C during testing) in terms of TVOSC production. Sludge samples from the field first stage were also evaluated.

A comparison of the TVOSC generation from the full-scale plant-digested sludge is compared to the laboratory-digested sludge in Figure 3-92. The "field thermo stage" denotes the sludge from the full-scale second stage that was operated at a higher temperature while the "field meso stage" denotes the full-scale sludge from the first stage. As indicated in the figure, there was little difference in the peak TVOSC between the field second stage and the laboratory second stage operated at a lower temperature. From this study, the WERF team concluded that the second stage mesophilic digestion practice at WLSSD, while not at a standard mesophilic temperature, is not a problem with regard to biosolids odor potential.

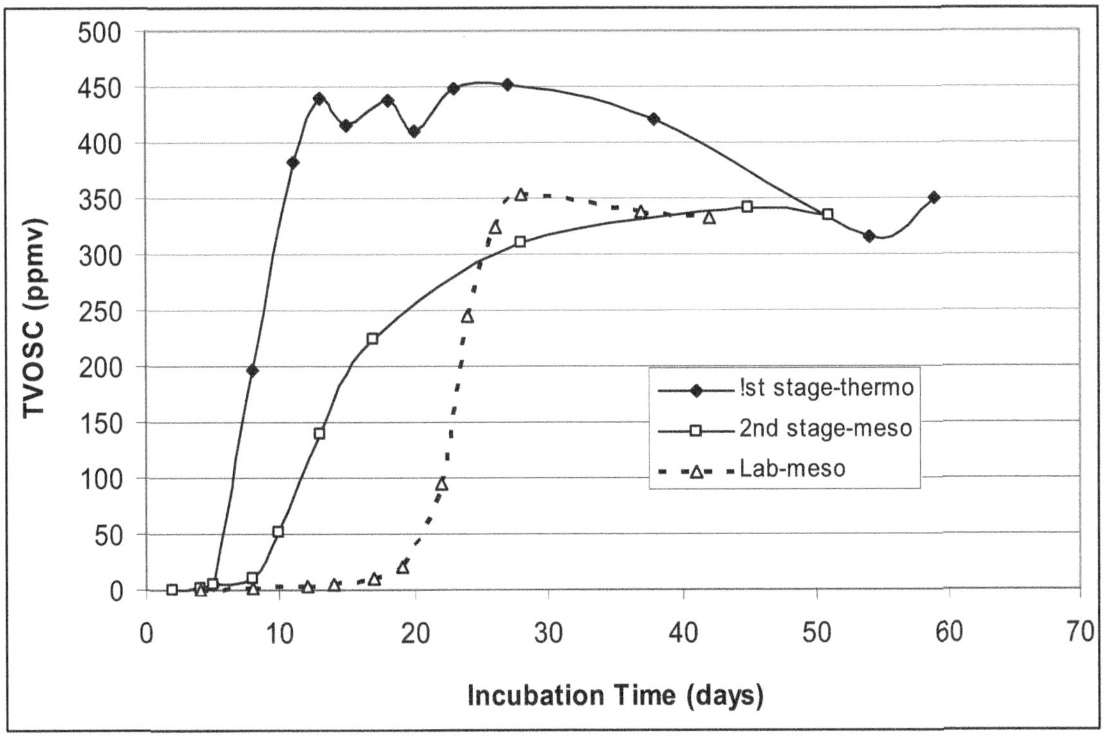

Figure 3-92. WLSSD Comparison of Headspace TVOSC of Field and Laboratory Digestion (BESA Added).

3.3.2 Dewatering Equipment Test Results

A series of full-scale trials was performed to examine the effect of dewatering processes on the production of odors. The trials included examination of the following:

- Different centrifuge types at two utilities
- Centrifuge operational parameters such as torque and bowl speed;
- Comparison of side-by-side belt filter press and centrifuge dewatering;
- Comparison of side-by-side centrifuge and Fournier rotary press.

The results from these tests are provided in the following sections.

3.3.2.1 Philadelphia Biosolids Recycling Center Centrifuge Tests

Full-scale side-by-side centrifuge testing on odor production was conducted at the Philadelphia Water Department Biosolids Recycling Center (BRC) in fall 2004. Samples were collected from six different centrifuges, all dewatering the same feed biosolids.

Effect of Solids Content on VOSC Production

The peak normalized headspace TVOSC concentration measured during dewatered cake storage is shown in Figure 3-93 for each biosolids sample, along with the cake solids achieved for each centrifuge. The "normalized" headspace TVOSC is defined as the peak TVOSC measured during storage divided by the mass of total solids in the headspace bottle. The purpose of this conversion is to normalize the results for the cake solids differences which result in different amounts of solids being placed in the headspace bottles, because during sampling the solids are weighed into the bottle on a wet solids basis.

The plot in Figure 3-93 shows that in general, higher cake solids correlated with higher TVOSC concentration, although not in every case. For example, a Centrysis modified Bird centrifuge (one of the medium-solids centrifuges) produced a cake of 30% solids, yet had a relatively low TVOSC production. It is interesting to note that since these are normalized results, the difference in cake solids alone does not explain the differences in headspace TVOSC. If cake solids was the only factor, the results would show equivalent peak, normalized TVOSC, but since the results are different, this suggests that shear or other factors may contribute.

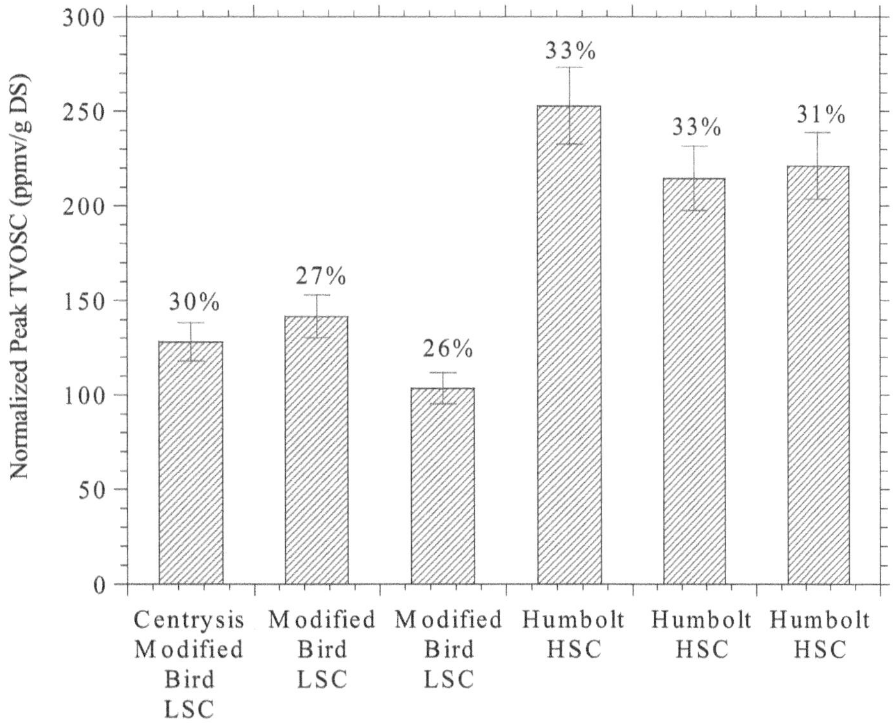

Figure 3-93. Normalized Peak TVOSC Concentration for Different Centrifuge Types at the Philadelphia Water Department BRC (Numbers on Bars Represent Percentage Cake Solids Content from the Machine on Day of Testing).

Effect of Centrifuge Torque on VOSC Production

Full-scale trials were performed at the BRC for the purpose of evaluating the impact of centrifuge torque on dewatering performance and TVOSC production. At the BRC, this trial was performed on a Humboldt high-solids centrifuge. The effect of increasing centrifuge torque on the cake solids and the normalized peak TVOSC production in the headspace of the dewatered cake is depicted in Figure 3-94.

The figure indicates that as torque on the centrifuge was increased, the cake solids increased proportionally from approximately 29% at the lowest torque setting to about 33% at the highest torque setting, a substantial increase in the cake solids. Similar results have been shown in previous research (Higgins et al., 2005). Additionally, the peak TVOSC from the cakes produced at different torque values increased with increasing torque. Therefore, based on this study, an increase in centrifuge torque is expected to produce more odorous cakes, although the benefit of greater cake solids would need to be considered as well.

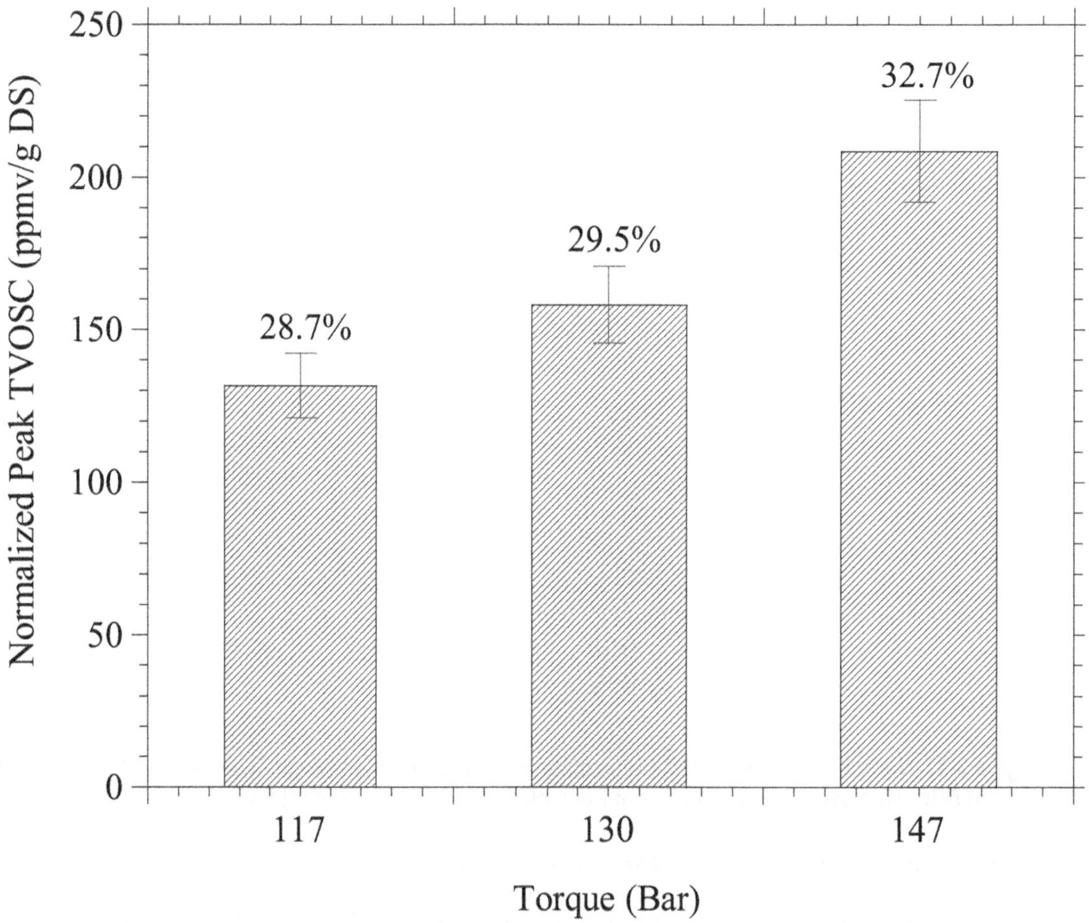

Figure 3-94. Average Peak TVOSCs for Different Torque Values of a High Solids Centrifuge at BRC.

The production of MT and DMS during cake storage is shown in Figure 3-95 and Figure 3-96, respectively. As shown in the figures, increased torque resulted in higher concentrations of these VOSCs and a delay in the time-to-peak value. A comparison of peak TVOSC values for the three torque settings is provided in Figure 3-97.

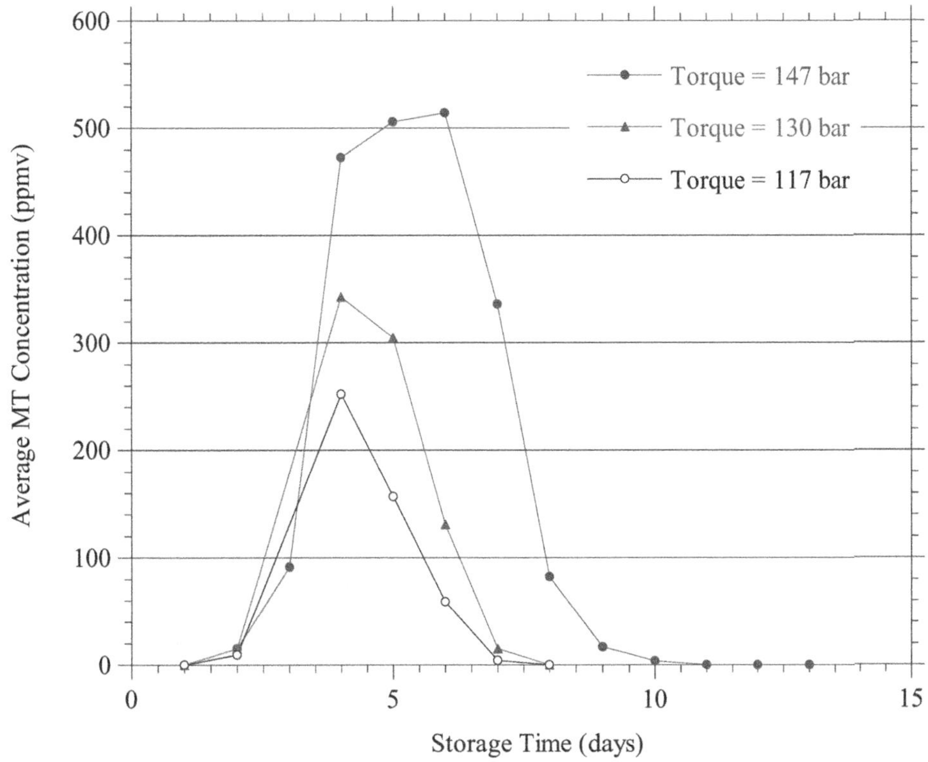

Figure 3-95. MT Headspace Concentrations During Cake Storage for Different Centrifuge Torque Values at BRC.

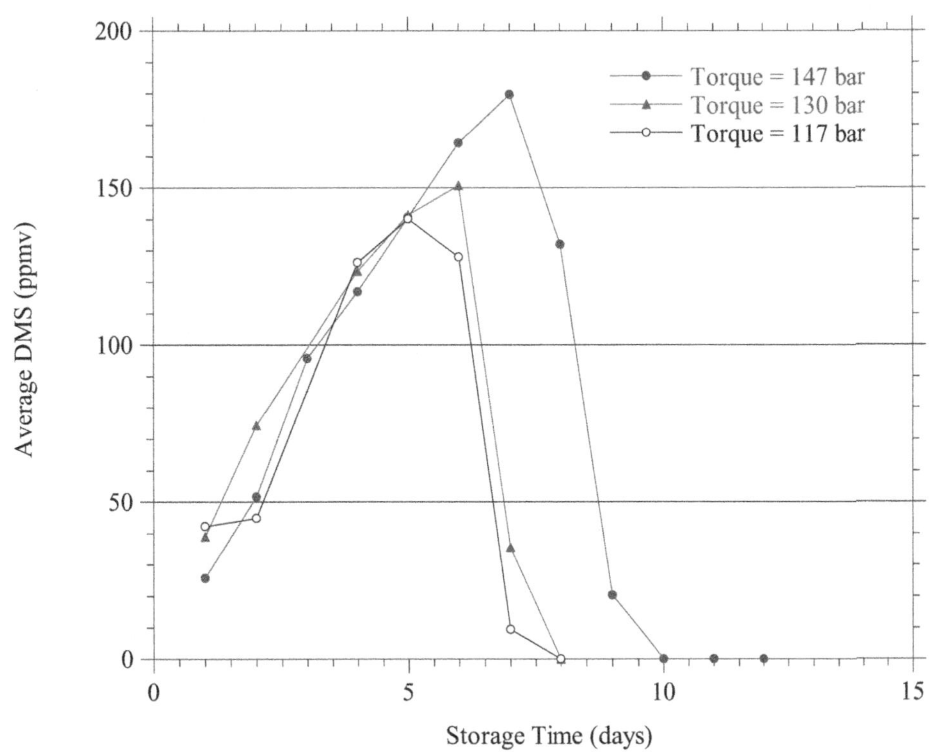

Figure 3-96. DMS Headspace Concentrations During Cake Storage for Different Centrifuge Torque Values at BRC.

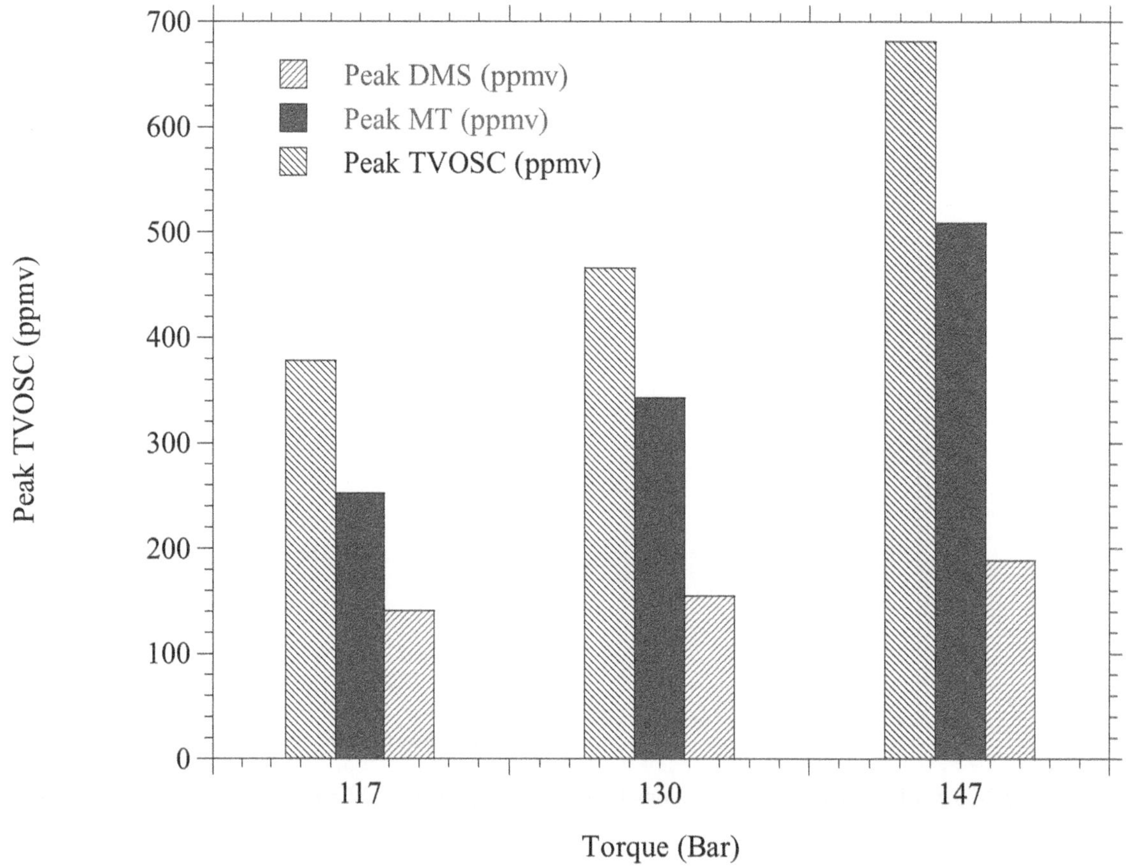

Figure 3-97. Peak Cake Headspace TVOSC Concentrations for Different Centrifuge Torque Values at BRC.

Figure 3-98 shows that in the BRC testing, the biosolids cakes produced at higher torque values had lower methane production. This could indicate a potential mechanism for greater VOSC production from higher torque settings on a centrifuge, as the methane decrease could be a result of greater shear during dewatering, which may inhibit methanogenic activity and increase VOSC production.

Another possible mechanism is the cakes produced at greater torque had a greater amount of available substrate for odor production. The bound protein concentration was the 26.9 mg/g dry solids for the highest torque setting, and 24.0 and 24.7 mg/g DS for the second and third highest torque setting respectively. This indicates that the highest torque had the greatest amount of easily extracted protein, while the lower two torque values had lower amounts of easily extracted protein.

Another measure of available substrate is BESA addition to the cake sample. BESA inhibits all methanogenic activity; therefore, after inhibition, differences in VOSC production can be attributed primarily to the available substrate. Peak MT, DMS, and TVOSC concentrations measured in BESA-amended cake samples are shown in Figure 3-99. When the methanogens were inhibited in all the cake samples with BESA, the highest torque was associated with the highest production in MT, DMS, and TVOSC.

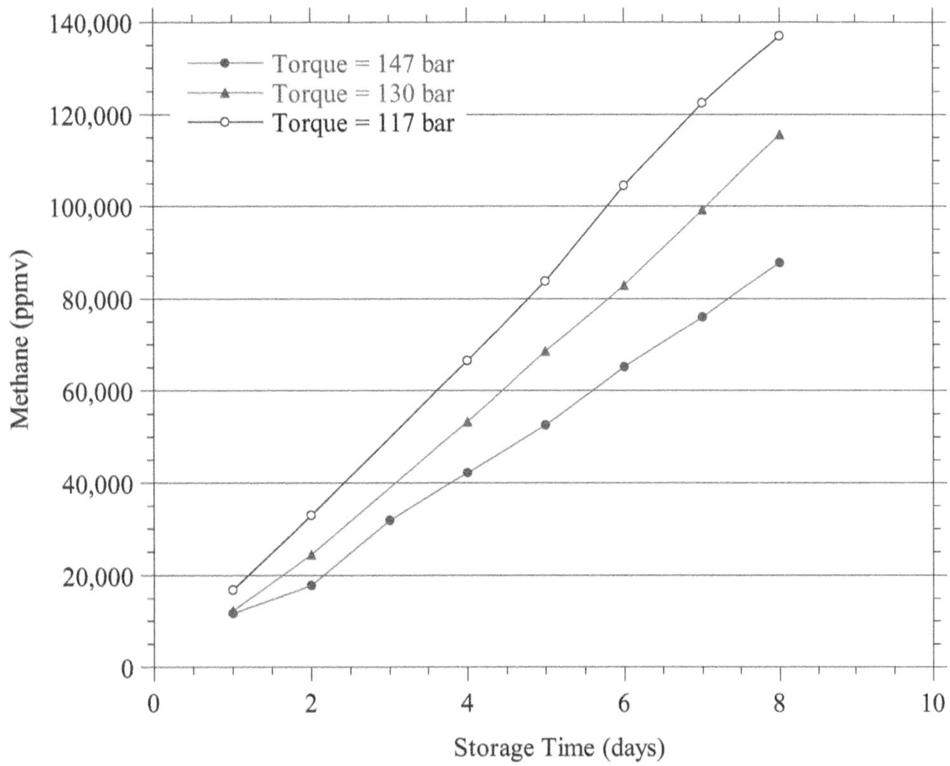

Figure 3-98. Methane Concentration During Cake Storage for Different Centrifuge Torque Values at BRC.

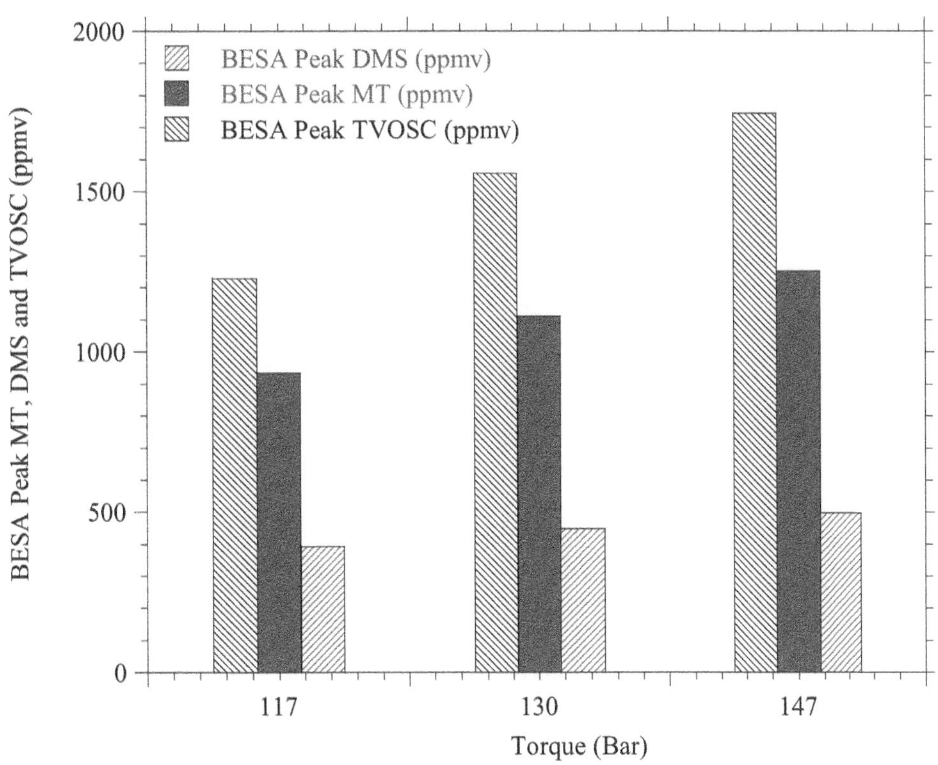

Figure 3-99. Peak VOSC Concentrations of BESA-Amended Cake for Different Centrifuge Torque Values at BRC.

Interestingly, the highest torque produced a 25% greater peak TVOSC in the BESA-amended cake than the lowest torque (Figure 3-99), compared to a 58% greater peak TVOSC in the highest versus lowest torque in the non-BESA amended cake. This observation suggests that both inhibition of methanogens and an increase in available substrate are important in production of greater VOSC concentrations in dewatered cake when using a high-solids centrifuge.

Effect of Centrifuge Bowl Speed (Centrifugal Force) on VOSC Production

Field trials were performed at the Philadelphia Water Department BRC to examine the impact of bowl speed or g-force on cake solids and headspace VOSC production from the dewatered cake. In these tests, the bowl speed was varied to obtain different g-values, and the dewatered cake was collected after operating under each condition for a period of time such that steady-state conditions had been achieved, typically one hour.

The cake solids and peak TVOSC for three different g-values at the BRC is shown in Figures 3-100 and 3-101, respectively. The figures indicate that bowl speed did not appreciably impact the cake solids content for the range of speeds that were tested. Similarly, the peak TVOSC concentration was not substantially impacted by varying bowl speed, although the highest speed did have the greatest peak TVOSC concentration of the three tested. The results suggest that at the BRC, the centrifuge bowl speed could be reduced to result in significantly reduced VOSC concentration without negatively impact cake solids.

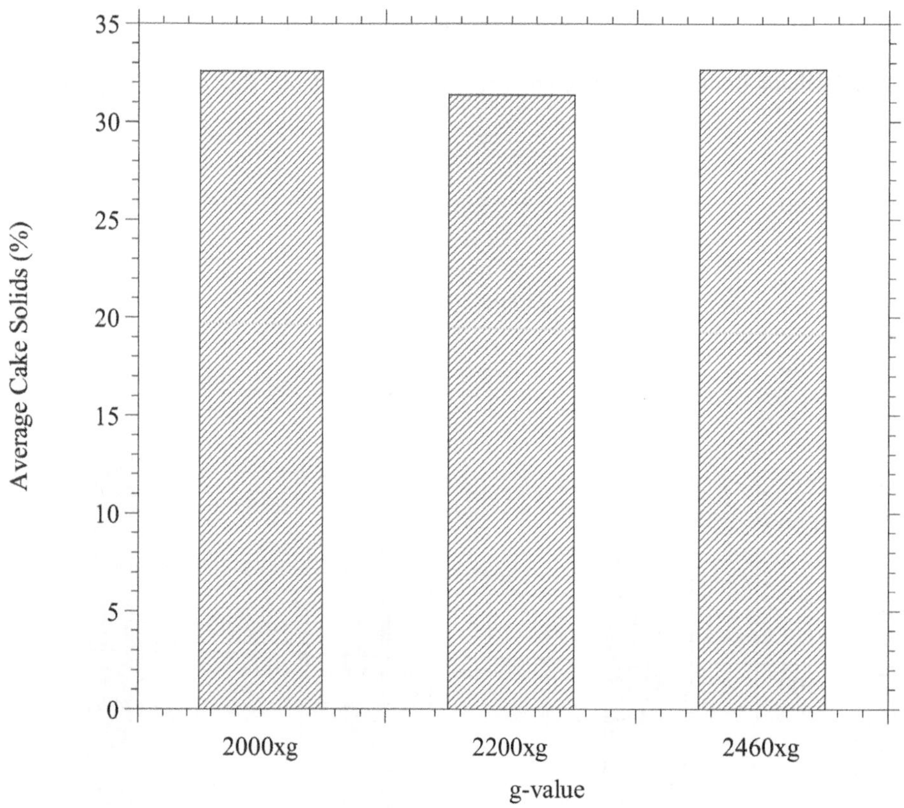

Figure 3-100. Effect of Increasing G-Force (as Produced by Centrifuge Bowl Speed) on BRC Cake Solids.

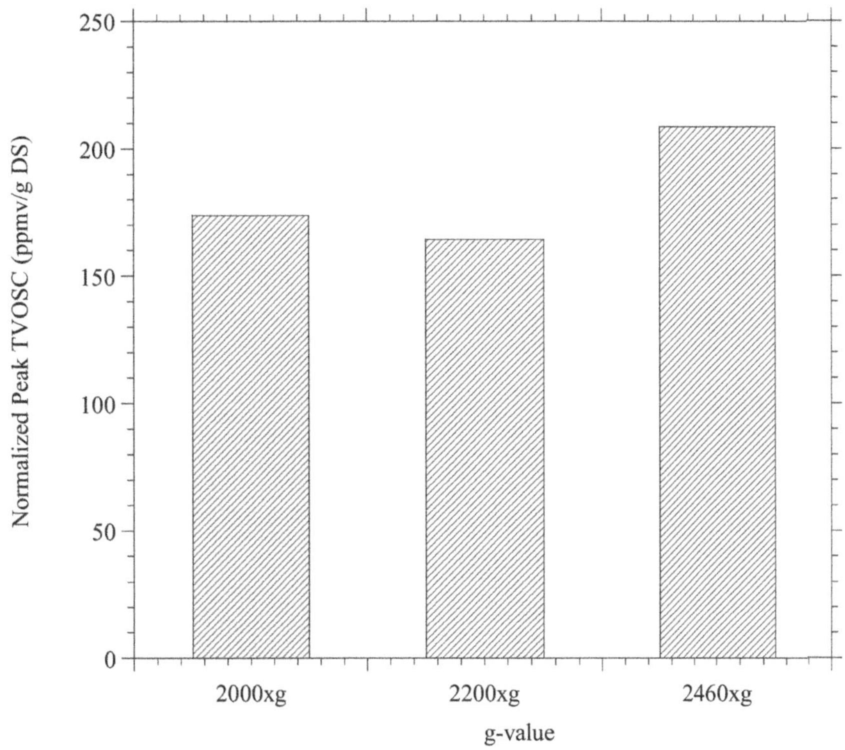

Figure 3-101. Effect of G-Force (as Produced by Centrifuge Bowl Speed) on BRC Cake Peak TVOSC Concentrations.

3.3.2.2 LACSD Joint Water Pollution Control Plant Dewatering Test Results

A series of trials similar to the Philadelphia BRC side-by-side dewatering tests were conducted at the LACSD Joint Water Pollution Control Plant (JWPCP) in the spring and fall of 2005 in order to examine the effect of centrifuge operational parameters on dewatering performance and VOSC production in the dewatered cake. The results from these tests are discussed in the following sections.

Effect of Solids Content on VOSC Production

Three different centrifuge models (Humboldt, Flottweg, and Bird-Humboldt) were used to produce varying solids content. The resulting peak TVOSC concentrations for the three centrifuges is shown in Figure 3-102. The figure shows the machine type, bowl speed (as measured by g force) and sludge feed rate in gallons per minute (gpm). Similar to the Philadelphia Water Department BRC results, higher cake solids for the different machines was associated with increased TVOSC production.

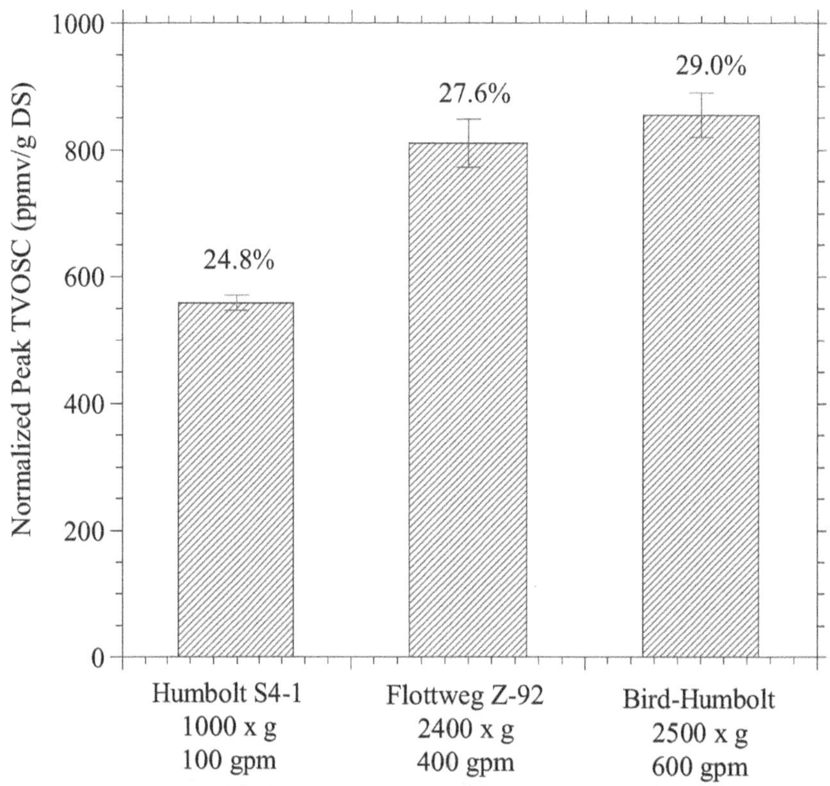

Figure 3-102. Peak TVOSC Concentration for Different Centrifuge Types at the JWPCP (Numbers on Bars Represent Cake Solids Content from the Machine on Day of Testing).

Effect of Centrifuge Torque on VOSC Production

Dewatering tests on varying torque settings were conducted at the JWPCP on two high-solids centrifuges: an Alfa Laval and a Humboldt type centrifuge. The effect of increasing torque on cake solids and peak TVOSC concentrations measured in the dewatered cake is depicted in Figures 3-103 and 3-104. Unlike the BRC results, the JWPCP Alfa Laval centrifuge cake solids did not have a consistent correlation with increasing torque (shown in Figure 3-103). However, the peak TVOSC concentration (normalized in these tests on a dry solids basis) increased as torque increased in a similar manner to the BRC.

Similar results for the impact of increasing torque on cake solids were observed for the Humboldt centrifuge (Figure 3-104), with a minimal correlation observed. The positive correlation for increasing torque and increasing normalized peak TVOSC concentration was observed but not as consistent as found with the Alfa Laval centrifuge. One outlying point was recorded at 2,100 pounds per square inch (psi) torque (see Figure 3-104).

Another experiment using the Alfa Laval centrifuge was performed to examine the impact of two polymer types (an emulsion and a Mannich polymer) on VOSC production while varying the torque settings, as depicted in Figure 3-105. Through the varying torque settings, the Mannich polymer produced cakes with lower normalized peak TVOSC concentrations. The bound protein and methane production were similar for both the Mannich and emulsion polymer (shown in Figure 3-106), and the differences in TVOSC may be a result of the lower cake solids associated with the Mannich polymer.

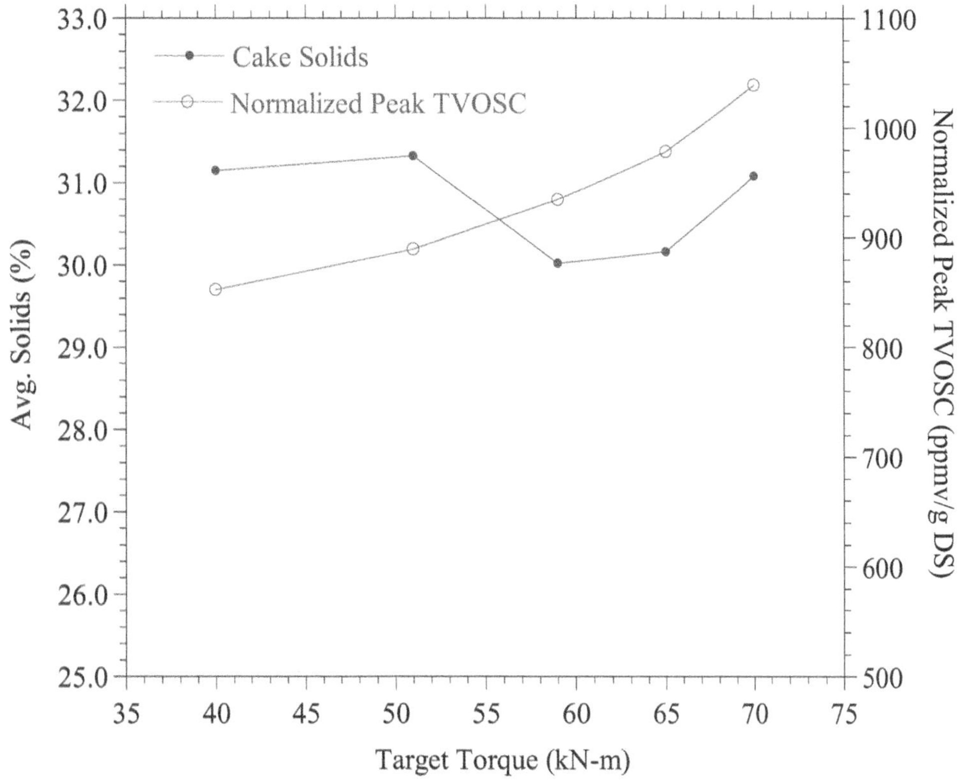

Figure 3-103. Effect of Torque on Cake Solids and Peak TVOSC Concentration for a JWPCP Alfa Laval Centrifuge.

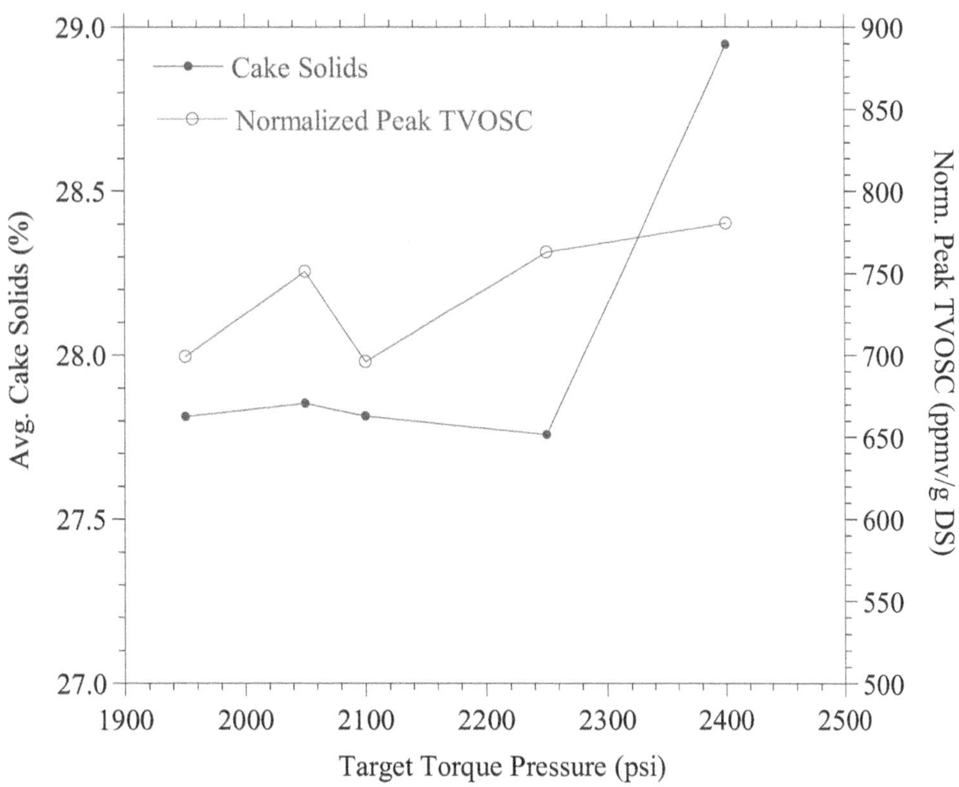

Figure 3-104. Effect of Torque on Cake Solids and Peak TVOSC for a JWPCP Humbolt Centrifuge.

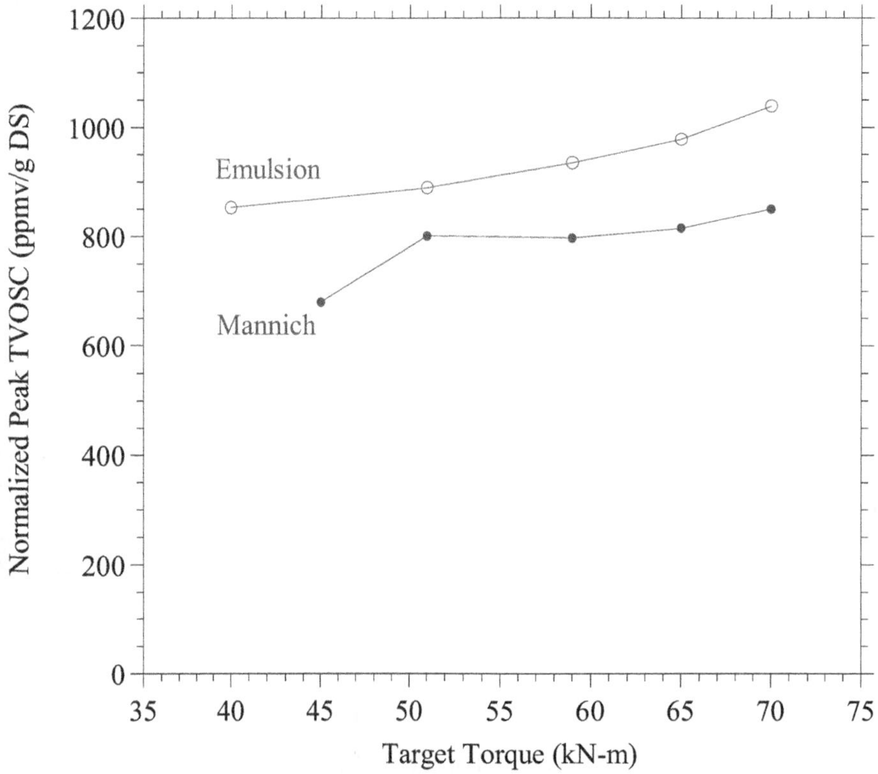

Figure 3-105. Effect of Different Polymer Types and Torque on Peak TVOSC for a JWPCP Alfa Laval Centrifuge.

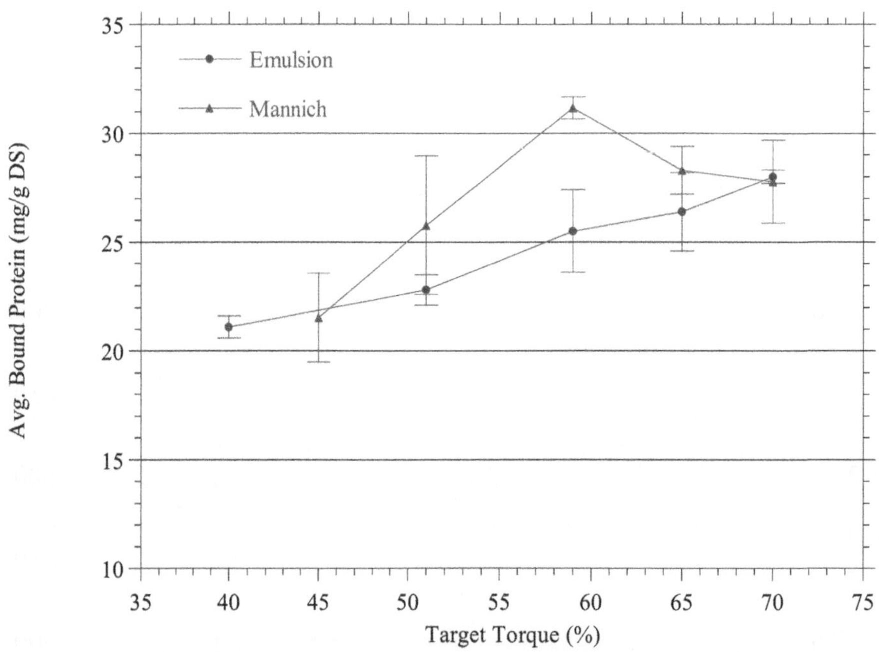

Figure 3-106. Effect of Different Polymer Types and Torque on Bound Protein for a JWPCP Alfa Laval Centrifuge.

Additionally, the WERF team theorized that the two tests with different polymers being separated by several weeks may have produced some variability in the process operation, possibly leading to the difference in TVOSC generation.

Effect of Centrifuge Bowl Speed or Centrifugal Force

A greater difference in cake solids and peak TVOSC with varying bowl speed was measured in the JWPCP testing than was measured at the BRC. Figures 3-107 and 3-108 show the variations in cake solids and TVOSC with bowl speed, respectively. The range of g-values was from 1,000g to 2,500g, which was a much larger range than was used in the BRC study. However, it should be noted that the g-values used at the JWPCP were for different machines whereas at the BRC the same machine was operated at different speeds. At the JWPCP, the lower speed machines (Humboldts operating at 1,000g) had lower cake solids content, and lower normalized peak TVOSC concentrations.

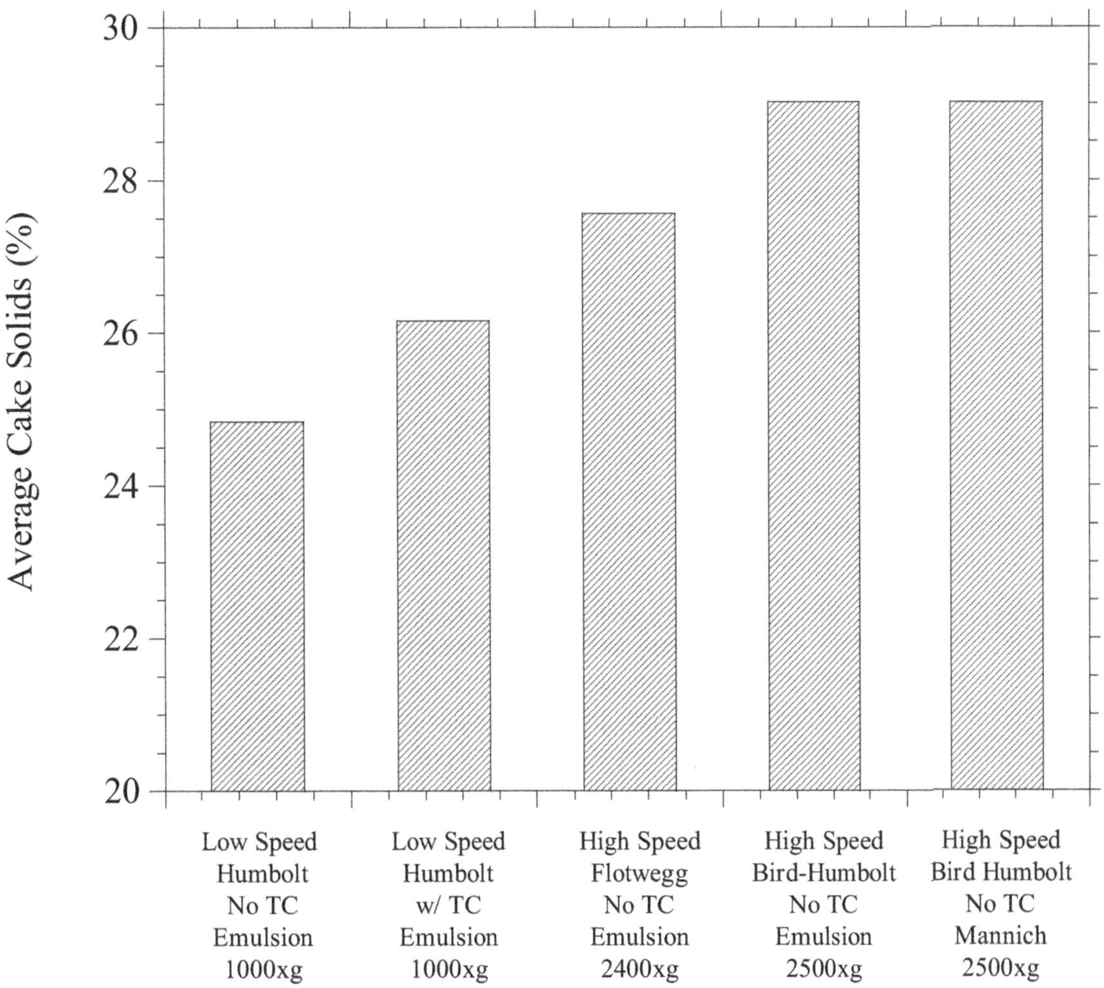

Figure 3-107. Effect of G-Force (as Produced by Centrifuge Bowl Speed) on JWPCP Cake Solids (Note: TC = Thickened Centrate).

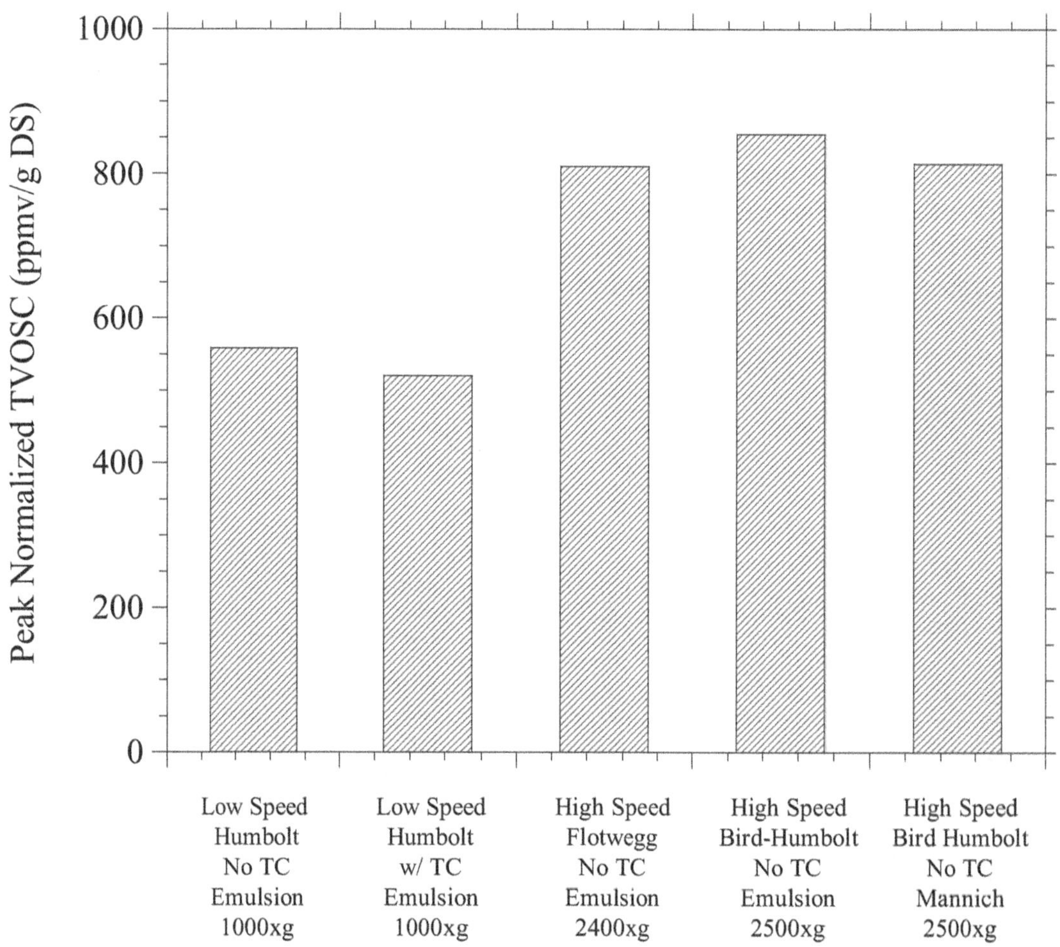

Figure 3-108. Effect of G-Force (as Produced by Centrifuge Bowl Speed) on JWPCP Cake Peak TVOSC Concentrations (Note: TC = Thickened Centrate).

3.3.2.3 King County Centrifuge and Belt Filter Press Comparison Test Results

In the dewatering tests at the King County West Lake WWTP, experiments were conducted on cake samples collected from a belt filter press and high-solids centrifuge that were both dewatering the same anaerobicly digested biosolids. Figure 3-109 depicts the cake solids and TVOSC profile during storage in the centrifuge and belt filter press (note that solids content in the centrifuge was significantly greater than the belt filter press). The belt filter press had no measurable TVOSC during storage, while the high-solids centrifuge had a peak concentration slightly greater than 1,000 ppmv. Previous research states that an important factor in increased TVOSC production is increased inhibition of methanogens, which are known to degrade odor-causing compounds (Higgins et al., 2006). Figure 3-110 indicates that methane production was much lower in the high-solids centrifuge than the belt filter press, confirming that this mechanism could be occurring in this test.

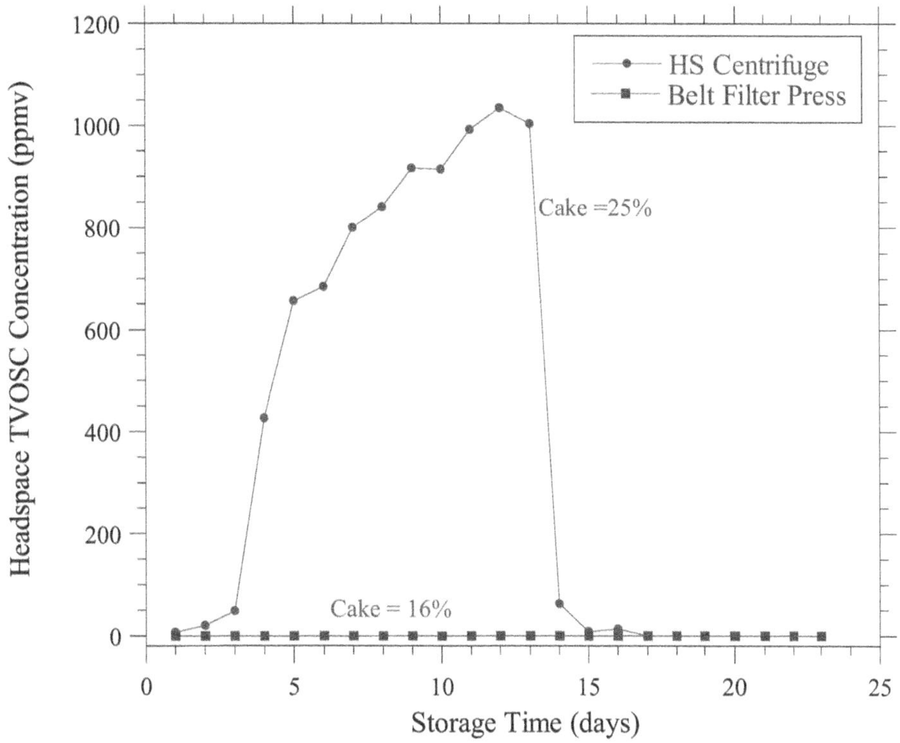

Figure 3-109. King County Biosolids Headspace TVOSC Concentration During Cake Storage.

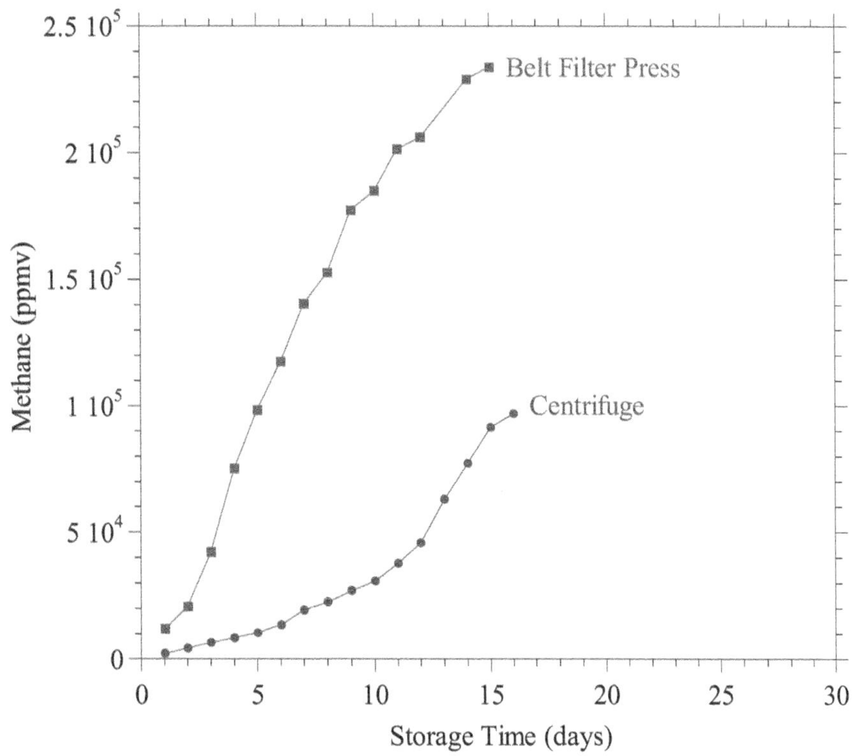

Figure 3-110. King County Biosolids Methane Concentration During Storage.

3.3.2.4 Joint Water Pollution Control Plant Centrifuge and Rotary Press Comparison

Another side-by-side trial was performed at the LACSD Joint Water Pollution Control Plant (JWPCP) to compare the production of odors from a Fournier rotary press two high-solids centrifuges (an Alfa Laval and a Bird-Humboldt). The TVOSC profile for each dewatering device is depicted in Figure 3-111. The cake from the rotary press produced much lower TVOSC concentrations than the high-solids centrifuges. The throughput of solids through the rotary press was also lower: the cake solids from the rotary press ranged from 16-22% solids depending on operating conditions. The results suggest that the rotary press imparts less shear during dewatering.

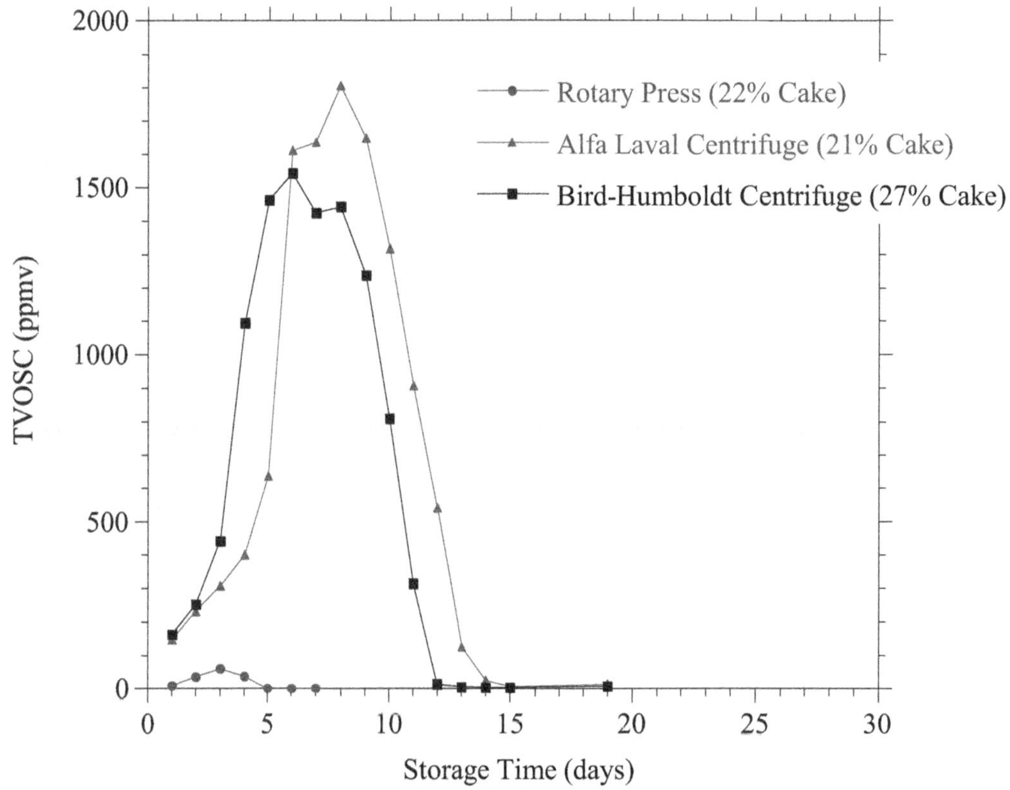

Figure 3-111. TVOSC Concentration During Storage of JWPCP Cake from Side-by-Side High Solids Centrifuges and a Fournier Rotary Press.

The methane production profile shown in Figure 3-112 demonstrates that the rotary press cake exhibited much greater methane production compared to the high-solids centrifuges. This could be a result of less shear during dewatering, as shear could inhibit methanogenic activity, possibly through greater exposure to air, and/or cell disruption. The greater methane production in the rotary press cake coincides with the reduction of TVOSC since the methanogens have been shown to degrade these odor-causing compounds (Higgins et al., 2006).

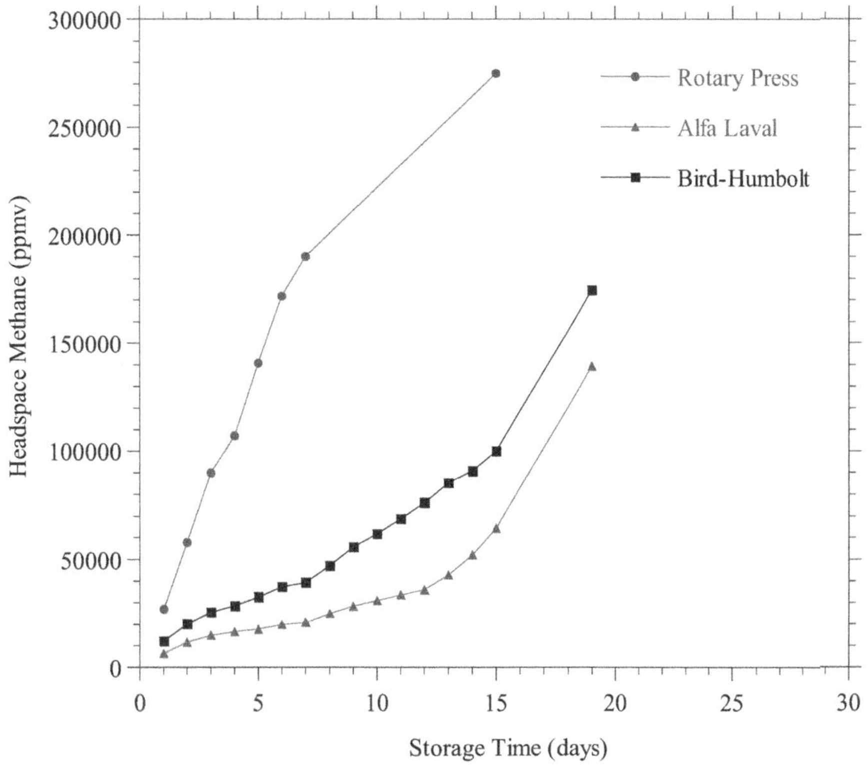

Figure 3-112. Methane Concentration During Storage of JWPCP Cake from Side-by-Side High Solids Centrifuges and a Fournier Rotary Press.

3.3.3 Combined MicroSludge and Dewatering Equipment Test Results

A set of field tests were conducted at the Los Angeles County's Joint Water Pollution Control Plant (JWPCP) plant to compare the effect of variations in digester operation and biosolids dewatering equipment on the odor potential of dewatered sludge cakes. At the time these tests were conducted, the Microsludge™ process, a patented process that uses chemical and pressure treatment to liquefy WAS, was being tested at one digester at the plant. In addition, the Andritz Company provided a high-solids centrifuge and a belt filter press pilot unit so that various operational characteristics of centrifuges could be studied to evaluate their effect on odors.

Details of the centrifuge operational variations are discussed in the following sections. These include a comparison of odor potential from sludge treated by the Microsludge™ process as a pretreatment step prior to conventional mesophillic, anaerobic digestion, centrifuge adjustments including the type of polymer, operational g-value and biosolids feed rate, and a comparison of results between the centrifuge and belt press.

3.3.3.1 MicroSludge WAS Pretreatment Test Results

The MicroSludge™ process is a patented chemical and pressure treatment that liquefies waste activated sludge (WAS) in order to improve the rate and extent of its degradation by anaerobic digestion. The MicroSludge process was operated at full-scale at the Los Angeles County Sanitation District's Joint Water Pollution Control Plant (JWPCP) from October 2005 until August 2006 to evaluate the performance of this technology for enhancing anaerobic digestion. Samples from the anaerobic digester receiving MicroSludge treated WAS were compared to samples from a control digester to determine the effect of the MicroSludge process on odor generation from centrifugally dewatered sludge cakes.

At the JWPCP, the WAS feed comprises 25% of the flow and 32% of the mass going to the digesters, so the fraction of WAS is not typical of most wastewater facilities. The increase in volatile solids destruction for the digester receiving MicroSludge processed material ranged between 3 and 6%. This suggests that degradation of the activated sludge fraction was enhanced by 9 to 18%.

Centrifuge Polymer Type and G-Value Comparisons

At the full-scale comparison of the MicroSludge™ process at the JWPCP in March 2006, the WERF team found that the MicroSludge™ process reduced cake headspace TVOSC concentrations by approximately 50%. Figure 3-113 depicts the TVOSC profile for the MicroSludge™ and non-MicroSludge™ (control) sample and Mannich polymer, and Figure 3-114 shows the same results for an emulsion polymer.

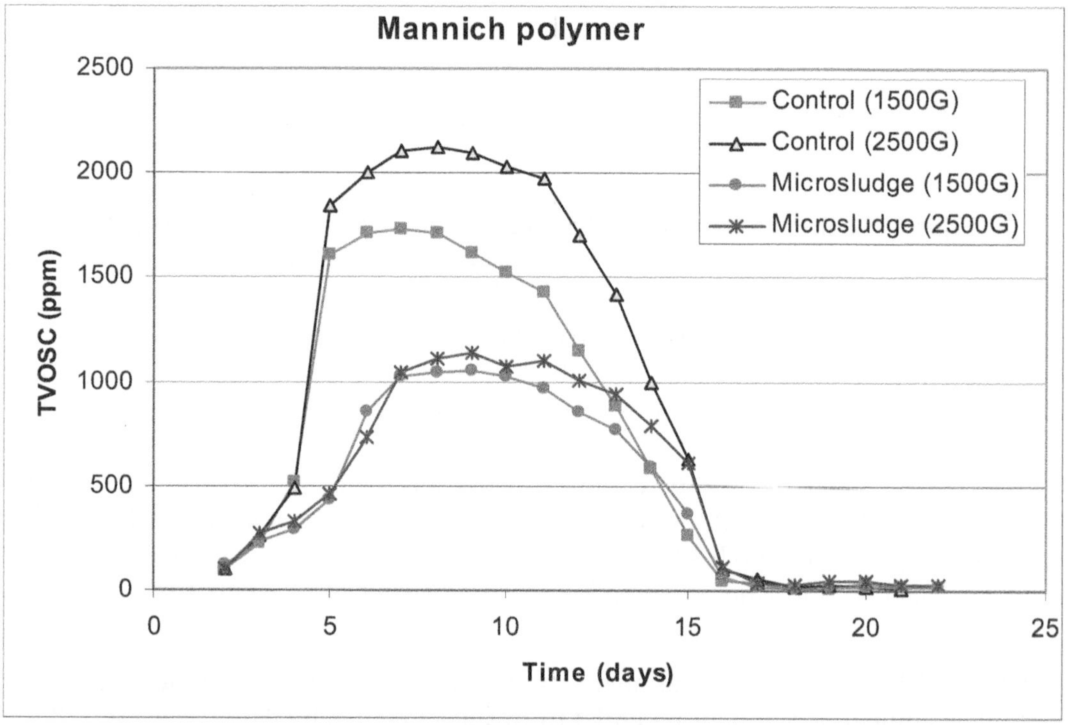

Figure 3-113. Comparison of JWPCP Cake Headspace TVOSC for Mannich Polymer with G=1500 and G=2500 using MicroSludge™ and Non-MicroSludge™ Digested Samples.

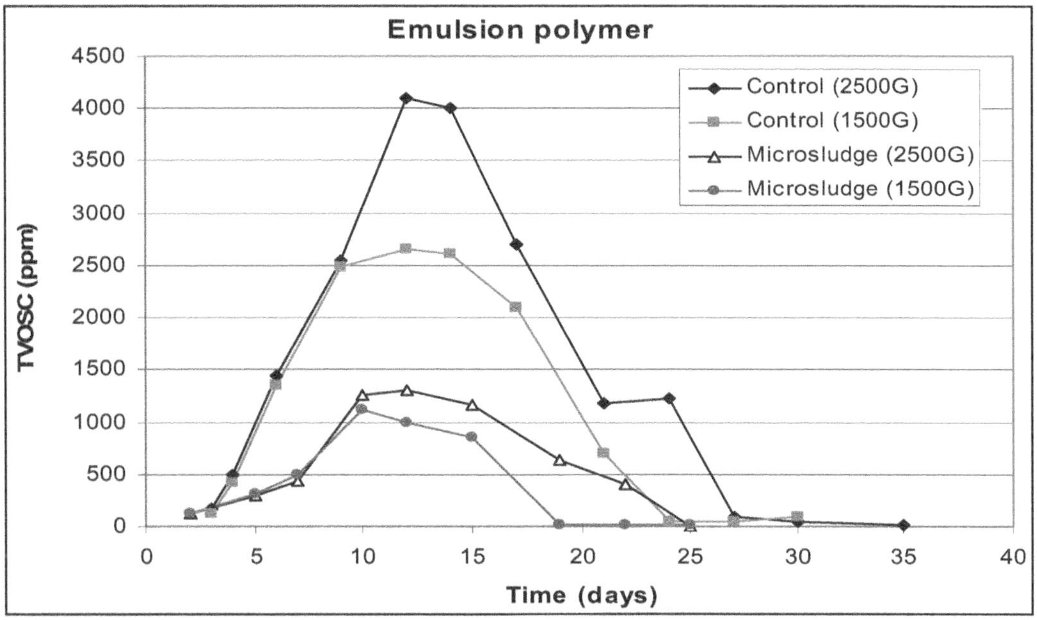

Figure 3-114. Comparison of JWPCP Cake Headspace TVOSC for the Emulsion Polymer with G=1500 and G=2500 using MicroSludge™ and Non-MicroSludge™ Digested Solids Samples.

The WERF team concluded from the LACSD JWPCP full-scale trials that the MicroSludge™ TVOSC generation was at levels about half that of the control digester. The difference in TVOSC generation between dewatered cakes produced by the MicroSludge™ enhanced and control digesters is greater than would be expected by the difference in volatile solids destruction alone. The results of these tests indicate that the MicroSludge™ process may assist digestion in destroying a larger proportion of odor precursors in biosolids.

Biosolids Feed Rate Studies

Variations in feed rate to the JWPCP centrifuges were studied in order to determine how feed rate influences TVOSC production in the dewatered biosolids cake. As part of this study, digested sludge that had been processed with the Microsludge process was compared to sludge from the control digester. In these tests, a Mannich polymer was used as a dewatering aid and a bowl speed of 2,500g was employed. TVOSC profiles from this study are depicted in Figure 3-115 (control digester) and Figure 3-116 (Microsludge digester).

In the control digester, the peak TVOSC was similar for both feed rates, falling in the range of 1,300 to 1,500 ppmv. For the Microsludge™ digester, the feed rate made a greater difference in the peak TVOSC concentration. At the 90 gpm feed rate, the peak TVOSC of the control digester sludge was approximately 40% higher than the Microsludge™ digester sludge. For the feed rate of 35 gpm, there was little difference between the sludges from the control and the Microsludge digesters in terms of TVOSC production.

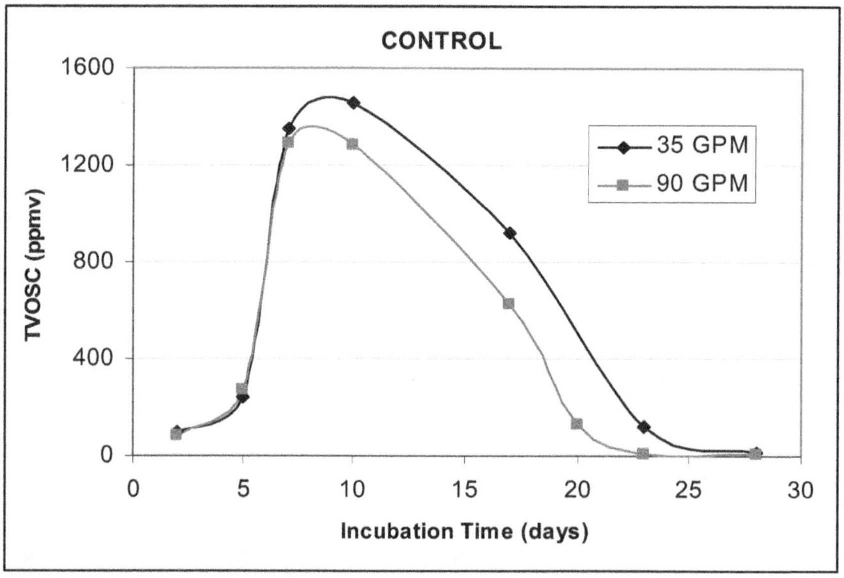

Figure 3-115. Comparison of JWPCP Cake Headspace TVOSC for a High-Solids Centrifuge at Two Different Feed Rates for Sludge from the Control Digester. Tests Were Conducted Using a Mannich Polymer at 2,500g.

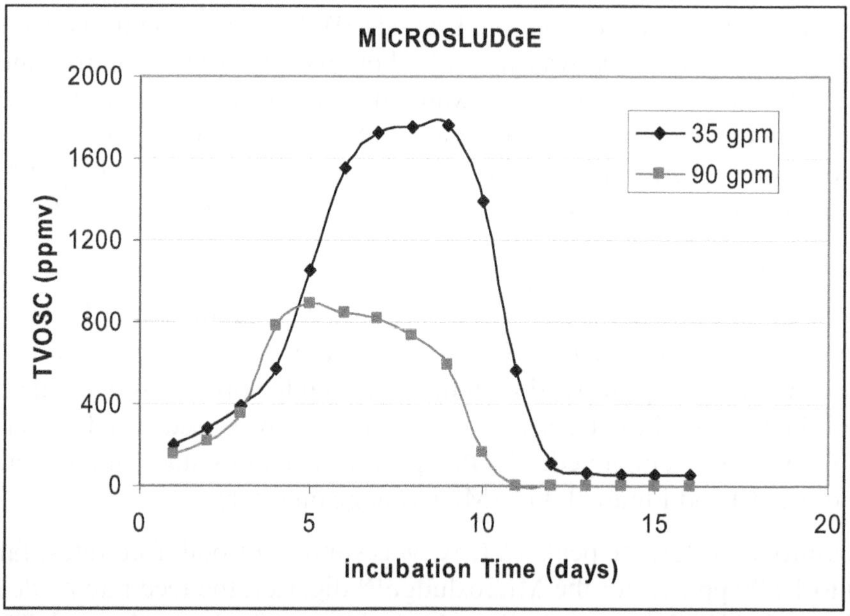

Figure 3-116. Comparison of JWPCP Cake Headspace TVOSC for a High-Solids Centrifuge at Two Different Feed Rates for Microsludge Processed Digested Sludge. Tests were Conducted Using a Mannich polymer at 2,500g.

These plots are consistent with data shown in Figures 3-113 and 3-114. At the feed rate of 90 gpm, the detention time in the centrifuge is less and therefore less shear is theorized to be imparted on the sludge, resulting in lower TVOSC. It appears that under lower flow and higher detention time conditions, there is less difference in TVOSC between the Microsludge processed and the control digested and dewatered biosolids.

Belt Filter Press Studies

A belt filter press was provided by Andritz to conduct a study comparing the TVOSC generation using a high-solids centrifuge with that using a belt filter press. As part of this study, TVOSC from digested sludge that had been processed with the Microsludge process was also compared to sludge from the control digester. As was found in the King County West Lake WWTP study (Section 3.3.2.3), the TVOSC from the belt filter press was much lower (all peaking at less than 300 ppmv) than that from cake generated by centrifugation, as seen in Figures 3-117 and 3-118. Note that the peak TVOSC from biosolids dewatered by centrifuges using the same sludge ranged from 1,300 to 1,800 ppmv (see Figure 3-115 and Figure 3-116). The difference in TVOSC profiles between the control and MicroSludge™ sludges is minimal, but the MicroSludge™ TVOSC measurements were slightly lower.

In these tests, cake solids content was varied by further pressing the sludge from the belt filter press using a laboratory press. If the data are re-plotted as the peak TVOSC versus cake solids (Figure 3-119), it can be seen that the TVOSC values for the belt filter press are much lower than for the centrifuge, even at the higher cake solids. All values were less than 300 ppmv, when centrifuge dewatered biosolids peak TVOSC ranged from 900 to over 4,000 ppmv. It can also be seen that the only differences between the control and MicroSludge™ digester TVOSC concentrations are due to the cake solids concentrations of the sample points.

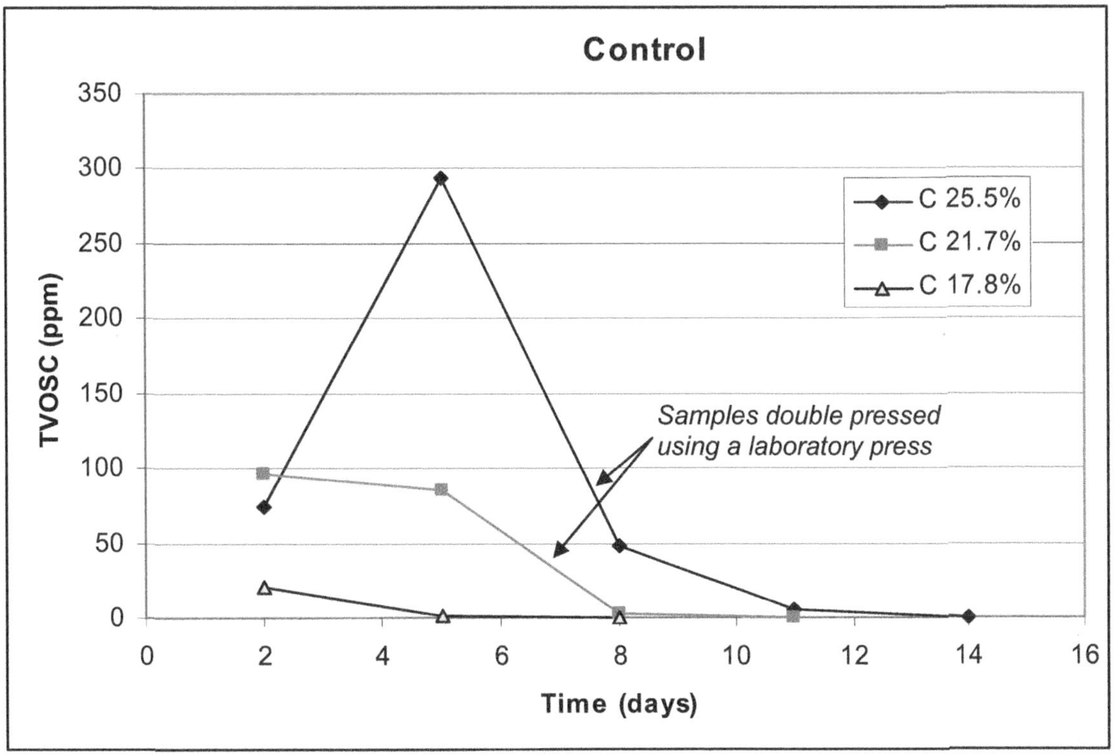

Figure 3-117. TVOSC from the Belt Filter Press using Biosolids from the JWPCP Control Digester.

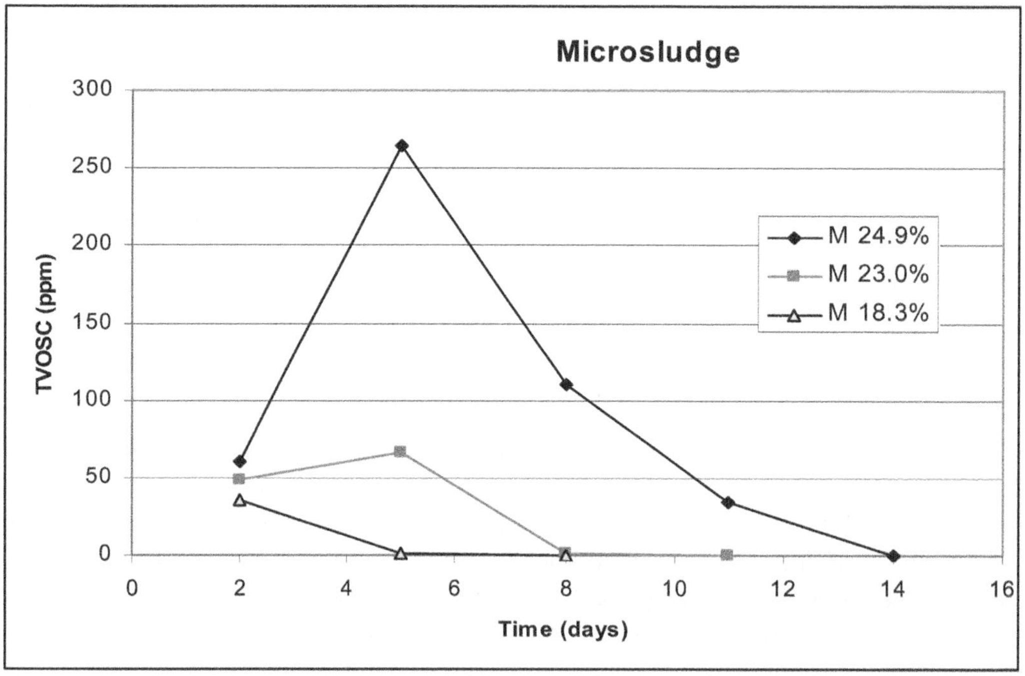

Figure 3-118. TVOSC from the Belt Filter Press Using Biosolids from the JWPCP MicroSludge™ Digester.

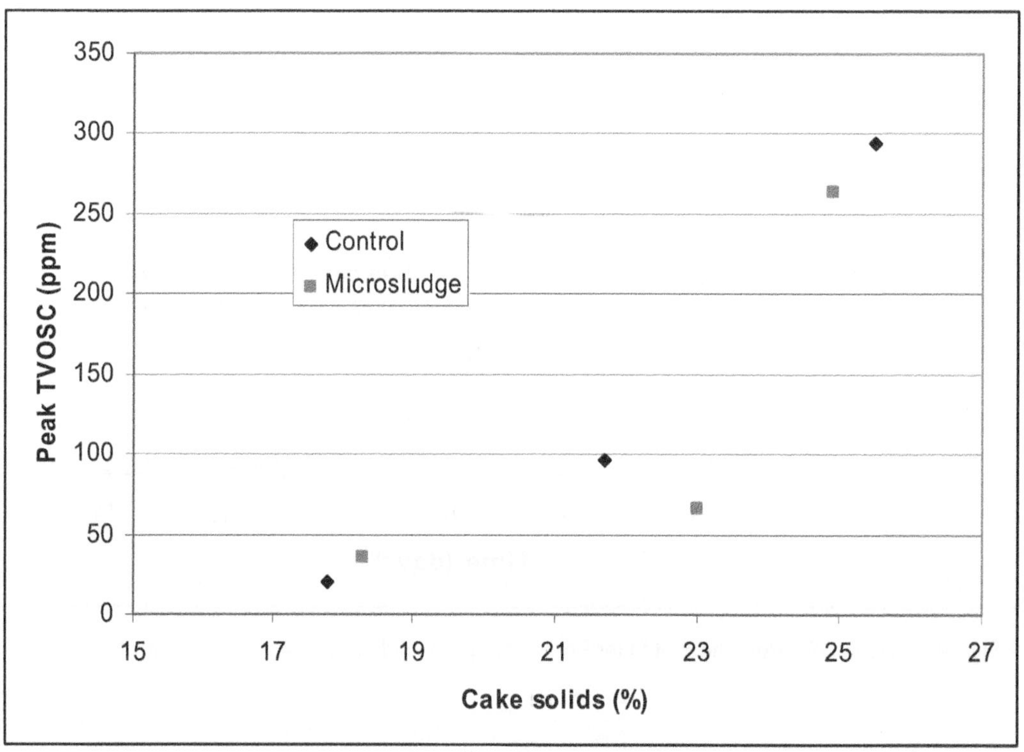

Figure 3-119. Effect of Cake Solids on TVOSC for JWPCP Dewatered Biosolids Using a Belt Filter Press.

Summary of Microsludge Effects on Cake TVOSC Production

In the belt filter press experiments at the JWPCP, the differences between the Microsludge digester and the control digester were relatively small. However, in the experiments conducted with high-solids centrifuges, the Microsludge-processed material produced lower headspace TVOSC than sludge from the control (non-Microsludge) digester.

The greatest differences between the Microsludge TVOSC and the control was when an emulsion polymer was used as a dewatering aid and under operational conditions that would create the highest shear (lower centrifuge feed rates). Under these conditions, the Microsludge-processed material generated TVOSC concentrations that were less than half those from the control digester. The benefits of Microsludge for VS reduction were in the range of an additional 3 to 6% VS reduction for the Microsludge unit. This suggests that something other than additional VS reduction accounted for the TVOSC reduction benefits of the Microsludge process.

Since the Microsludge system processed only 25% of the sludge flow (the WAS fraction), the reduction of peak TVOSC by more than 50% suggests that the sludge homogenization technology may alter the degradability of the precursors of the organic sulfur compounds, reducing their buildup in the headspace. Data for methanethiol, dimethyl sulfide and methane for the control and Microsludge-processed sludges are depicted in Figures 3-120 through Figure 3-122. A Mannich polymer was used as a dewatering aid for all of these tests, and bowl speed was varied as indicated in the plots.

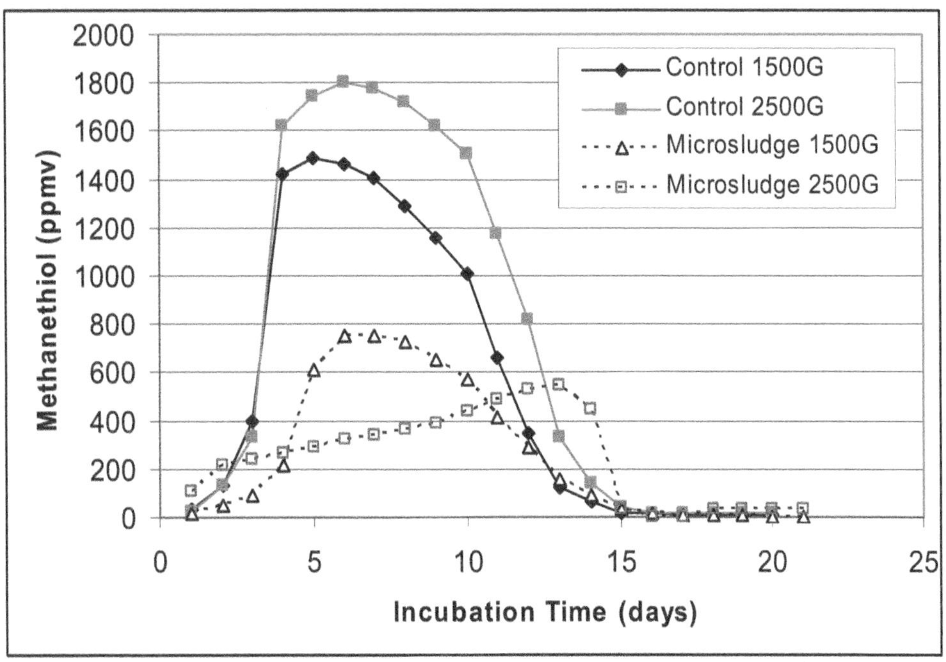

Figure 3-120. Comparison of Headspace MT Concentration Using MicroSludge™ and Control Digester Samples of JWPCP Biosolids.

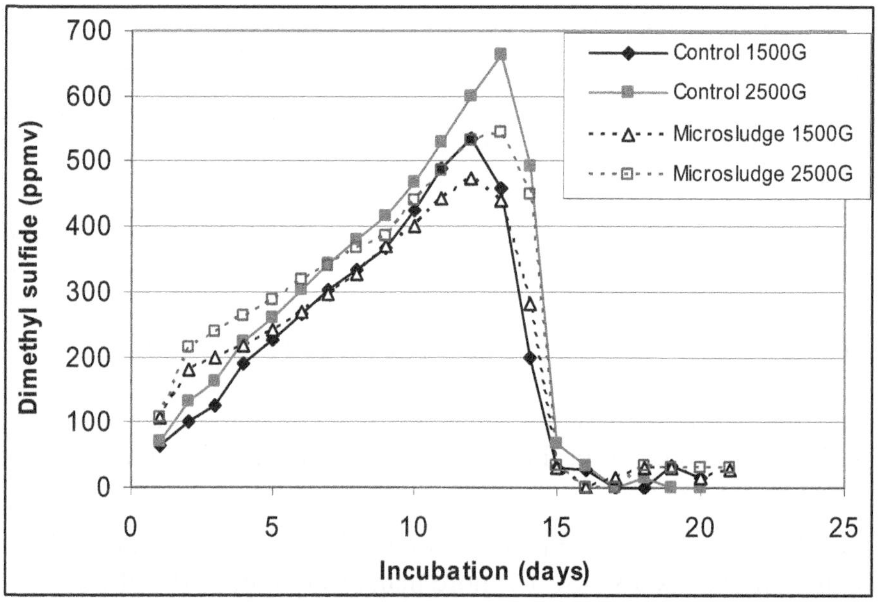

Figure 3-121. Comparison of Headspace DMS Concentration Using MicroSludge™ and Control Digester Samples of JWPCP Biosolids.

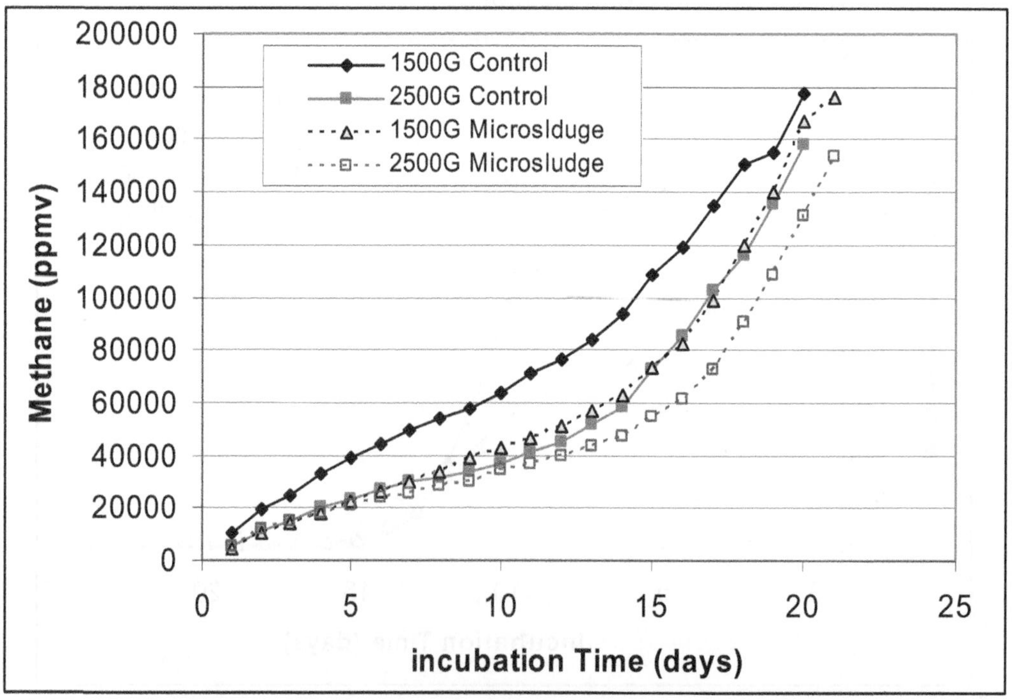

Figure 3-122. Comparison of Methane Production using MicroSludge™ and Control Digester Samples of JWPCP Biosolids.

The data in Figure 3-120 shows that the main difference in MT profiles between the Microsludge processed material and the control is that significantly less MT accumulates in the headspace for the Microsludge processed and dewatered sludge. However, the dimethyl sulfide concentrations were similar (Figure 3-1215) as were the concentrations of dimethyl disulfide (data not shown). In addition, the methane that accumulates (Figure 3-122) is similar but slightly higher in the control systems.

It appears that the benefit of the Microsludge process for odor reduction is that it creates a microbial environment that results in rapid conversion of proteins to sulfide so that less TVOSC, particularly MT, accumulates in the dewatered cake headspace. It may be that the high degree of solubilization of sludge solids that occurs with the Microsludge process greatly accelerates the degradation process, so a lower amount of intermediates accumulate.

3.3.3.2 Centrifuge Adjustments Test Results

In this portion of the JWPCP testing, the WERF team conducted full-scale experiments on varying bowl speed and feed rate of centrifuges to identify effects on odors from the dewatered biosolids cake.

The effect of centrifuge bowl speed was studied for the JWPCP Digester 3 feed (the control digester) and feed from the Microsludge treated digester. Figure 3-123 shows the TVOSC production from control digester cakes produced at two bowl speeds: 1,500g and 2,500g. A Mannich polymer was used as a dewatering aid for all samples. Solids content is indicated on each TVOSC profile, and error bars represent one standard deviation. Consistent with previous results (Figure 3-107, also for the JWPCP sludge), the greater g-value correlated with an increase in the TVOSC concentrations, in addition to an increase in the cake solids from 24% to 26%.

The corresponding graph showing results from the same bowl speed test but instead using the Microsludge digester feed is shown in Figure 3-124. A Mannich polymer was used as a dewatering aid for all samples. One observation made by the WERF team was that the cake solids content was slightly greater when using the Microsludge-treated sludge as centrifuge feed.

Interestingly, there was much less difference in the Microsludge digested sludge TVOSC profile when compared to the dewatered sludge from the control digester, in that the peak TVOSC concentrations were almost the same. This may suggest that the Microsludge feed was less susceptible to shear or had lesser bioavailable TVOSC precursors compared to the control digester that received no pre-treatment before digestion.

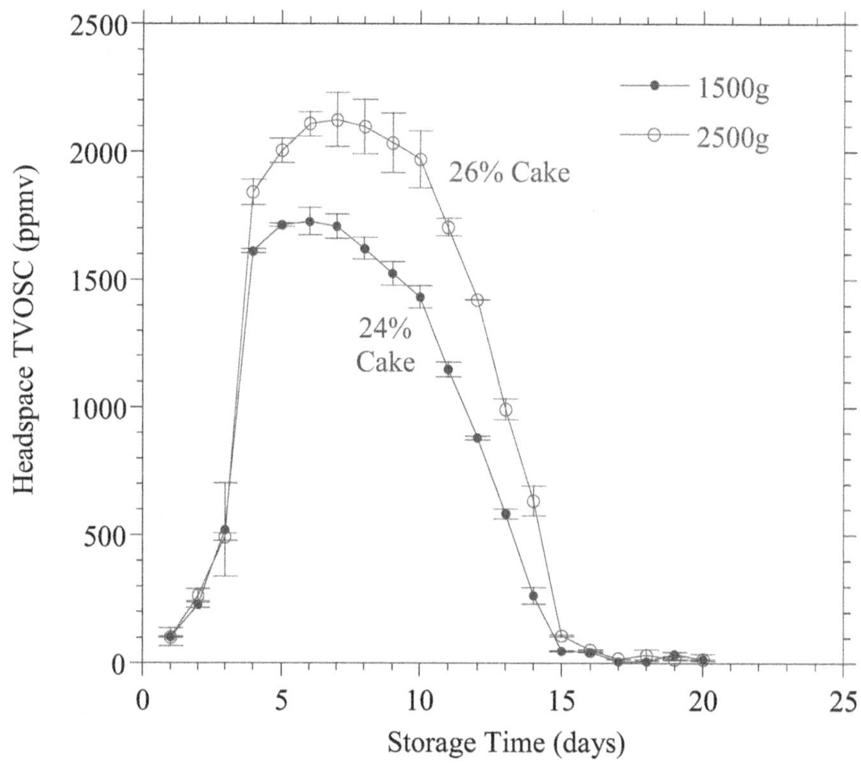

Figure 3-123. Comparison of JWPCP Biosolids Headspace TVOSC Using the Control Digester as Centrifuge Feed.

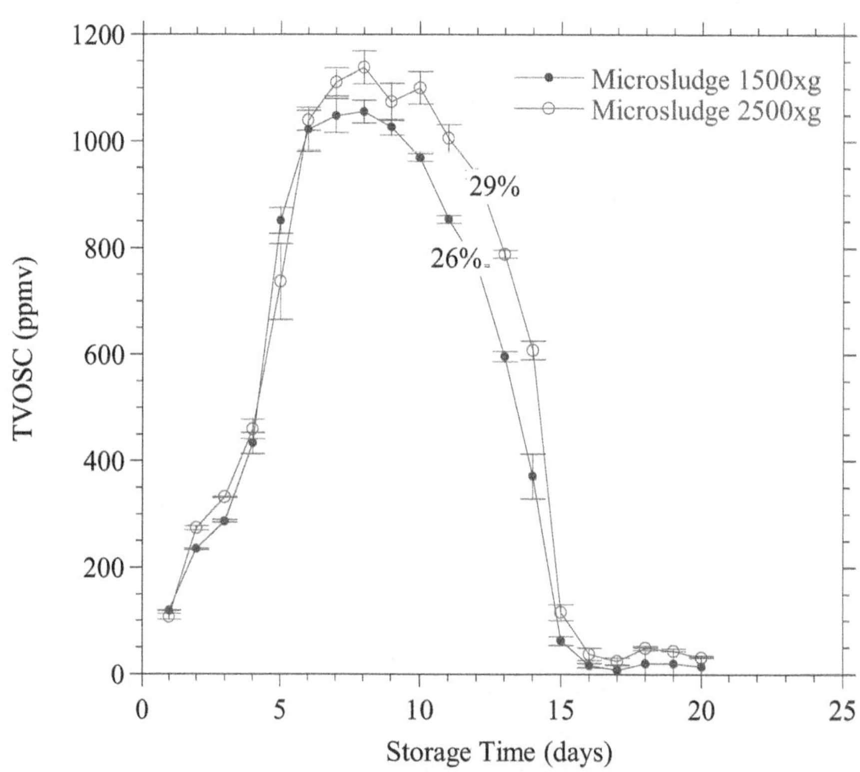

Figure 3-124. Comparison of JWPCP Biosolids TVOSC Using the Microsludge Digester as Centrifuge Feed.

3.3.3.3 Centrifuge Scroll Test Results

At the JWPCP using the Andritz centrifuge, the WERF team collected samples from several different points along the scroll of the centrifuge to determine the impact of scrolling on TVOSC production. The sample locations are shown in Figure 3-125 where locations A, B, and C, represent samples that are progressively further down the scroll, and location D is actually at the centrifuge discharge port. Between the sampling points C and D, the centrifuge has an adjustable weir plate, or disk, through which centrifuged solids are extruded before passing through the centrifuge discharge port. The internal pressure, or torque, of the centrifuge is set by adjusting this weir disk. A substantial amount of shear is exerted on the cake biosolids as they are extruded through this constricted disk opening on the way to the discharge port. The additional shear as this location contributed to a higher TVOSC profile at location D even though the solids content was lower than at location C.

The TVOSC concentration profile during cake storage for each sample location is shown in Figure 3-126, along with the cake solids content measured for each sample. As the sample collection point moved down the scroll, the resultant cake produced higher TVOSC concentrations. The cake solids also increased progressively for the first three locations (A, B, and C), but the final cake solids content being discharged was slightly lower than what was measured for location C. The results of the scroll testing demonstrate the importance of shear in impacting TVOSC profiles in the dewatered cake, and indicate that odor production is more than just a cake-solids phenomenon.

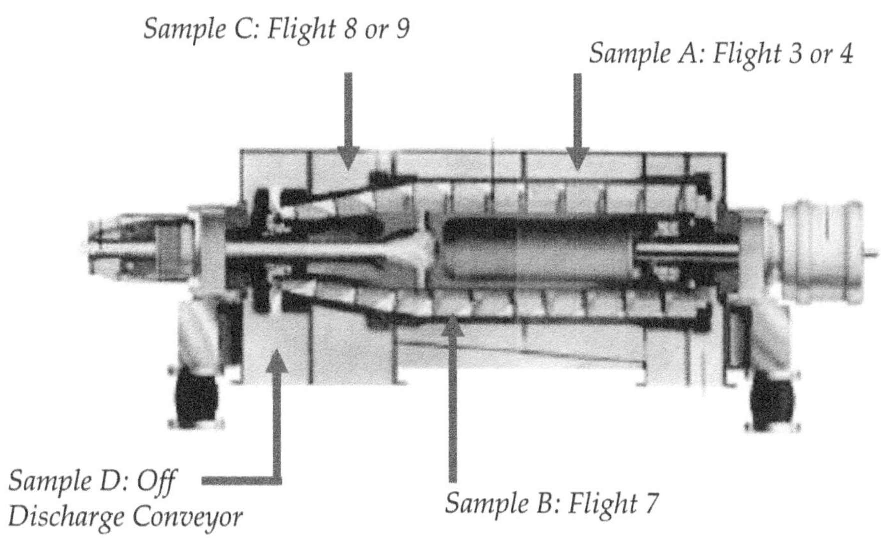

Figure 3-125. Cake Sample Locations along the Centrifuge Scroll.

3.3.3.4 Centrifuge and Belt Filter Press Comparison

Another side-by-side comparison of a belt filter press and a centrifuge was conducted at the JWPCP, in which the dewatering equipment feed was from the control digester. The TVOSC production profile for the test is shown in Figure 3-127, along with the respective cake solids concentrations. Very similar to previous testing, the belt filter press cake had little TVOSC production, while the centrifuge had peak concentrations near 1800 ppmv.

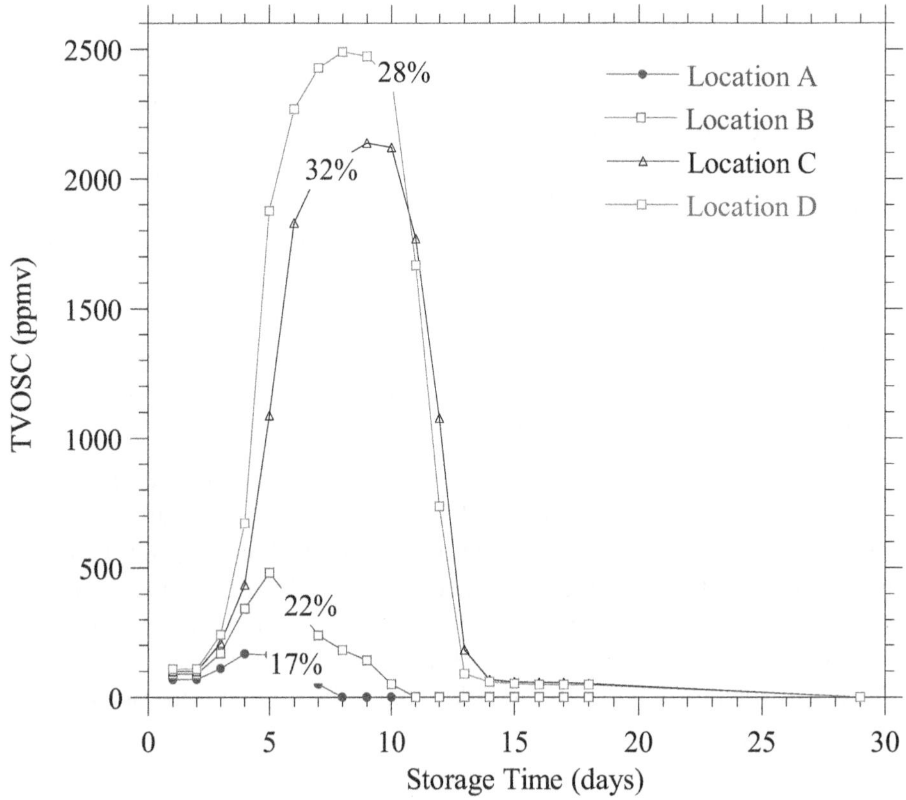

Figure 3-126. Effect of Scrolling on TVOSC Production, with Cake Solids Percentage Shown.

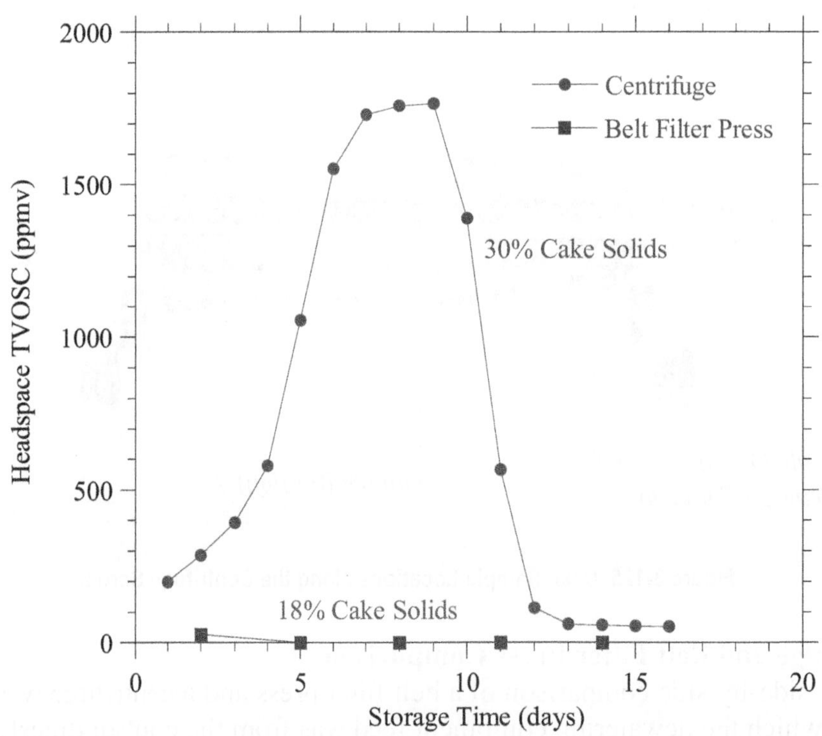

Figure 3-127. Comparison of Belt Filter Press and Centrifuge Cake Headspace TVOSC from Field Trial at the JWPCP.

WERF

3.3.4 Ferric Chloride and Lime Addition Tests Results

Research has shown the calcium and iron play a role in binding of biopolymeric material. Therefore, the WERF team hypothesized that lime and/or ferric chloride ($FeCl_3$) addition may reduce odorant production through protein binding making it less bioavailable. This decrease in bioavailability may then reduce odor production since many odor-causing compounds are produced by the biodegradation of protein.

Several trials were performed in the field at the Charlotte Mecklenburg Utilities (CMU) McAlpine Creek Wastewater Facility WWMF to examine the impact of low-dose lime addition and ferric chloride addition on TVOSC production in the dewatered biosolids cake. The McAlpine Creek WWMF employs mesophilic anaerobic digestion and dewatering with high-solids centrifuges. The addition points for lime and ferric chloride are shown in Figure 3-128. Ferric chloride was added prior to the centrifuge and prior to polymer addition, while lime was added to the dewatered cake (lime was mixed with the cake in the screw conveyors).

All samples were taken after the screw conveyors. At this facility the screw conveyors consist of three distinct parts: 1) horizontal portion collecting the centrifuge cake, 2) vertical portion taking the cake to the top of the storage silo, and 3) horizontal portion dropping the cake into the silo. In a separate previous study, the detention time between the cake drop off from the centrifuges and the top of the conveyors was determined to be one half hour.

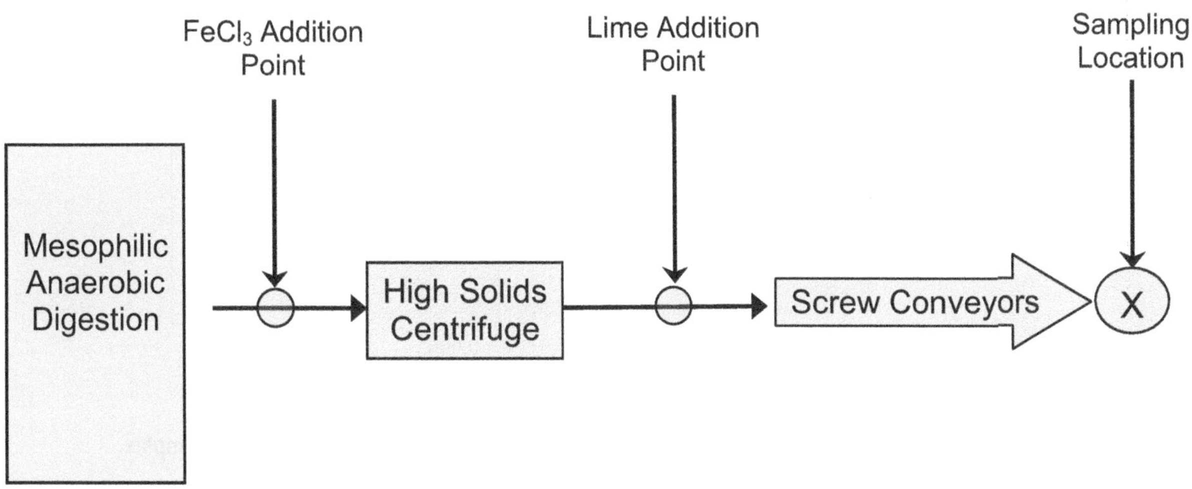

Figure 3-128. Schematic of the CMU McAlpine Creek WWMF Chemical Addition and Sampling Locations.

In these chemical addition tests the goal was to keep all variables constant except one. For example, to study the effect of ferric chloride addition, the lime dosage was kept constant while the ferric chloride dosage was varied. After a change was made, samples were collected once the system had been allowed to reach steady-state condition. Collected samples were shipped to Bucknell University for analysis as described in Chapter 2.0. The samples were collected at the top of the vertical screw portion, after the dewatered and chemical amended cake samples have gone through the screw conveyor. This allowed mixing time for the chemical and the cake. It should be noted that previous studies has shown that the screw conveyor type used at this facility led to increased odor when cake from the top of the screw conveyors was compared side-by-side with the cake collected from the first horizontal portion (Erdal et al., 2003)

3.3.4.1 Ferric Chloride Addition with and without Lime Dosing Test Results

The WERF team first conducted tests on ferric chloride addition to the dewatering process feed sludge. Ferric chloride was first tested without lime addition, and then with a constant lime dose.

Ferric Chloride Addition without Lime Addition

A trial was first performed to examine the addition of iron only to the dewatering feed. Ferric chloride was added at a dosage range from 0 to 3% as Fe on a dry weight basis. These four tests were conducted time with no lime addition. The peak TVOSC concentration measured for the different iron dosages are shown in Figure 3-129.

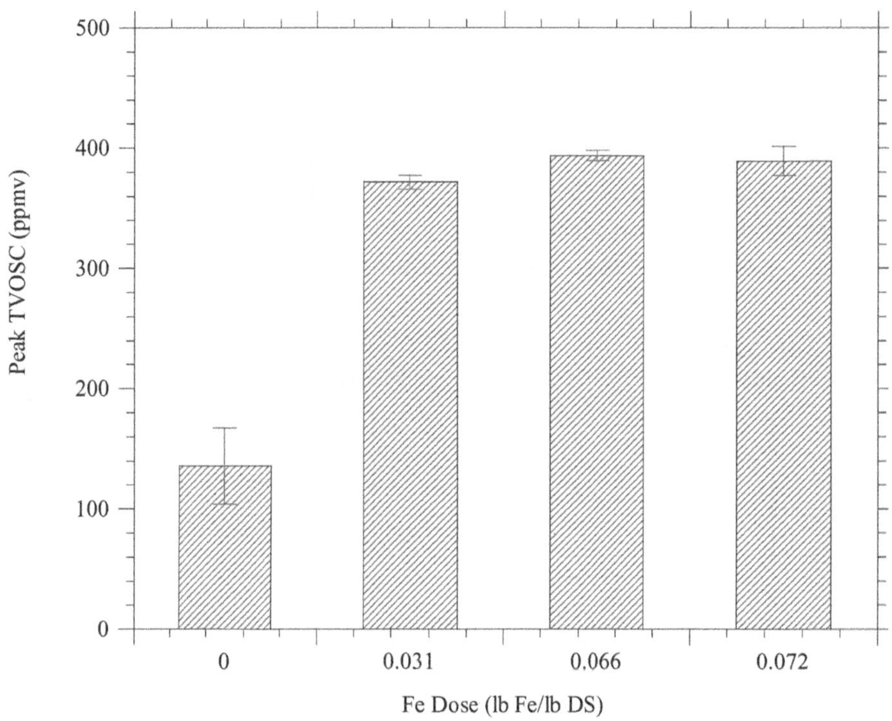

Figure 3-129. Effect of Ferric Chloride Dosage on Peak TVOSC Concentration from Cake Samples.

The results indicated that iron addition actually increased the TVOSC production compared to the control. This increase may have been due to the associated decrease in pH and inhibition of methanogenic activity as shown in Figures 3-130 and 3-131, respectively.

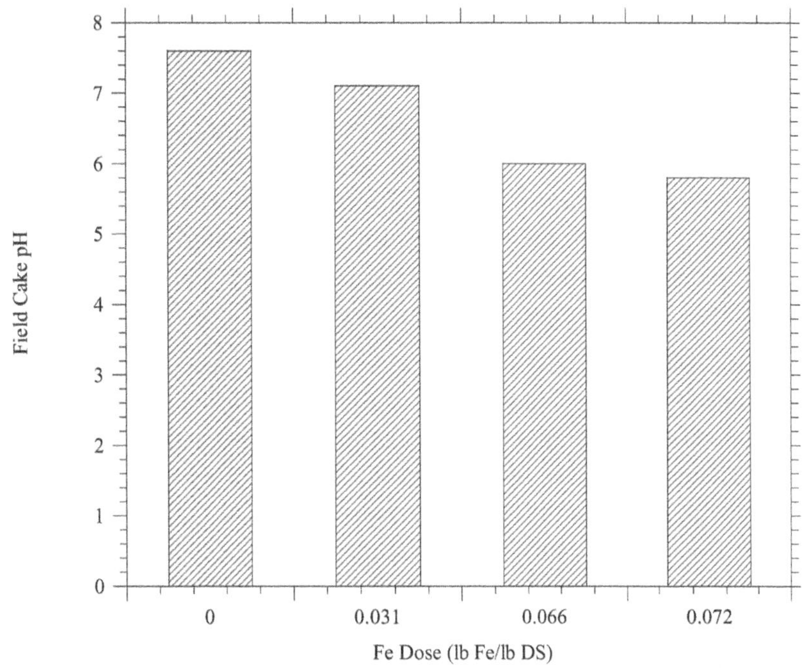

Figure 3-130. Effect of Ferric Chloride Dosage on Cake pH as Measured in the Field.

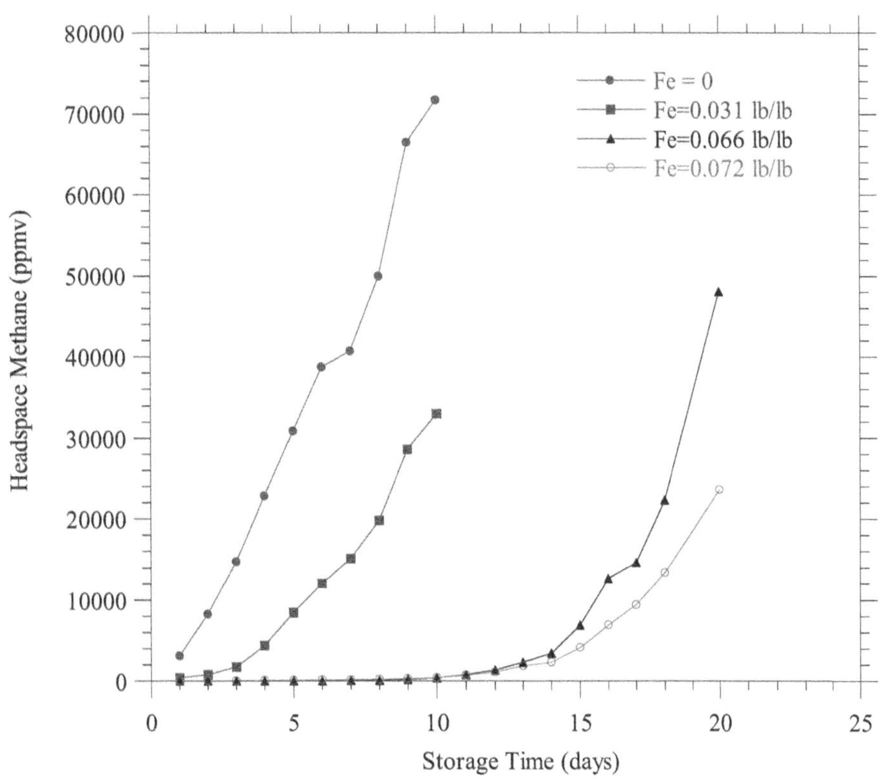

Figure 3-131. Effect of Ferric Chloride Dosage on Methane Production.

Interestingly, the control sample had relatively low odorant production compared to many plants sampled as part of the WERF studies. The increase in pH may have led to imbalances in microbial kinetics that caused an increase in odorant production.

Ferric Chloride Addition with Constant Lime Dosing

In the next set of tests, ferric chloride was added at a dosage range from 0-3% as Fe on a dry weight basis while the lime dosage was kept constant at 5%. The TVOSC profile for the different dosages is shown in Figure 3-132 and the peak TVOSC concentration for each sample is shown in Figure 3-133. As the ferric chloride dosage was increased for the first two dosages, the peak TVOSC *increased.* Only at the highest dosage of 3% as Fe on a dry weight basis was the TVOSC concentration less than the control sample. The bound protein content measured in the different cake samples is shown in Figure 3-134. Similarly, the bound cake protein content also increased for the first two dosages, but decreased at the highest dosage.

The content of different metals for each cake sample was also measured, and the iron content for the different ferric chloride dosages is shown in Figure 3-135. The trend shows that as the ferric chloride dosage was increased, the iron in the cake increased proportionally. Therefore, the iron measurements are likely reliable. The increase in iron in the cake was much less than would have been predicted if the entire ferric chloride dosage resulted in all the iron being deposited in the cake. For example, at the highest dosage of 0.03 kg Fe/kg dry solids (DS), the cake iron concentration increased by 9.61 mg Fe/g DS, which means only about 32% of the Fe dosage was deposited in the cake. The remainder presumably remained in solution and was discharged in the centrate.

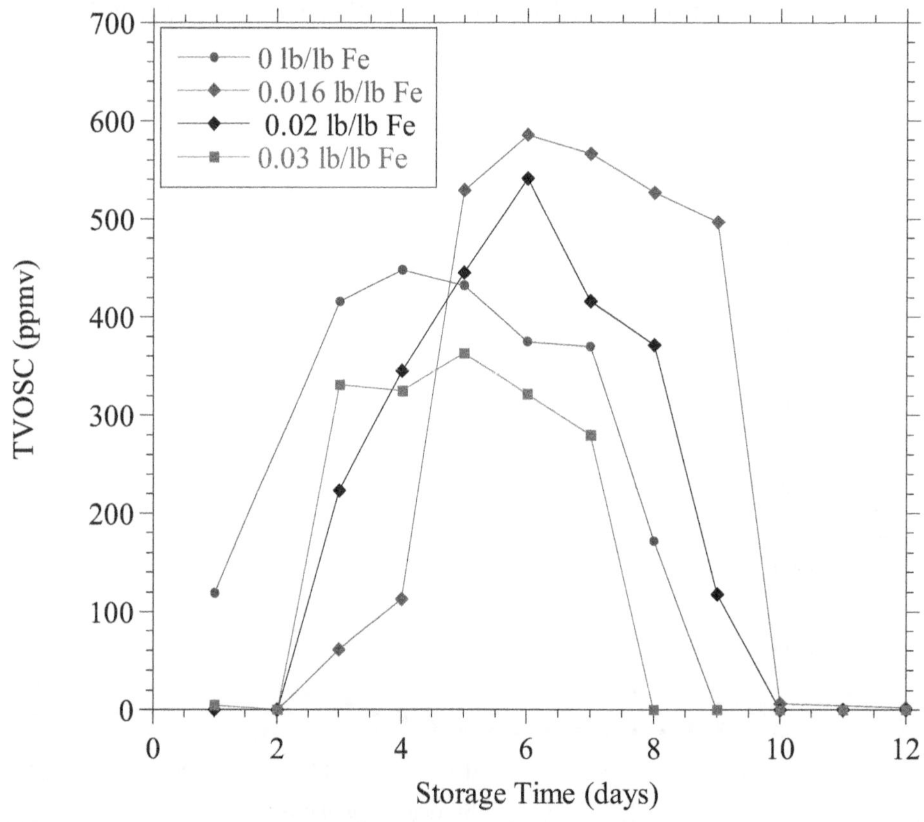

Figure 3-132. Production of Headspace TVOSC from Cakes Produced With Different Ferric Chloride Dosages.

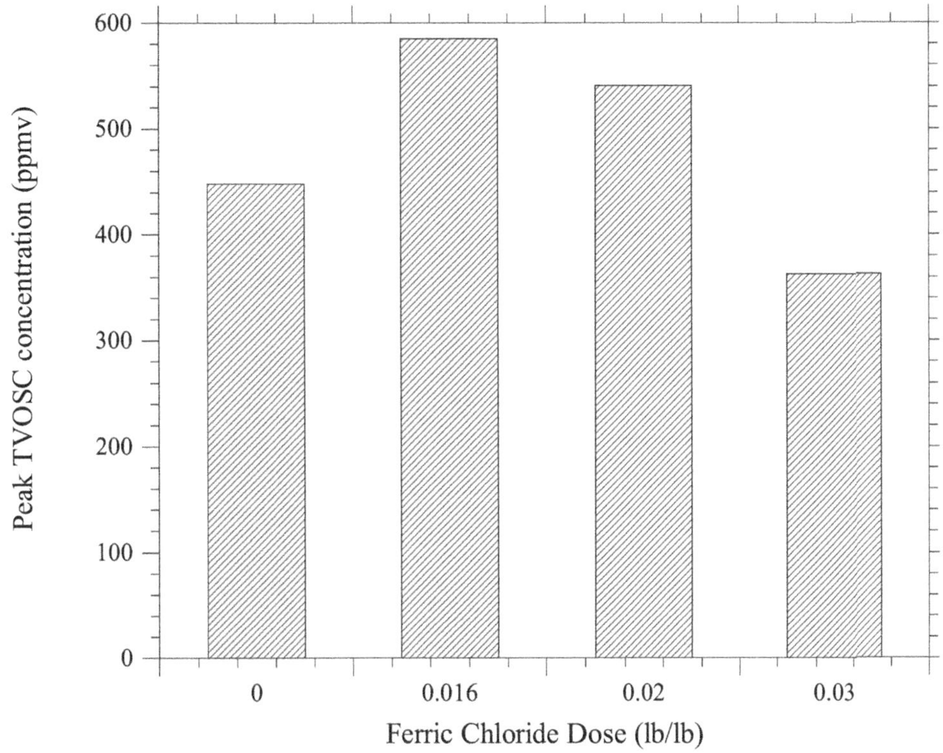

Figure 3-133. Peak TVOSC Concentration From Cakes Produced With Different Ferric Chloride Dosages.

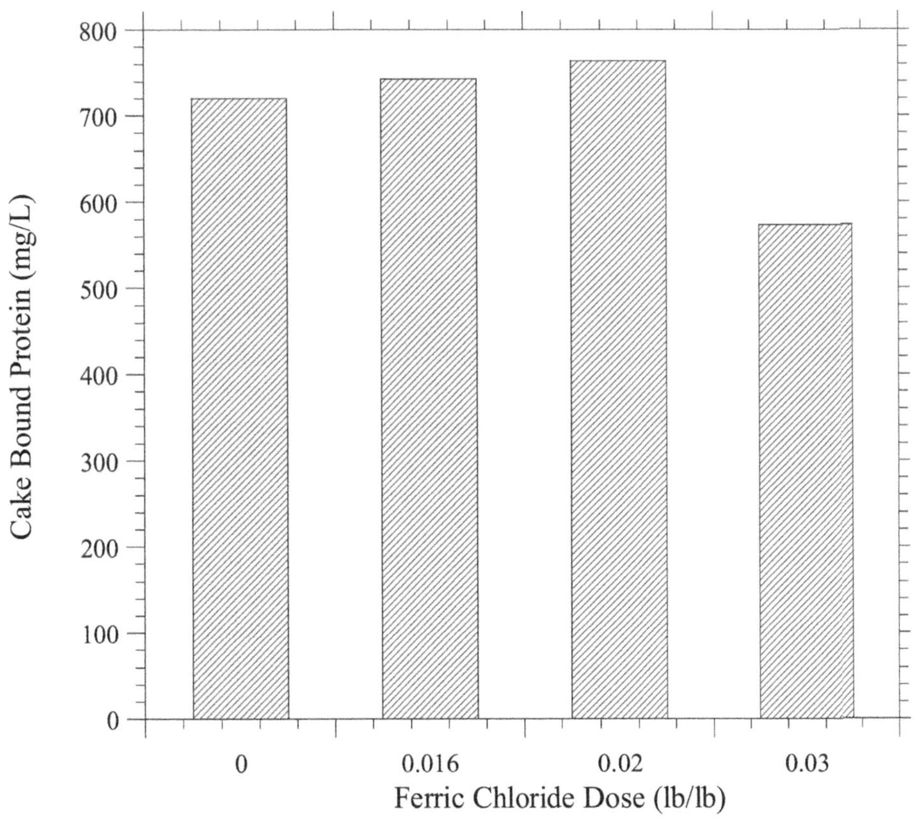

Figure 3-134. Bound Cake Protein Content Measured for the Different Ferric Chloride Dosages.

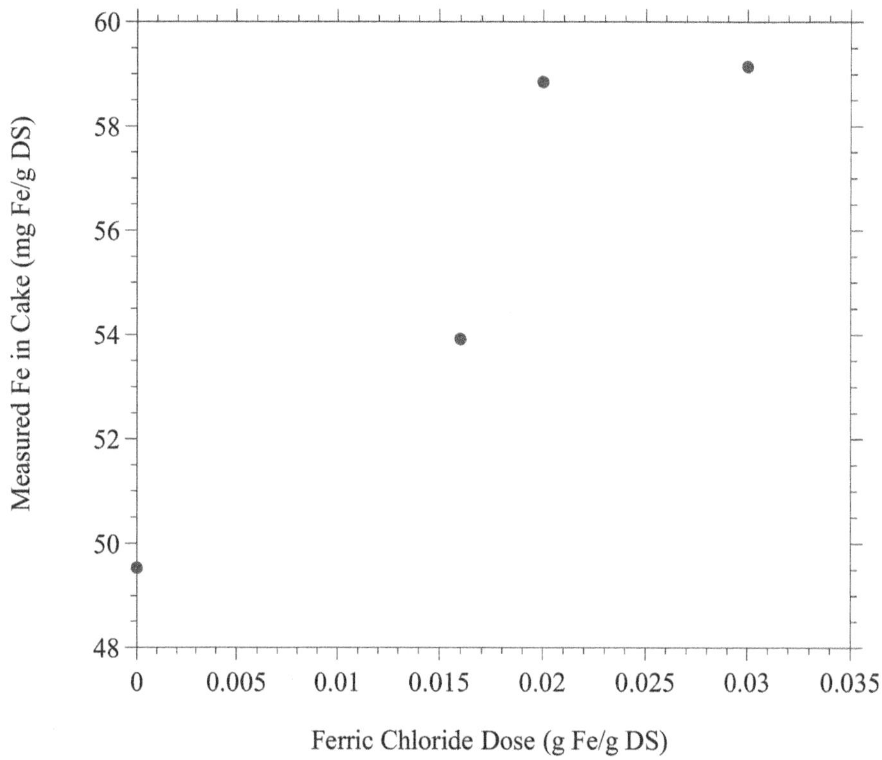

Figure 3-135. Iron Content Measured in Cakes for the Different Ferric Chloride Dosages.

Since the iron dosage did not directly correlate with the peak TVOSC concentration, the WERF team examined other variables that might affect TVOSC. The peak TVOSC as a function of cake pH is plotted in Figure 3-136. The figure indicates that a direct positive correlation exists between the cake pH and peak TVOSC. However, the change in pH was not correlated to the iron dosage as shown in Figure 3-137. Since ferric chloride addition would tend to decrease the pH of the solution, increasing the dosage should result in lower pH values.

As shown in Figure 3-137, the sample with no ferric chloride actually had a lower pH than the first two ferric chloride dosed samples. This may suggest a possible problem with the dosing of chemicals. Lime was also being added to the cake at a dose of 5%, therefore the concentration of lime may be impacting pH as well, although during this trial the lime dosage was ideally being held constant. There might have been an improper mixing of lime into the cake in these experiments, hence the sample may not have represented limed cake, thus affecting cake pH.

The calcium content of the cake is a good indicator of the lime dose, and in this testing the calcium content measured is directly related to the cake pH as shown in Figure 3-138. Unfortunately, the calcium content varied during the trial, suggesting that the lime dosage was not constant. However, the data does provide some useful insight into the role of pH in odor production. The results indicate pH is a very important factor in controlling the reactions which produce VOSCs.

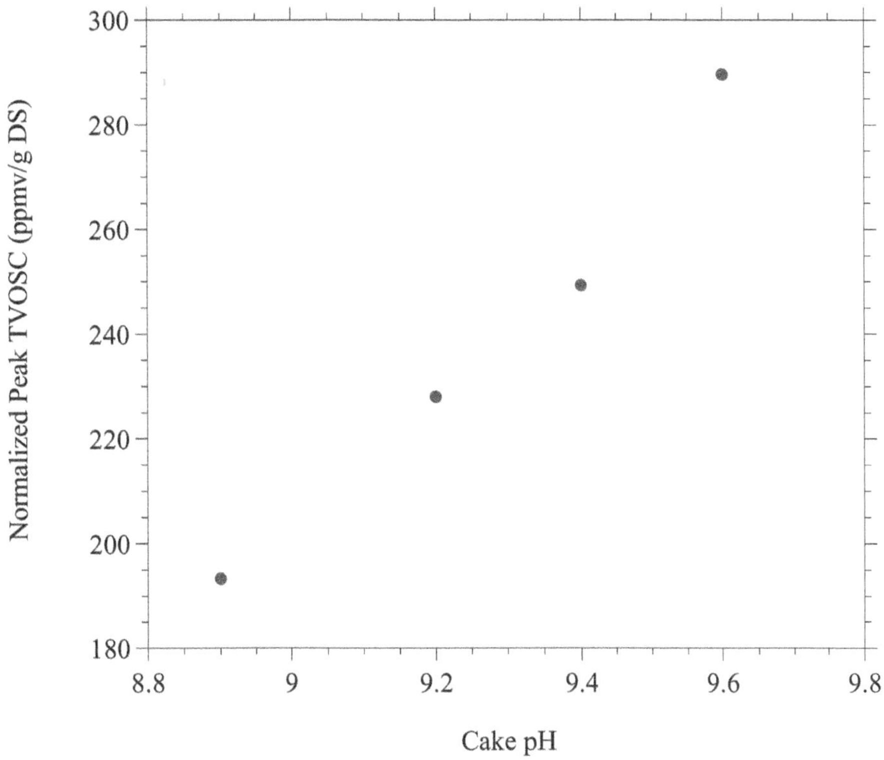

Figure 3-136. Relationship Between Peak TVOSC Concentration and Cake pH (Ferric Chloride Addition Tests).

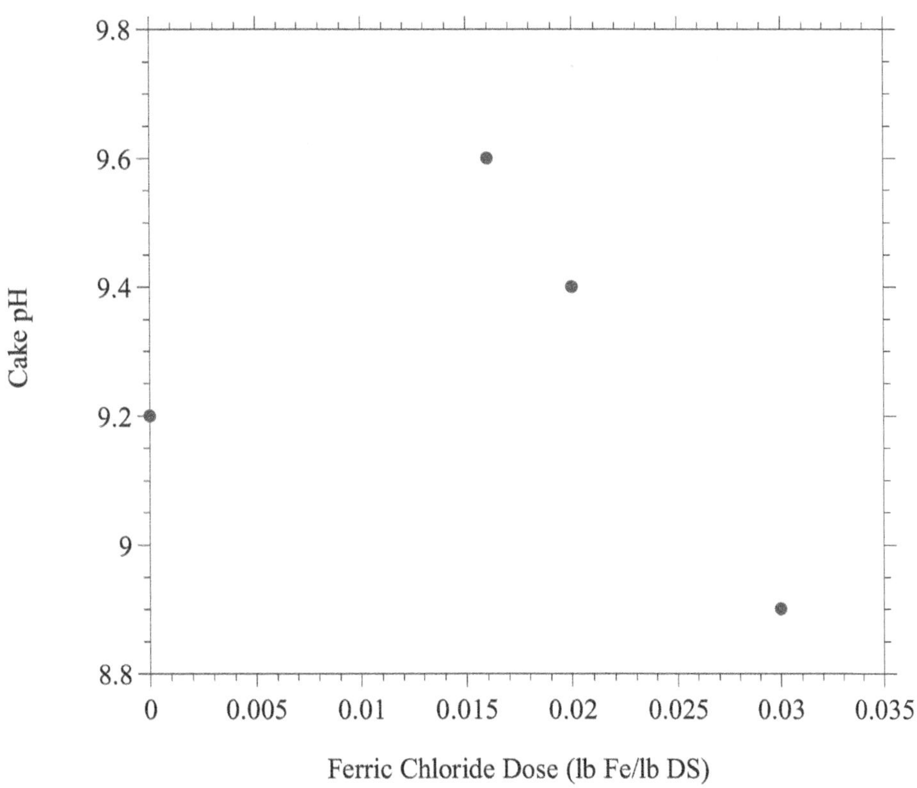

Figure 3-137. Relationship Between Ferric Chloride Dosage and Cake pH (Ferric Chloride Addition Tests).

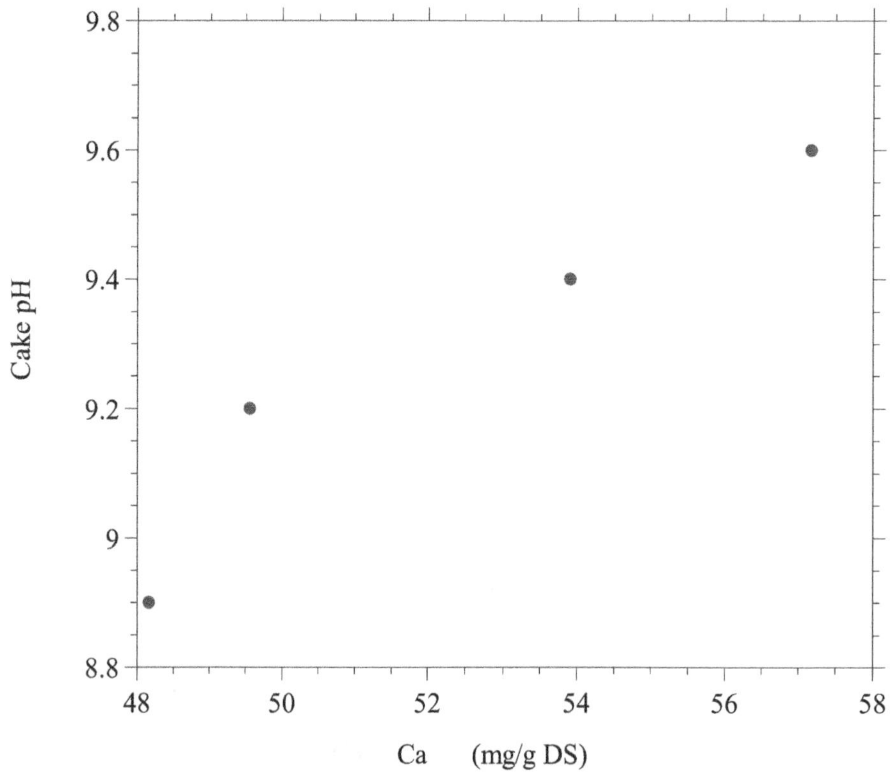

Figure 3-138. Relationship Between Calcium Content of the Cake and the Cake pH (Ferric Chloride Addition Tests).

A second full-scale trial was performed to further examine the impact of ferric chloride and lime addition on odorant production. This trial examined a greater range of ferric chloride dosages: from 3-9% as Fe on a dry weight basis. The results were very similar to the first trial. The WERF team concluded that the addition of ferric chloride clearly did not clearly reduce VOSCs and the addition of low dosages of lime significantly increased the TVOSC concentrations produced by the cake. The main parameter that seemed to correlate with TVOSC was the cake pH.

3.3.4.2 Lime Addition with Ferric Chloride Dosing Test Results

During the same trial, the WERF team conducted a study that examined the effect of hydrated lime addition on the production of TVOSC in the cake headspace. The lime dosage ranged from 0-5%, and the lime was added directly to the cake (mixed in a screw conveyor). Cake samples were collected downstream of the screw conveyor. The ferric chloride dose was held constant for this trial at 0.016 lb Fe/lb DS.

Figure 3-139 shows the headspace TVOSC production profile during cake storage. Contrary to initial hypotheses by the WERF team, low-dose lime addition had a detrimental effect on odor concentrations produced by the cake samples, as the figure shows that as the lime dosage increased, the headspace TVOSC concentrations also increased. The peak TVOSC concentrations measured during cake storage for each lime dose are depicted in Figure 3-140. Peak TVOSC also increased with increasing lime dose. At the 0% lime dose (constant FeCl3 dose) the TVOSC was very low, not leaving much room for improvement.

This trend is likely due to the increased pH associated with the lime addition. The relationship between pH and peak TVOSC for all the ferric chloride and lime additions is depicted in Figure 3-141. This data clearly shows a trend for the impact of pH on TVOSC production in the cake headspace. Specifically, pH effects were negligible up to a pH of approximately 8 (this determined from extrapolation of the data). Above 8, as the pH increased, the TVOSC concentrations increase significantly and with a strong correlation. This appears to be due to inhibition of methanogens that degrade organic sulfur compounds. As the lime dosage increased, methanogenic activity decreased, as indicated in Figure 3-142. In other studies where some, it was shown that cake pH decreases and stabilized over time under low level lime addition (Erdal et al., 2004). The cake pH profile over the storage period was not monitored in this study so as to not disturb the headspace vials.

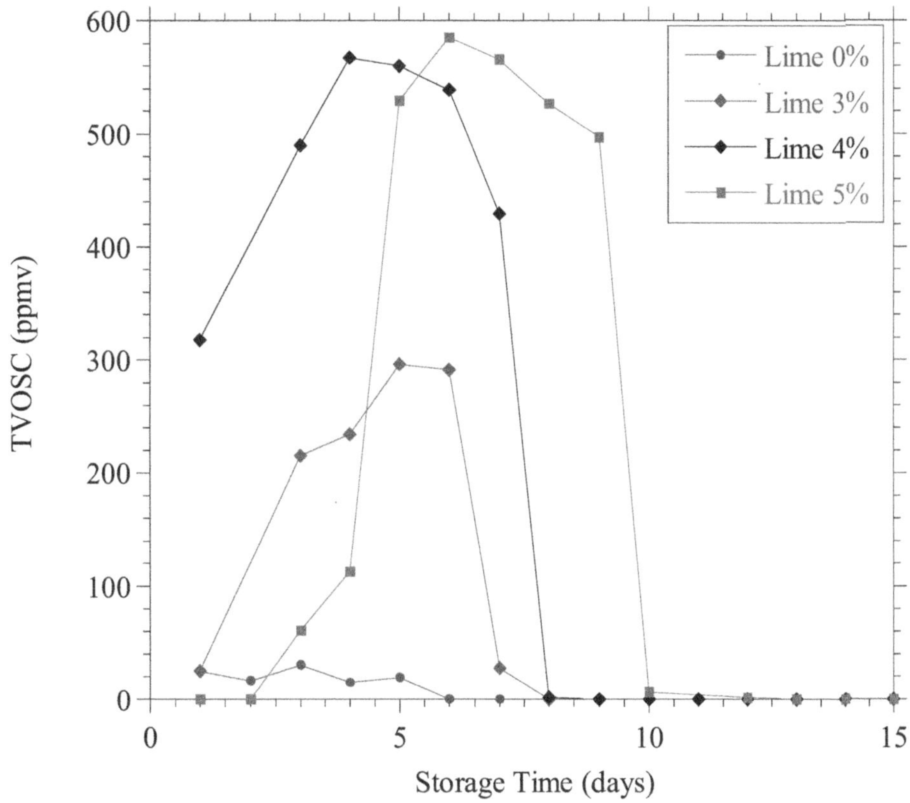

Figure 3-139. TVOSC Production Profile for JWPCP Cakes with Different Lime Dosages ($FeCl_3$ Dose Held Constant).

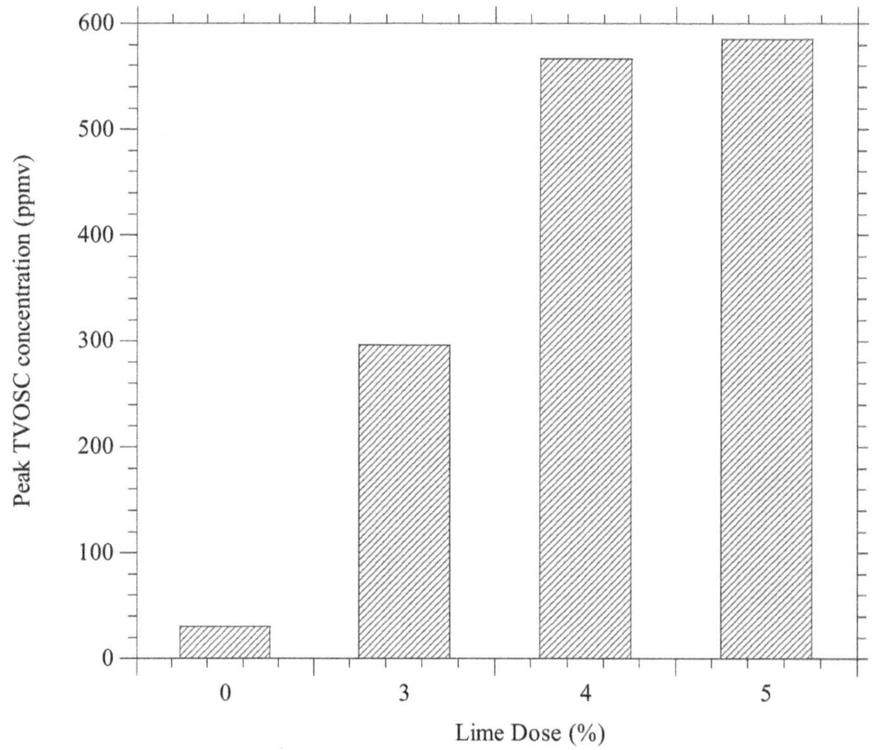

Figure 3-140. Peak TVOSC Concentration for Cakes with Different Lime Dosages ($FeCl_3$ Dose Held Constant).

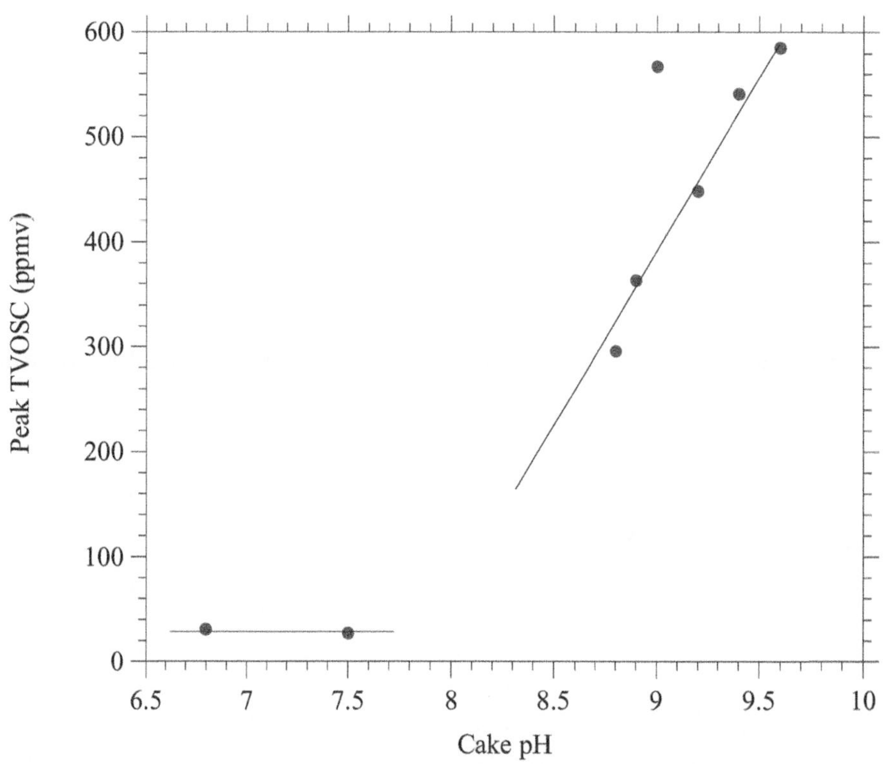

Figure 3-141. Peak TVOSC Concentration as a Function of Cake pH for JWPCP Lime Addition Testing.

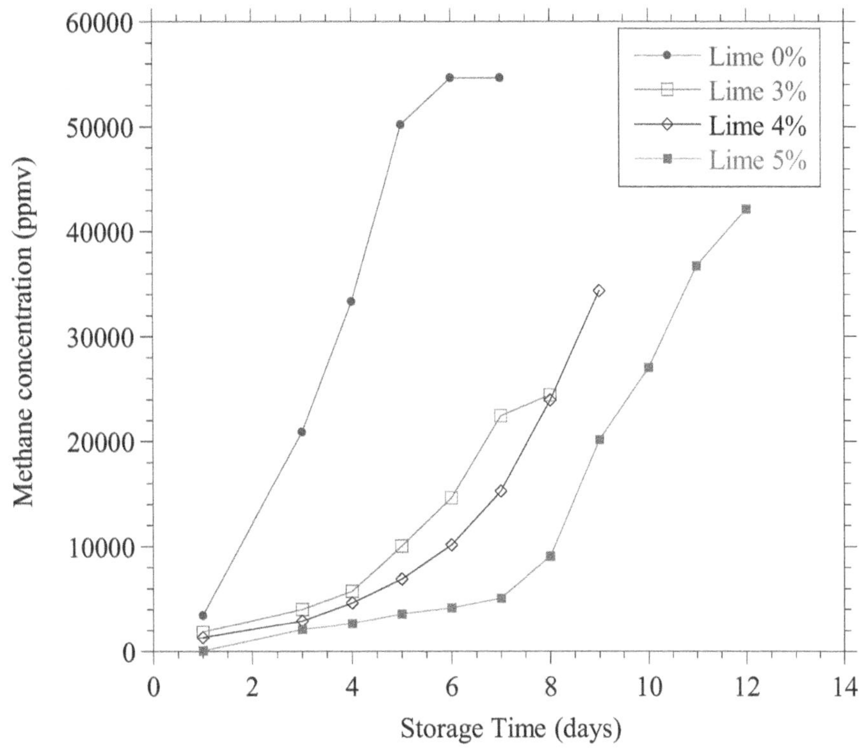

Figure 3-142. Methane Production by JWPCP Cakes with Different Lime Dosages ($FeCl_3$ Dose Held Constant).

3.3.5 Alum Addition Test Results

The WERF team conducted full-scale alum addition testing at the Philadelphia Water Department Biosolids Recycling Center (BRC). Results of laboratory testing (as detailed in Section 3.2.4) indicated that alum addition during dewatering significantly reduced headspace VOSC production in several different dewatered cake samples. As a result, a full-scale investigation was performed at the BRC to determine if alum addition would reduce the production of odorous compounds under field conditions. A Specific Test Protocol was used for testing the odor reduction properties of aluminum, added as aluminum sulfate (alum) at the feed point of a high-solids dewatering centrifuge at the BRC, in conjunction with a polymer dose that was optimized for each accompanying alum dose.

The general process and sampling schematic is shown in Figure 3-143, and the detailed sampling and testing protocol for the alum addition testing is located Section 2.4.2.2.

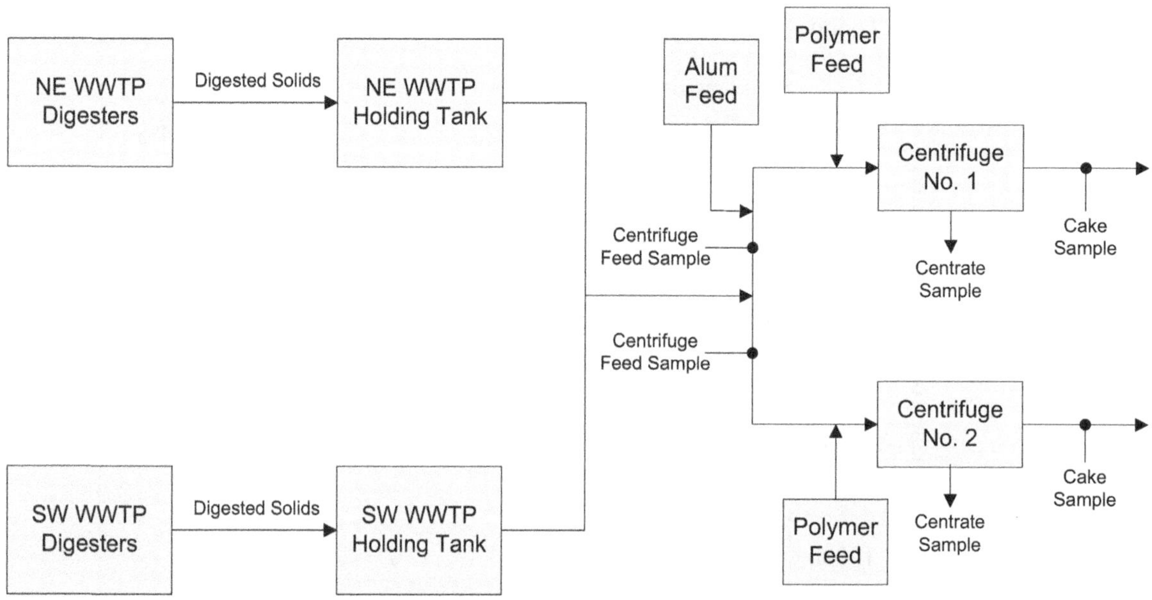

Figure 3-143. BRC Alum Full-Scale Test Schematic, Showing Test and Control Centrifuges.

A summary of the testing protocol is as follows:

1. The alum feed pumps were calibrated at the start of the testing. Prior to feeding alum, the centrifuge was stabilized with the feed and polymer dose using the BRC standard operating procedures.
2. A control sample was collected without alum.
3. Alum was added to the feed upstream of polymer addition. The polymer dose was adjusted to produce a good centrate and cake quality based on visual inspection.
4. After steady-state conditions in the centrifuge were achieved, samples of centrifuge feed, cake and centrate were collected. The pH and temperature of the samples were measured immediately, followed by placing the samples in their appropriate containers and preparing them for shipping.

In the alum addition testing, two trials were performed: one with the BRC Northeast (NE) WWTP digester feed and one with the Southwest (SW) digester feed. The results from the two trials are provided in the following sections.

3.3.5.1 Test Results for Test Run 1 (Northeast WWTP Feed Sludge)

For the first trial, the target alum dosages were 0, 0.5, 1, 2, and 4% as Al on a dry weight basis. Table 3-5 provides a summary of the cake and centrate parameters for the trial. Following are observations made by the WERF team during the testing and on the results:

- For most of the alum dosages, the cake solids remained between 30-34%. The 0.5% as Al dosage, which had a very low cake solids content of about 23%, was the one exception to this observation.
- The control pH for the feed and centrate both start around neutral (7 and 7.3 respectively). As alum dose was increased above 1% as Al, the pH of the feed and

the centrate decreases. At the highest alum dosage of 4% Al, the feed and centrate pH were down to 4.3 and 4.2, respectively.

- For the two highest alum dosages of 2% and 4% as Al, the WERF team observed that the centrate clarity was poor. This observation indicates a lack of good solids capture by the centrifuge.
- The cake aluminum concentration increased proportionally with the alum dosage, and at the two highest alum dosages, the centrate aluminum concentration was very high. This indicated that the aluminum was remaining soluble, likely due to the low pH.
- The optimum polymer dosage was decreased for the first three alum dosages, and the centrate and cake quality generally remained good (except for the 0.5% dosage that produced the 23.1% solids cake, which appears to be an anomaly). At the highest dosage of 4%, the polymer flow had to be increased to the original levels, although the centrate quality remained marginal. The results indicate that at lower dosages (up to about 1-2%), alum addition may result in a decrease in polymer demand.

Table 3-5. Summary of NE WWTP Testing Cake and Centrate data with Varying Polymer and Alum Addition.

Sample #	Target Al Dose (% as Al on dry wt. basis)	Avg. Cake Solids (%)	Feed pH	Centrate pH	Cake Avg Alumium (mg/g DS)	Centrate Avg Alumium (mg/L)	Cake Avg Fe (mg/g DS)	Polymer Flow (gpm)
7	0	33.6	7	7.3	10.95	0.3	80.9	13
8	0.5	23.1	6.9	7.5	16.87	1.5	84.3	11
9	1	30.0	6.8	7.3	20.99	2.5	84.0	9
10	2	30.0	6.2	6.8	28.54	139.4	80.7	8
11	4	32.2	4.3	4.2	33.11	439.6	68.0	13
12	0	34.6	7.3	7.6	21.38	0.5	78.3	13

Bucknell University conducted TVOSC testing on the alum addition samples as detailed in Section 2.2 General Protocols. The effect of alum addition on the cake headspace TVOSC production profile is depicted in Figure 3-144 and Figure 3-145 depicts a summary of the peak TVOSC concentrations measured during storage.

The figures indicate that at the 0.5 and 1% as Al dosages, the overall TVOSC production profile and the peak TVOSC concentrations were lower than the control, but as the dosage was increased beyond those levels, the TVOSC production also increased. At the highest dosage of 4% as Al, the cake headspace TVOSC profile and the peak TVOSC were greater than the control. Interestingly, for the 2 and 4% as Al dosage, the TVOSC production profile during storage was delayed with respect to the time that transpired to reach the peak TVOSC (see Figure 3-146). The WERF team did not expect this time-to-peak TVOSC shift, which became more dramatic with increasing alum dosage.

Also of note is that the second control sample, taken after the alum feed was stopped and the centrifuge was allowed to return to steady-state conditions, had a lower TVOSC profile and peak TVOSC than the first control that was taken at the start of the testing (see Figures 3-144 and 3-145). However, the cake Al data (Table 3-5) shows that the Al concentration was about twice that of the control (21.38 versus 10.95 mg/g DS). This suggests that not all of the alum had washed out of the system prior to sampling, and that the second 0% as Al dose did not serve well as a second control. The result shows there is a balance between the positive effects of alum associated with protein binding, and the negative effects associated with a substantial decrease in the pH.

Figure 3-146 shows the effect of cake aluminum concentration on the peak headspace TVOSC concentrations in the NE WWTP cake. Concentrations up to around 21 mg/g DS result in a decrease in the TVOSC concentration as compared to the control, but in general as the cake aluminum concentration increases, the peak TVOSC concentrations also increase in a similar manner as shown in Table 3-5, Figure 3-144 and Figure 3-145. The low pH associated with the higher dosages is the likely cause for the increase in TVOSC.

Figure 3-147 shows the effect of the alum dosages on methane production. The figure shows the resulting inhibition of methanogens with increasing alum dosage, which was consistent through the testing in this trial. Higher alum dosages were associated with a decrease in methane production (indicative of an inhibited methanogen population), which is generally detrimental to the TVOSC production.

Therefore, the net impact on odor production is the combined effect of methangen inhibition which would tend to increase TVOSCs, and the reduction in bioavailable substrate, which would tend to decrease TVOSCs. The relative impacts of these competing effects will determine the net increase or decrease in TVOSC production. To fully realize odor reduction in the future, the focus would need to be on maximizing the reduction in substrate bioavailability, while minimizing the negative impact on methanogens.

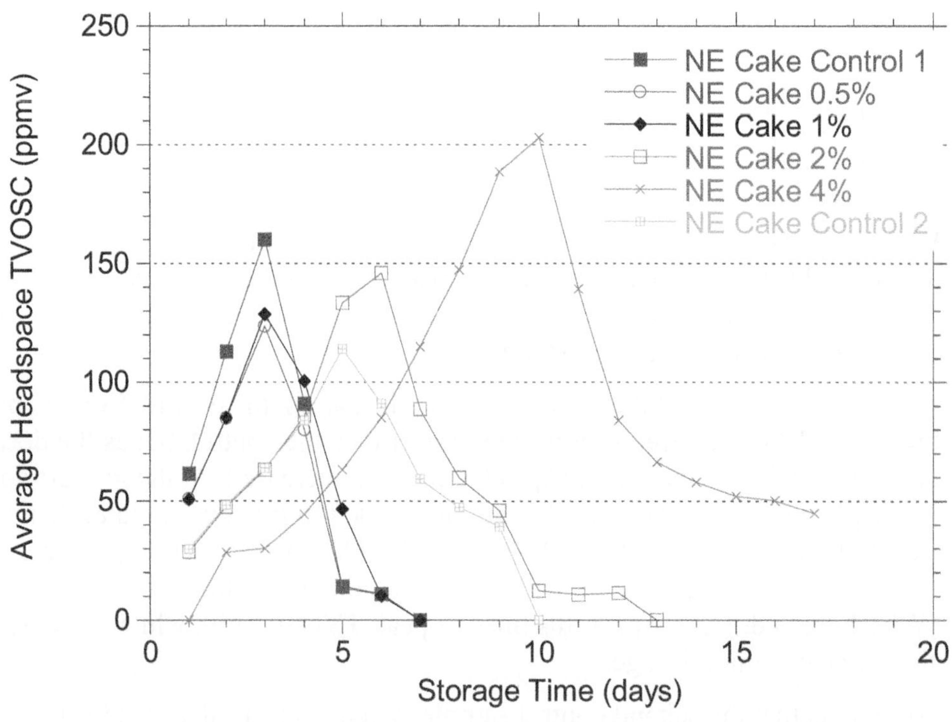

Figure 3-144. Effect of Alum Addition on the BRC Cake Headspace TVOSC Production Profile.

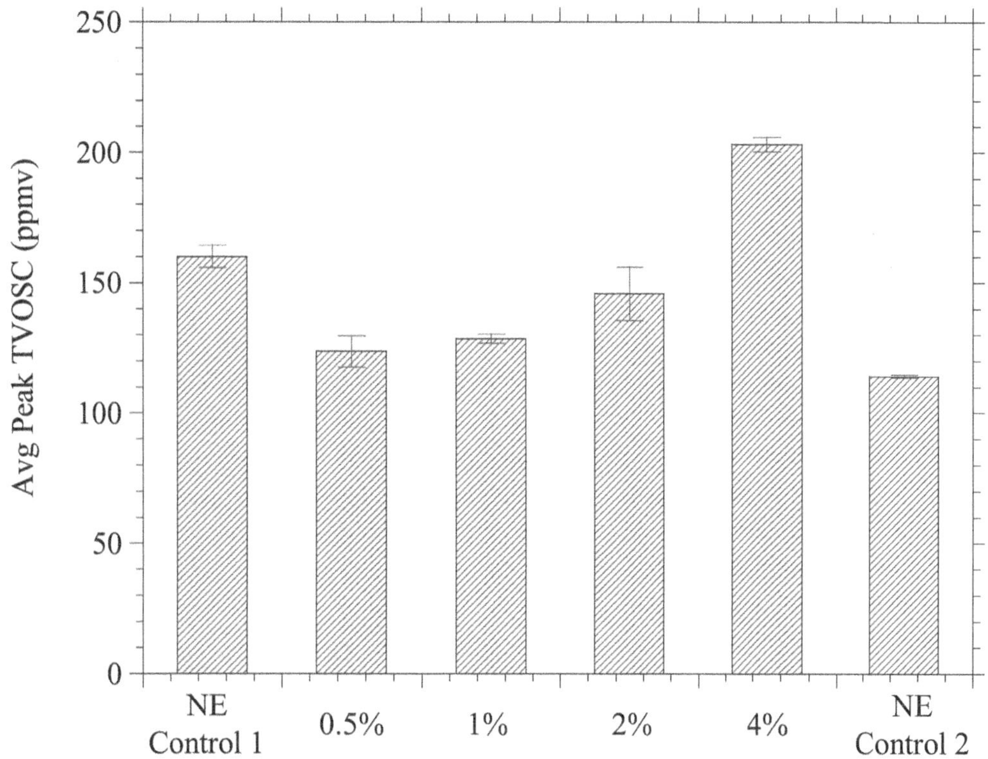

Figure 3-145. Peak TVOSC Concentrations Measured for Different Alum Dosages with the BRC NE WWTP Feed.

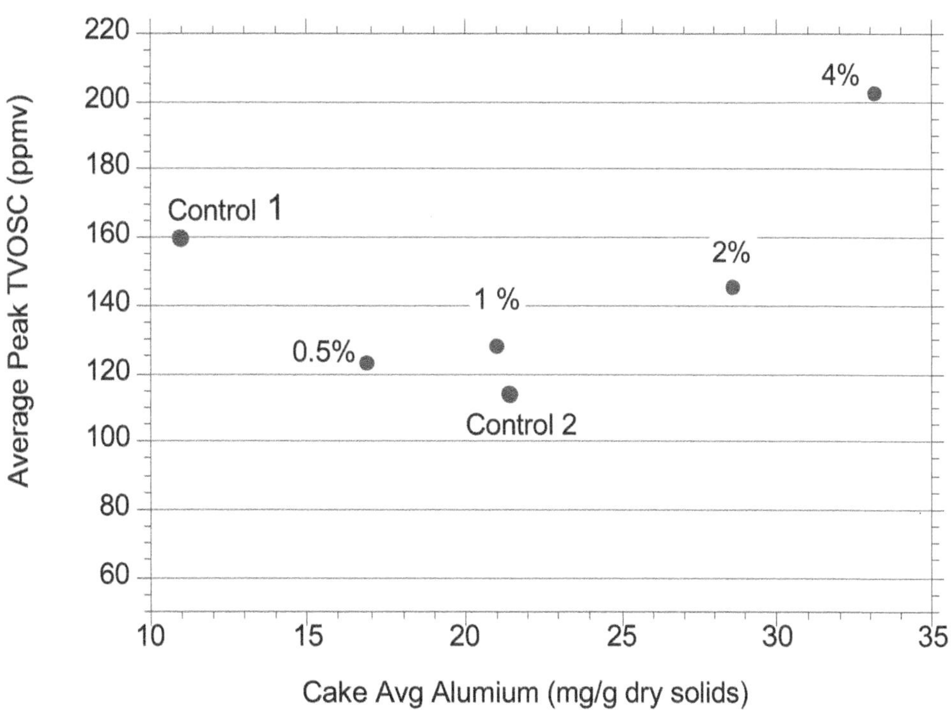

Figure 3-146. Relationship Between Peak TVOSC and Cake Al Concentration in NE WWTP Samples.

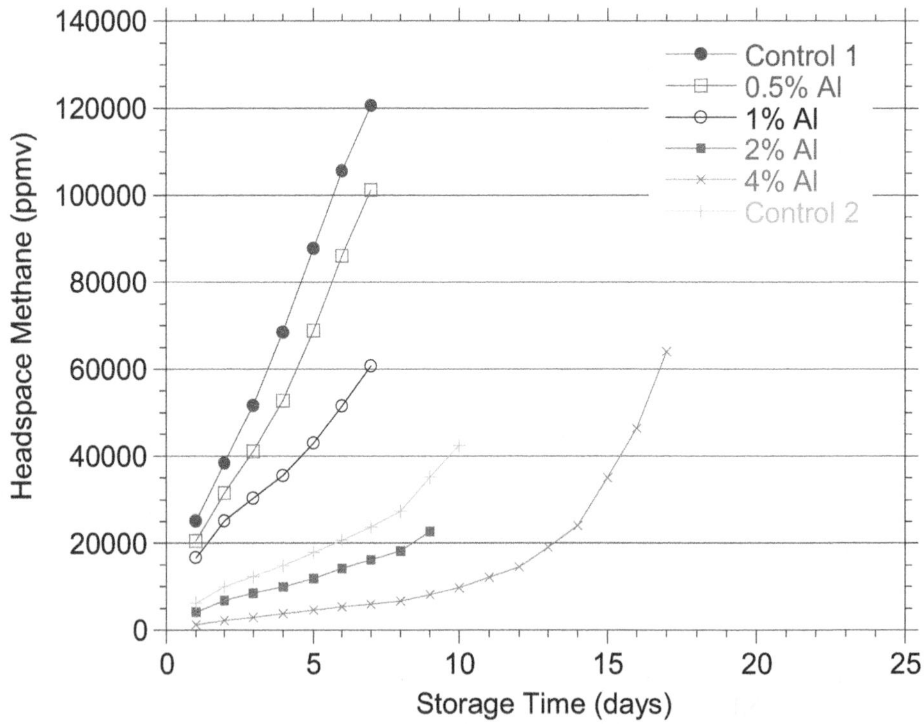

Figure 3-147. Effect of Alum Addition on Methane Production in the NE WWTP Cake.

3.3.5.2 Test Results for Test Run 2 (Southwest WWTP Feed Sludge)

The SW WWTP feed was utilized in the second Philadelphia BRC trial, which was conducted the morning after the first trial using the NE WWTP feed sludge. A feed solids measurement was only taken at the beginning and end of the trial and a centrate solids percentage was calculated for this trial. Otherwise, the procedure followed by the WERF team was identical to that which was conducted for Test Run 1, with alum dosages ranging from 0% to 4% as Al and the polymer dose optimized for each trial.

Table 3-6 provides a summary of the cake and centrate parameters for the trial. Following are observations made by the WERF team during the testing and on the results:

- As the alum dose as Al increased, the pH of the feed and centrate decreased proportionally, similar to the results of Trial 1. The highest alum dose (4% as Al) resulted in a feed pH of 3.9 and a centrate pH of 4.2.
- As the alum dosage increased, the centrate quality generally declined, as measured by the centrate solids concentration.
- At the highest alum dosage of 4%, the centrate had the highest solids content, about 1.1%, representing a relatively low capture rate.

Table 3-6. Summary of SW WWTP Testing Cake and Centrate data with Varying Polymer and Alum Addition.

Sample Name	Feed pH	Feed Temp (°C)	Feed Solids (%)	Target Al Dose (% Al dw)	Polymer Dose (gpm)	Centrate Solids (%)	Centrate pH
PWD Cake Control 1	7.5	32.7	2.35%	0	8	0.130%	7.7
PWD SW Cake 2	6.6	32.8		0.5	7	0.181%	7.3
PWD SW Cake 3	6.3	34.4		1	6	0.238%	6.9
PWD SW Cake 4	5.6	34.1		2	6	0.568%	6.3
PWD SW Cake 5	3.9	34.2		4	9	1.077%	4.2
PWD SW Cake 6	7.0	34.5	2.03%	0	8	0.113%	7.3

For the second trial, Bucknell University reported the cake headspace TVOSC results in terms of the two most prominent individual components: methane thiol (MT) and dimethyl sulfide (DMS). The effect of alum addition on the production of these components, the peak TVOSC, and methane are shown in Figure 3-148 through Figure 3-151. The second trial had more encouraging results with respect to the effectiveness of alum addition on TVOSC production. Increasing alum dose up to 4% as Al correlated with lower peak TVOSC and depressed MT and DMS headspace profiles. Similar to Trial 1 using the NE WWTP sludge, there was a time-to-peak TVOSC shift for increasing alum dosages in the individual components MT and DMS.

Figure 3-152 depicts a plot of increasing alum dosage on the optimal polymer dose determined for dewatering. The figure shows that for increasing alum dosage up to and including 2% as Al, polymer flow requirements decrease. However, polymer requirements for the 4% as Al dose were higher than the first control, equal to the final control.

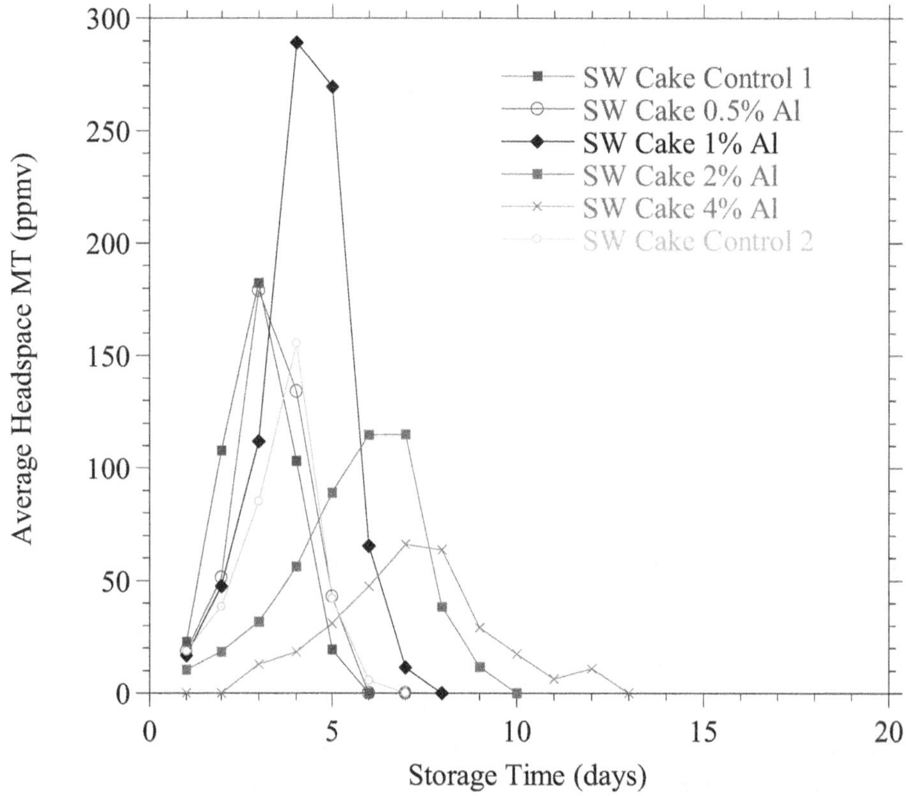

Figure 3-148. Effect of Alum Addition to SW WWTP Feed on the Production of Headspace Methane Thiol.

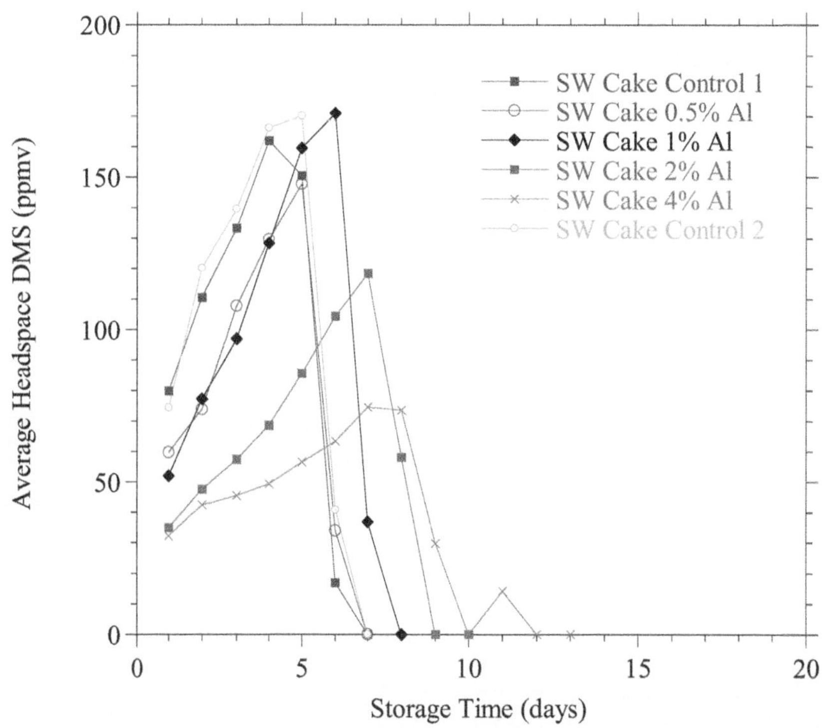

Figure 3-149. Effect of Alum Addition to SW WWTP Feed on the Production of Headspace Dimethyl Sulfide.

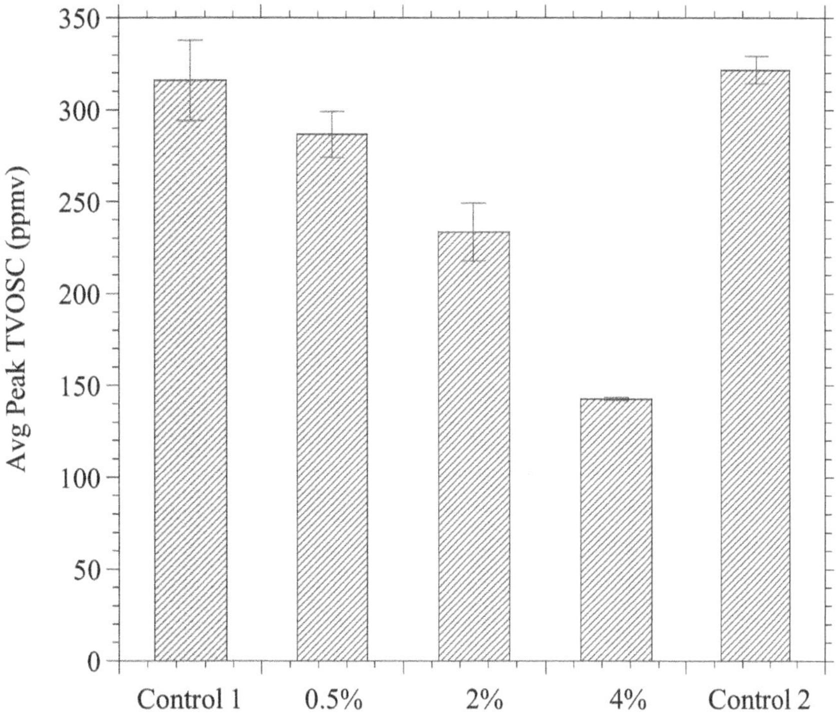

Figure 3-150. Effect of Alum Addition to SW WWTP Feed on the Production of Peak Headspace TVOSC.

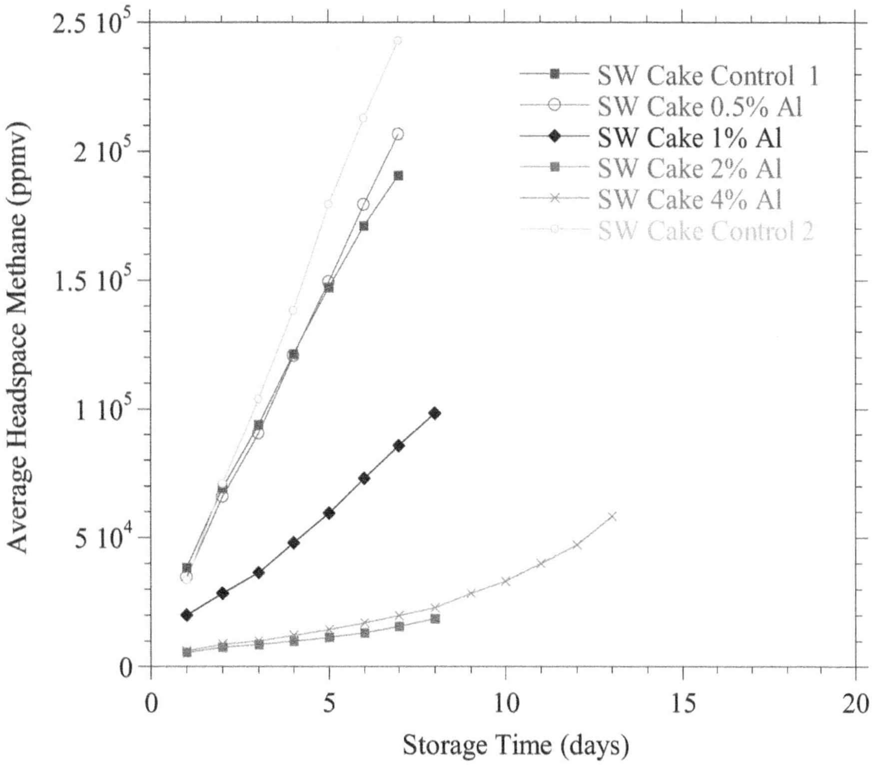

Figure 3-151. Effect of Alum Addition to SW WWTP Feed on Methane Production.

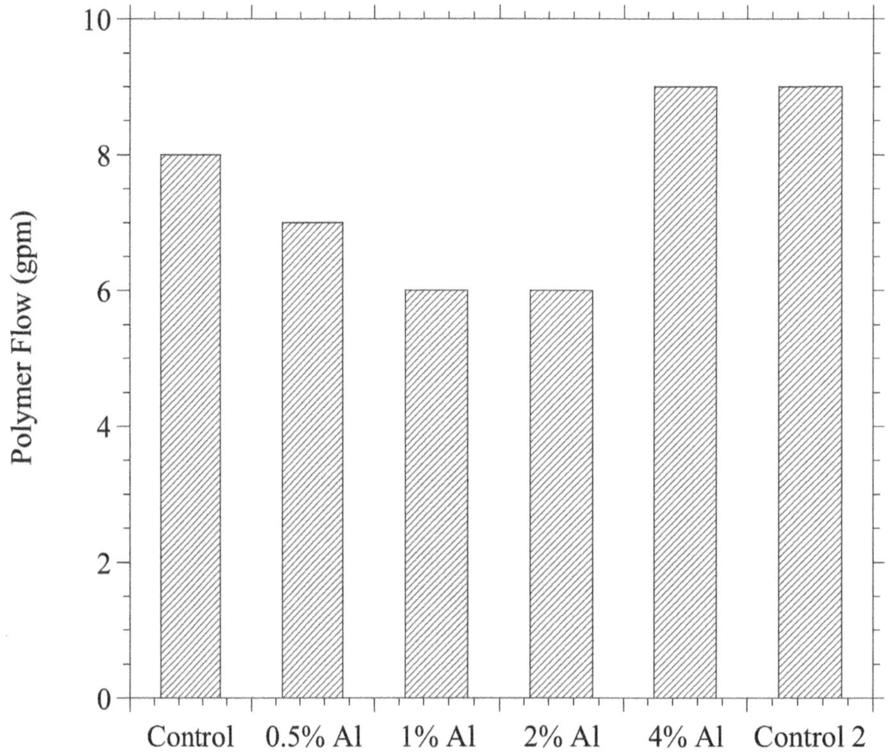

Figure 3-152. Effect of Alum Addition to SW WWTP Feed on the Optimum Polymer Dose.

3.4 Quality Assurance/Quality Control Test Results

During the course of the project, it was important for the WERF team to assure the quality of data that was being collected, especially the headspace TVOSC data by the Bucknell University and Virginia Tech laboratories. In addition, both Bucknell University and Virginia Tech developed centrifuge simulation procedures, which used different approaches. Therefore, these methods needed to be verified to determine if they produce cakes with similar odor profiles as full-scale centrifuges and similar odor profiles to each other. This Quality Assurance/Quality Control (QA/QC) study was performed with the following objectives:

1. Perform and inter-laboratory study to compare TVOSC analytical results using the same cake sample. The three laboratories that were tested were Bucknell University, Virginia Tech, and LACSD. These three laboratories conducted TVOSC analysis over the course of the Phase 3 work.
2. Compare the Bucknell University and Virginia Tech centrifuge simulation processes to results from full-scale centrifuge cakes.
3. Compare the Bucknell University and Virginia Tech centrifuge simulation processes to each other.

3.4.1 Laboratory Comparison of TVOSC Analysis Methods

In this comparison, a cake sample was collected from a high-solids centrifuge at the LACSD Joint Water Pollution Control Plant (JWPCP), split up into even quantities, and sent to three laboratories: 1) to Bucknell University via overnight shipping in a cooler with no ice, 2) to

Virginia Tech via overnight shipping in a cooler with no ice, and 3) to the LACSD laboratory located at the JWPCP (no shipping, as the sample remained on site).

Upon arrival at each laboratory, cake samples were placed in headspace bottles and analyzed for TVOSC concentration during cake storage. The TVOSC concentrations measured during cake storage by each lab is shown in Figure 3-153. The results between Virginia Tech (VT) and Bucknell compared well overall. The TVOSC profiles produced by the LACSD laboratory were significantly lower, and showed a delay in the time to reach the peak TVOSC concentration.

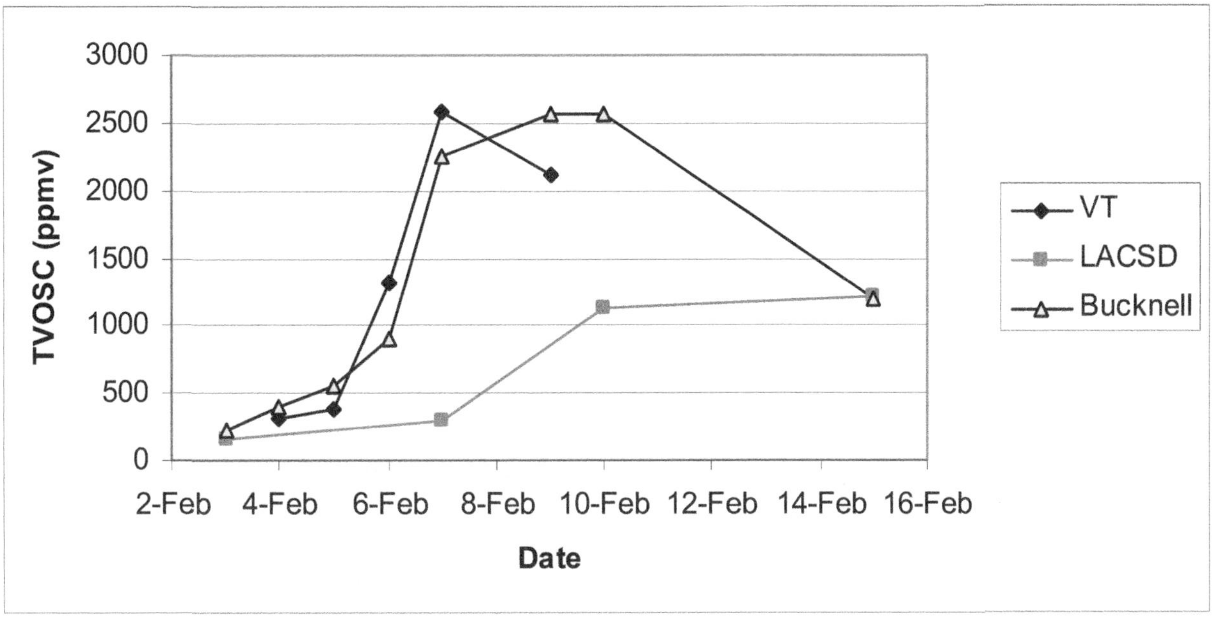

Figure 3-153. JWPCP Cake TVOSC Production Profiles Measured at Three Laboratories.

The WERF team theorized that cooling or heating during shipment across the country to Virginia Tech and Bucknell might have affected the cake quality. If this was occurring, then a smaller sample size would be more susceptible to temperature changes. Therefore, the WERF team conducted a second trial to determine if the sample size made a difference in TVOSC measurement. A small and a large plastic bag containing JWPCP biosolids were both shipped to the three laboratories for TVOSC analysis. The results from this trial are shown in Figure 3-154. No substantial difference was measured for the different sample sizes or the two laboratories. The results from Bucknell had a slightly higher peak TVOSC, and the degradation or disappearance of the TVOSC was faster in the Bucknell results, but overall the data does confirm that the Bucknell and Virginia Tech headspace TVOSC analyses were in satisfactory agreement.

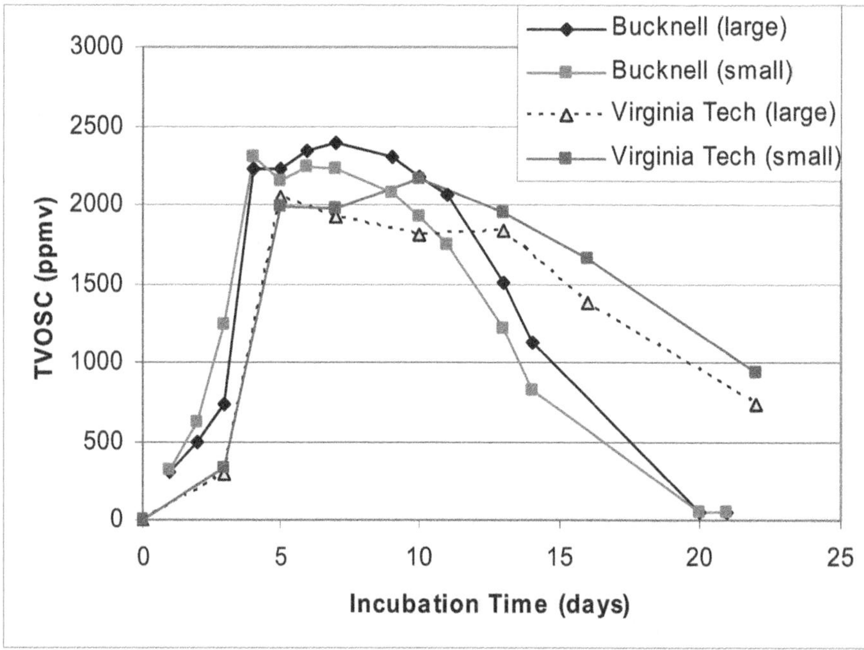

Figure 3-154. LACSD JWPCP Dewatered Cake Testing Between Virginia Tech and Bucknell Laboratories (Sludge was Sent in Small and Large Bags).

3.4.2 Comparison of High-Solids Centrifuge Simulation Procedures.

As discussed in Chapter 2.0, Bucknell and Virginia Tech have developed laboratory methods which produce a cake that has a similar TVOSC production profile as cakes generated from field high-solids centrifuges. To verify the accuracy of these methods, a set of experiments was designed in which JWPCP digested sludge was shipped to Bucknell and Virginia Tech for dewatering. A field cake sample was also shipped to each lab in order to compare the laboratory procedure with the cake generated by the JWPCP centrifuge.

The comparison between the Bucknell method and the LACSD field cake is shown in Figure 3-155. The Bucknell method can change the number of passes through the device to increase the amount of shear that is imparted to the cake, with more passes resulting in greater shear and greater TVOSC production. As shown in the figure, the peak TVOSC of the field cake was around 2,500 ppmv, while the Bucknell laboratory procedure with 20 passes through the centrifuge was around 2,300 ppmv. Therefore, the Bucknell laboratory method compared well with the field cake sample with respect to TVOSC generation.

Similar results were found in Figure 3-156 for the Virginia Tech dewatering and TVOSC analysis process, although the peak TVOSC measured with the laboratory procedure was greater than what was measured with the field cake. A comparison of the Bucknell and Virginia Tech laboratory procedures is shown in Figure 3-157. Feed sludge to the laboratory centrifuges came from the Microsludge digester (process described in Section 3.3.3). The two methods compared adequately overall with respect to TVOSC profile and peak TVOSC, though the sludge processed at Virginia Tech had a slightly higher peak TVOSC and a longer time-to-peak (20 days versus 8 days).

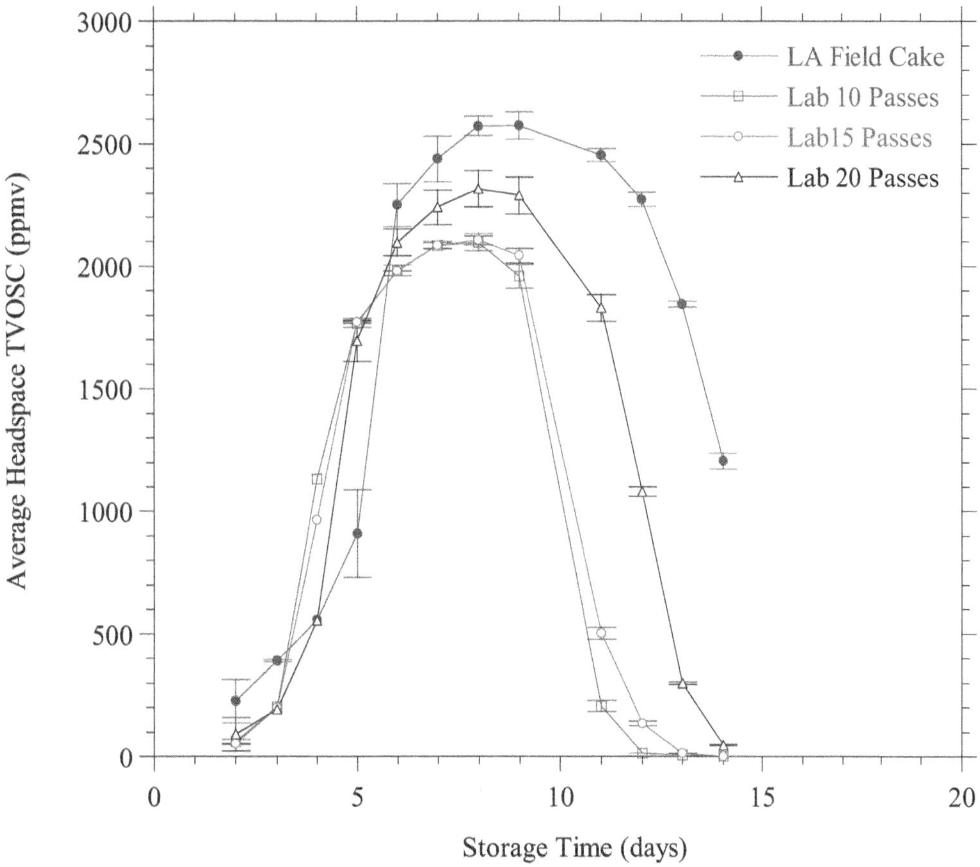

Figure 3-155. TVOSC Profile for Bucknell Dewatered Cake Versus JWPCP Field Cake.

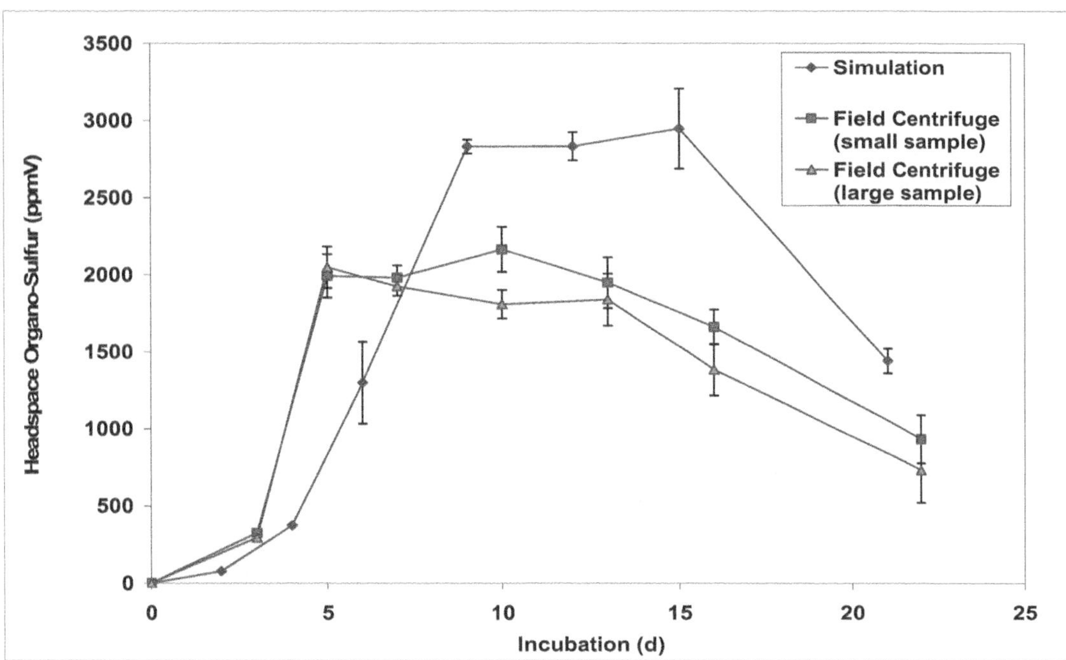

Figure 3-156. TVOSC Profile for Virginia Tech ("Simulation") Dewatered Cake Versus JWPCP Field Cake.

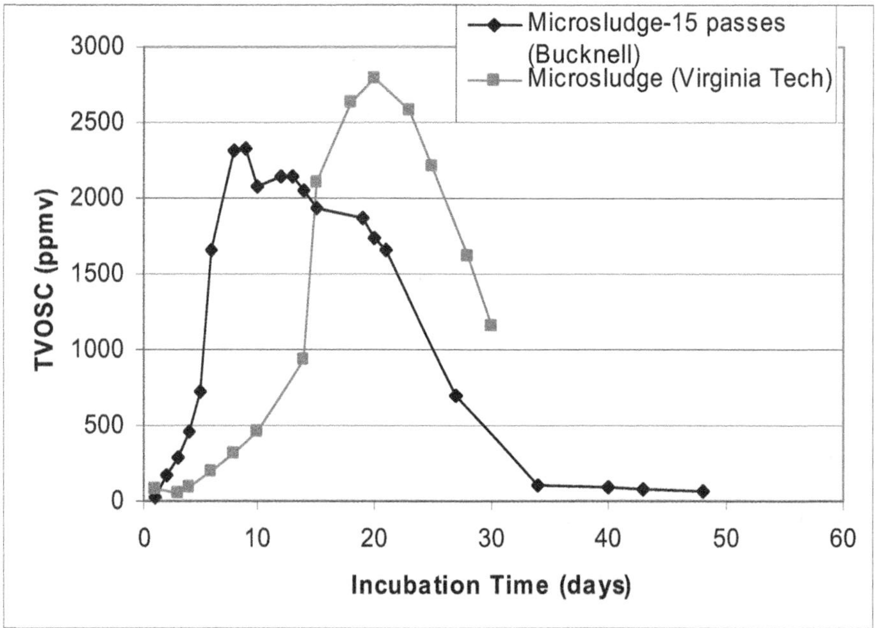

Figure 3-157. LACSD JWPCP Dewatered Cake (Microsludge Digester) Testing Between Virginia Tech and Bucknell Laboratories.

The following conclusions were made by the WERF team associated with this QA/QC testing at the three laboratories used in the Phase 3 research:

- Headspace TVOSC results from the Bucknell University and Virginia Tech laboratories compared adequately for analysis on the same field cake.
- The LACSD laboratory TVOSC results did not compare well with the Virginia Tech or Bucknell TVOSC results, and the difference does not appear to be due to the sample size used in shipping.
- The centrifuge simulation process worked sufficiently at both Virginia Tech and Bucknell laboratories to produce a cake that had a representative TVOSC production profile and peak TVOSC when compared to cake samples produced in JWPCP full-scale centrifuges.

CHAPTER 4.0

RESEARCH FINDINGS AND CONCLUSIONS

The purpose of this chapter is to summarize the research findings of this project and discuss the pertinent conclusions related to those findings. This chapter refers heavily to the results discussed in Chapter 3.0, but organizes the research findings by research subject area rather than by research experiment or full-scale trial. The findings discussed in this chapter cover the following research areas that were investigated in this phase of the project:

- Section 4.1 – The relationship between odors and total, volatile organic sulfur compounds (TVOSC)
- Section 4.2 – The effects of the metal cations as iron and aluminum on biosolids cake odors
- Section 4.3 – The effects of anaerobic digestion and various enhancements to the anaerobic digestion process on biosolids cake odors
- Section 4.4 – The effects of dewatering on biosolids cake odors, with an emphasis on the effects of high-solids centrifuges
- Section 4.5 – The effects of chemical addition on biosolids cake odors, with an emphasis on aluminum and iron addition after digestion
- Section 4.6 – Summary of findings and conclusions

This chapter summarizes our findings by subject area so they can be more readily applied to wastewater treatment and biosolids handling facilities. The findings are therefore discussed in a more general sense than Chapter 3.0, which included specific and detailed results supporting the findings summarized in this chapter.

Chapter 5.0 will conclude this report by presenting recommendations for practical applications of these findings as well as recommendations for further research in the various subject areas.

4.1 Relationship between Odors from Digested Biosolids and Headspace TVOSC Concentration

As noted previously in this report, the findings of this research agree with the previously established principle that there is a strong correlation between odors produced by biosolids from anaerobic digestion and the concentrations of volatile organic sulfur compounds in biosolids headspace samples, which are expressed in this report as TVOSC. The correlation between odor level and TVOSC is important because the required procedures for odor measurements are difficult and costly, whereas TVOSC can be analyzed in laboratories from biosolids field samples with a high degree of precision and reproducibility.

The relationship between odors and TVOSC from the headspace of anaerobically-digested biosolids was established in the previous research phase from biosolids samples taken at eleven different WWTPs across North America, as shown in Figure 4-1. The letter designations F, G, and I represent headspace measurements taken from digested biosolids, fresh cake biosolids, and stored cake biosolids, respectively, from the 11 WWTPs sampled.

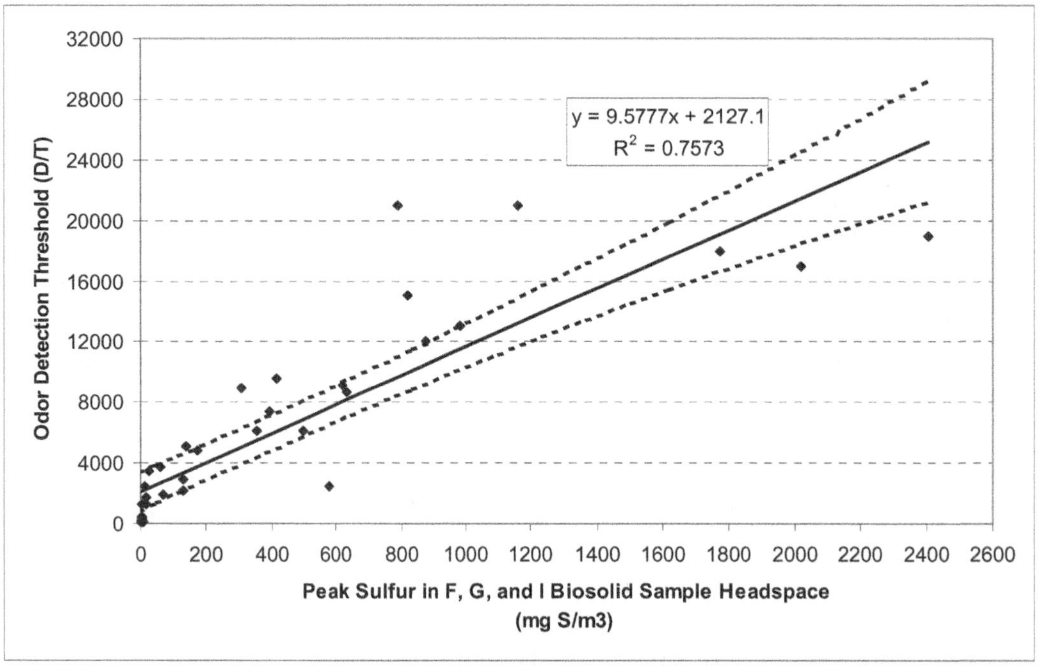

Figure 4-1. Odor Detection Threshold (DT) in Odor Units (D/T) Versus Volatile Total Sulfur. Dashed Lines Represent 95% Confidence Interval), from Phase 2 Research Results.

This relationship showed a correlation coefficient of 0.75 based on a total of 33 headspace samples that were analyzed in the laboratory as well as by a certified odor panel. It is a strong correlation for such a diverse data set from different WWTPs, consisting of at least three different sample locations within each of the 11 WWTPs.

The correlation between odor level and TVOSC is expected to be stronger when all samples are taken from an individual WWTP, and this proved to be the case with respect to samples analyzed at the LACSD's JWPCP in early 2006, as shown in Figure 4-2. This set of five samples from different locations in the JWPCP solids processing train showed a wide range of odor levels. When the odor levels were plotted against corresponding TVOSC concentrations, the resulting correlation coefficient from five samples was 0.93.

It has also been found by comparison of the data from Figure 4-2 with the results of other research projects (Webster at al, 2006), that the relationship between odors and TVOSC also exists with headspace samples taken from flux chambers placed on biosolids storage piles. The flux chamber samples were less concentrated than bottle-headspace samples (by about three orders of magnitude), which was expected due to the continuous movement of air through the chamber. Flux chamber samples would more closely represent odors detected in the field, but as shown in Figure 4-3, the flux chamber and headspace samples exhibit similar trends with respect to odor levels and TVOSC, showing a correlation coefficient of 0.98 on a log-log scale. This is an important finding for predicting odor impacts of biosolids based on laboratory analyses of TVOSC.

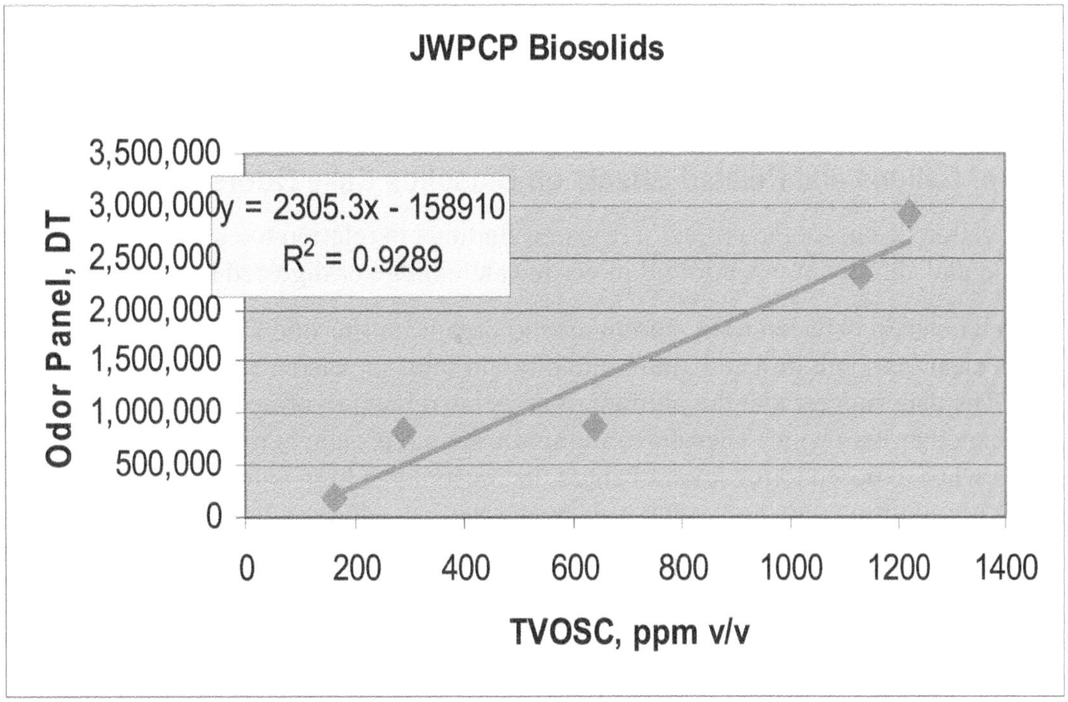

Figure 4-2. Odor Detection Threshold (DT) in Odor Units (D/T) Versus TVOSC from LACSD Biosolids Cake Samples.

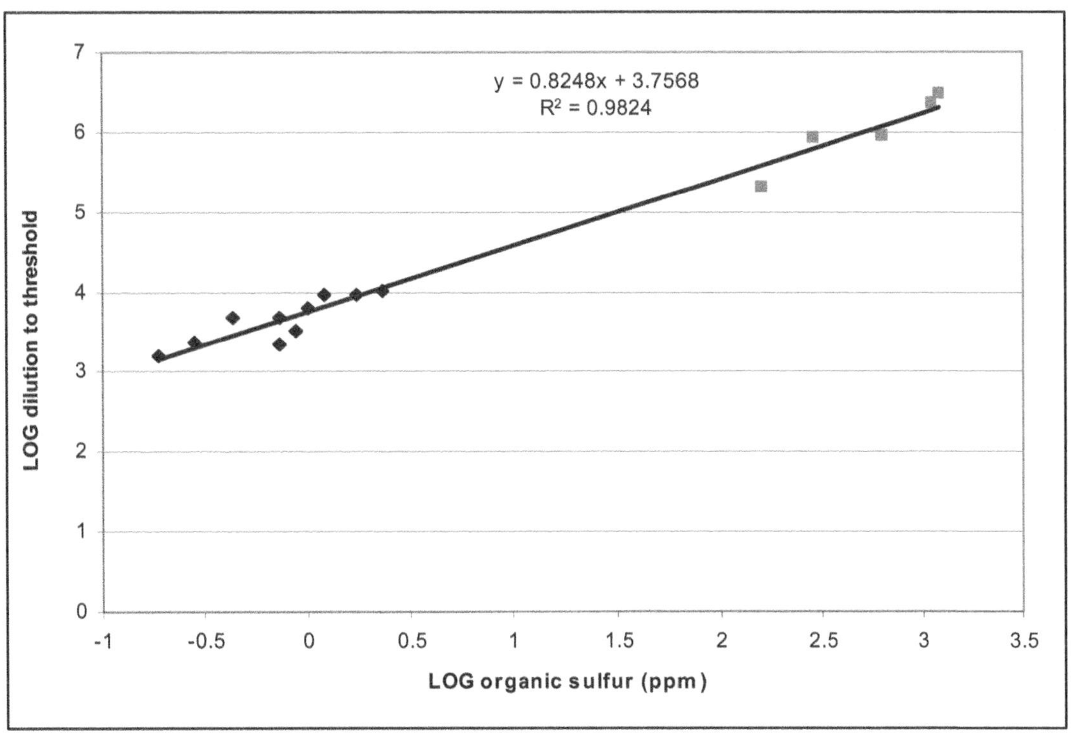

Figure 4-3. Odor Detection Threshold (DT) in Odor Units (D/T) Versus TVOSC from LACSD (Right Side of Line) Compared to Biosolids Cake Odors from Flux Chamber Samples (Webster et al., Left Side of Line).

The strong relationship between biosolids odor levels and headspace TVOSC that has appeared repeatedly in this research is based on biosolids that have undergone anaerobic digestion and mechanical dewatering. These relationships may not necessarily apply to biosolids that undergo other stabilization processes besides anaerobic digestion, but they are viable for the research summarized in this report, which has focused on anaerobic digestion and mechanical biosolids dewatering.

4.2 Role of Cations and Related Effects on Biosolids Cake Odors

This section discusses the project's research findings in relation to cations, focusing on the role of iron and aluminum when found in wastewater sludge or digested biosolids.

The relationship between iron, aluminum and organic sulfur odors in sludge is complex. It is clear that both iron and aluminum bind proteins in wastewater and biosolids to some extent. The data suggest that the aluminum-associated bond to protein is stronger than the iron-associated protein bond. Therefore, aluminum-bound protein is more difficult to degrade, even when exposed to the level of shear that occurs in a high solids centrifuge. It follows that when aluminum is high in the wastewater and raw sludges, both the sludge digestibility (VS destruction is generally lower) and odor generation are impacted (odors from TVOSC are generally lower as well), at least within the pH ranges of 6-8 that are normally seen in wastewater treatment and biosolids processing.

Iron also binds proteins, but apparently to a lesser degree than the bond between protein and aluminum. When iron is reduced in an anaerobic digester, some of the associated protein is released and degraded during digestion, reducing the odor potential of the digested biosolids. However, not all of the iron-associated protein is released and degraded. It is not clear why this occurs, but it appears that iron reduction is slow during anaerobic digestion, and the association of iron and organic matter can reduce the rate of chemical reduction. Some iron-associated protein remains in biosolids after anaerobic digestion, and it appears that higher concentrations of iron leave more iron-associated proteins after digestion. Some of these iron-associated proteins are then released in a dewatering centrifuge as a result of shear, causing higher odors in dewatered cake, as detailed further in the following discussion.

4.2.1 Effects of Iron

Iron is a complex molecule that plays many different roles in nature. Iron can exist in either an oxidized or a reduced form. Iron can be incorporated into biological materials within cells, can bind to organic and inorganic materials outside of cells and can precipitate as a solid in either the oxidized or reduced form. The role of iron with regard to its impact on odors in wastewater and sludge is equally complex. In this project and other related research, iron was observed at different times to be associated with both increased and decreased odor potential in dewatered biosolids. These observations and proposed mechanisms, along with research gaps, are discussed in this section.

4.2.1.1 Iron in Raw Wastewater

When iron is added to wastewater or sludge, it binds with the free sulfide ion and decreases the headspace or off-gas concentration of H_2S, no matter what form of iron is in the wastewater or sludge. Iron readily precipitates with available sulfide ions to form insoluble ferric sulfide (FeS). Iron salts are used successfully in this manner to control H_2S in wastewater collection systems and treatment facilities, including anaerobic digesters. There is little need to further investigate this well-known chemical property of iron, although it is important that this role be understood in reducing H_2S, especially when used to control the H_2S in digester gas.

With respect to the iron present in wastewater and sludges prior to digestion, the research team found in this research that higher concentrations of iron correlate with higher TVOSC concentrations in the headspaces of dewatered biosolids. This finding was somewhat unexpected, since iron addition post-digestion (such as to the feed of a dewatering centrifuge to facilitate capture of phosphorus in biosolids) has led to decreased TVOSC concentrations from biosolids. One possible explanation is that iron in biosolids is associated primarily with organic materials, both inside and outside cell bodies in the biosolids. When this organic material enters an anaerobic digester, most of the iron is reduced, and some of the organic material bound to iron is then degraded. However, a significant portion of the organic material associated with iron remains does not degrade until after digestion. When that organic material is exposed to shear in a centrifuge, bioavailable proteins are released or solubilized, and the proteins become substrate for bacteria that emit TVOSC when the dewatered cake is incubated or stored.

The research team found that low iron concentrations in wastewater and sludge prior to digestion correlated with lower TVOSC concentrations in the headspace of dewatered biosolids samples. One possible explanation is that when iron in wastewater is low or when the concentration of aluminum is high, less organic matter is associated with iron and therefore, less organic material is released in the digested biosolids and available to be sheared by centrifugation. So the resulting odors from dewatered biosolids are lower than for sludges containing higher iron concentrations prior to digestion.

Iron is a good scavenger for protein; so much of the organic matter associated with iron has the potential to create odors. When iron in wastewater is not associated with organic matter, such as the iron-based residual solids from water treatment found in some biosolids, the presence of iron did seem to increase the odor potential from biosolids.

4.2.1.2 Iron Added to Digester Feed Sludge

Other studies have shown that when iron is added directly to the feed sludge entering an anaerobic digester, VS destruction increases and TVOSC generation from dewatered biosolids cakes decreases. It is clear that iron contained in raw wastewater and sludges functions differently from iron added at the WWTP, either upstream or downstream of anaerobic digesters. The varying influence of iron, depending on its chemical state or its application point in a process, reflects the variety of roles or functions that iron can take in wastewater. Iron entering a digester in the feed sludge is predominantly associated with organic matter. Iron fed directly to a digester enters as an inorganic ion, which is rapidly reduced and combines with a variety of organic and inorganic materials including phosphorus, sulfide and organic matter.

If iron binds organic compounds and makes them less available for digestion, why would adding iron to digester feed sludge appear to improve the anaerobic digestion process? Some practitioners have observed that iron addition upstream of digestion appears to improve the percent VS destruction during digestion.

Little research has been conducted to date that attempts to investigate the mechanisms by which iron addition to a digester feed sludge reduces biological inhibition or increases digester VS destruction. Even less research has been conducted to measure the effects on TVOSC production from digested and dewatered biosolids. More study in this area would be beneficial for utilities trying to manage odor generation from digested, dewatered biosolids.

4.2.1.3 Iron Added to a Dewatering Centrifuge

Iron is sometimes added to the feed stream of dewatering centrifuges for sludge conditioning or phosphorus removal. When added in this manner at different WWTPs, iron has been found to increase TVOSC from the dewatered cakes in some cases and decrease TVOSC in other cases. Iron can combine with some organic sulfur compounds, thereby reducing their volatility (as odorants) and resulting in lower TVOSC concentrations. However, the binding properties of iron to organic sulfur compounds are generally poor for iron in its reduced form, so the amount of TVOSC reduction is limited.

When iron is added to a dewatering centrifuge in higher concentrations, two effects can occur. More iron can bind more organic sulfur compounds and thereby reduce headspace TVOSC concentrations, but more iron also tends to reduce the pH of the biosolids. The effects of a lower pH are a weakening of the bond between iron and organic sulfur compounds, and an increase in the inhibition of methanogens. Both of these effects tend to increase TVOSC in the headspace of dewatered cakes. Therefore, the post-digestion addition of iron to centrifuge feed appears to have limited value for reducing biosolids odors from TVOSC.

The role of iron in increasing or reducing odors from dewatered biosolids needs further study. If the apparent correlation between increasing concentrations of iron in wastewater and higher odors from anaerobically digested and dewatered biosolids proves to be valid, this research has provided the first set of data linking biosolids odors to a specific wastewater constituent.

4.2.2 Effects of Aluminum Compared to Iron

Based on preliminary indications that aluminum addition may be more effective at biosolids cake odor reduction than iron addition, it appears that aluminum forms a tighter bond with organic sulfur compounds than does iron, especially under the reduced conditions found in an anaerobic digester. As a result, the organic material bound with aluminum does not release as easily as iron-bound material when the biosolids are sheared in a dewatering centrifuge.

The roles of both iron and aluminum as conditioning chemicals for centrifugal dewatering need to be further investigated and compared. It appears based on the results of this project that the post-digestion addition of aluminum to biosolids can reduce the TVOSC emissions from dewatered biosolids as long as the pH can be controlled, while the post-digestion addition of iron to biosolids has shown very mixed results.

4.2.3 Iron or Aluminum Addition Practiced for Phosphorus Removal

The practice of feeding iron to a treatment plant either in the primary or secondary system for reduction of phosphorus in effluent is becoming more common due to increasingly stringent effluent phosphorus limits. The effects of iron fed in this manner on downstream biosolids odors are unclear and have not been well researched. It is apparent that some of the added iron forms chemical precipitants that may help to bind organic sulfur and reduce biosolids odor potential, but the remainder of the added iron is likely to be incorporated into activated sludges, thereby reducing digestion efficiency and increasing the odor potential of dewatered biosolids.

Feeding aluminum to the primary or secondary system to bind phosphorus in biosolids appears to have a more beneficial effect on biosolids odors based on our research. The body of research under this project indicates that aluminum addition; with its propensity to form chemical precipitants that effectively bind organic sulfur as well as phosphorus, reduces the odor potential of dewatered biosolids. The net effect of competing influences from the addition of iron or aluminum will no doubt depend on the characteristics of the wastewater and treatment processes employed at each WWTP.

The choice of whether to use aluminum or iron salts for phosphorus reduction, the valence state of the iron that is used (there is only one valence state for aluminum), and the locations where iron or aluminum are fed, can all be expected to influence VS reduction in digesters as well as the odor quality of digested and dewatered biosolids. The practices of feeding iron or aluminum to a wastewater treatment process for nutrient removal therefore require further study with respect to their impacts on the odor characteristics of dewatered biosolids.

4.2.4 Summary Effects of Iron and Aluminum on Biosolids Odors

The research team has drawn the following conclusions from our study of cations and related effects on biosolids cake odors, with an emphasis on the effects of iron and aluminum:

- In general, an increase in the iron concentration in sludge or biosolids results in higher TVOSC concentrations in the dewatered biosolids headspace, especially if iron is added to the treatment process prior to or during digestion.
- VS reduction during digestion shows no consistent correlation with TVOSC emissions from biosolids. RBA after digestion appears to be weakly correlated to TVOSC emissions, the trend being that lower RBA did result in lower TVOSC emissions. RBA may be a useful parameter for predicting the extent of TVOSC odors from digested and dewatered biosolids.
- Higher concentrations of aluminum in raw sludge or digested biosolids tend to result in lower TVOSC emissions from dewatered biosolids. The relationship between iron, aluminum and TVOSC generation may provide a useful tool for predicting if a raw sludge is likely to generate odors after anaerobic digestion and dewatering by high-solids centrifuges.

4.3 Effects of Anaerobic Digestion

This section discusses the effects of mesophilic anaerobic digestion on biosolids cake odors, as well as the effects of several advanced or enhanced digestion processes that were investigated when opportunities arose for comparison.

4.3.1 Effects of Mesophilic Digestion SRT

The research team performed a controlled experiment to determine the effect of anaerobic digester solids retention time (SRT) on TVOSC emissions from dewatered biosolids cake. A typical blend of primary sludge and WAS was digested in laboratory-scale anaerobic digesters and operated at SRTs ranging from 10 days to 40 days. Digested biosolids were tested for TVOSC generation using a laboratory-scale centrifuge. EPS extraction studies were also used to improve our understanding of TVOSC generation, and to better understand what fractions of EPS may be related to TVOSC production. The detailed results of this study are presented in Chapter 3.0.

Findings based on results of the experiments on mesophilic digestion SRT are summarized as follows:

- VS destruction during digestion of the same feed sludge under identical conditions (except for SRT) tends to increase as the SRT during digestion increases. This trend was corroborated by the finding of lower concentrations of proteins in solution from samples taken from digesters operated at higher SRTs.
- RBA results showed a reasonable correlation with VS reduction, and also correlated well with headspace TVOSC concentration. Lower RBA consistently resulted in lower TVOSC concentrations in the cake biosolids headspace. RBA is measured by digesting biosolids samples for an additional 40 days after the initial digestion period ends (U.S. EPA, 2003).
- In general, TVOSC emissions from biosolids cake increase with increasing cake solids concentrations. The data suggest that solids concentration itself influences the generation of TVOSC from dewatered biosolids, and that there may be an optimum solids-concentration range with respect to TVOSC generation.
- Reductions in peak TVOSC concentrations correlate well with higher VS destruction, although not as strong a correlation as with RBA. In most cases, higher VS and lower RBA destruction resulted in lower TVOSC.

The maximum extent of TVOSC reduction corresponding with longer digestion SRTs appears to be about 30% lower than the peak value of TVOSC, which is generally insufficient to reduce biosolids cake odors to acceptable levels, even as a result of extended 30- to 40-day digestion SRTs. Some odor reduction can generally be achieved with SRTs beyond the normal range of 15-20 days, but it is unlikely that many WWTPs have enough digestion capacity to provide 30 days of anaerobic digestion SRT, since most digestion systems are designed for only 15-20 days SRT.

In summary, the reduction of cake odors by implementing longer SRTs during digestion may not always be a cost-effective or sufficient solution to the biosolids odor problem, although it might be effective in combination with other appropriate measures as described in this report.

4.3.2 Effects of a Pre-Digestion Conditioning Process

Chapter 3.0 describes the results of the sampling and analyses conducted as part of trial to test the odor impacts of a pre-digestion enhancement process (the MicroSludge™ Process, in this case). The researchers found that the MicroSludge™ Process reduced TVOSC by approximately 50% as compared with conventional, mesophilic-anaerobic digestion, when run under identical conditions in two trials that were identical in procedure, except for the type of polymer used. An emulsion (oil-based) polymer was used in the first trial, and a mannich (water-based) polymer was used in the second trial.

The MicroSludge™ Process prior to mesophilic anaerobic digestion reduced the TVOSC profiles as compared with the control samples under both medium-speed (1500 G) and high-speed (2500 G) centrifuge conditions, and using both the emulsion-based and mannich-based polymers. In this experiment, the mannich-based polymer resulted in substantially lower TVOSC emissions than the emulsion-based polymer. The MicroSludge™ Process resulted in similar odor profiles for the emulsion-based and mannich-based polymer trials, and for both the medium-speed and high-speed centrifuge runs.

The results of MicroSludge™ also suggest that there was an asymptotic lower limit of headspace TVOSC concentration that the MicroSludge™-enhanced digestion process enabled, since the TVOSC headspace concentrations peaked at approximately 1000 ppmv during all four trials that were run on biosolids produced by the MicroSludge™-enhanced digestion process.

In conclusion, it appears that the MicroSludge™ process may be capable of reducing TVOSC generation by 50% compared with single-stage, mesophilic anaerobic digestion. The difference in TVOSC generation between dewatered cakes produced by the MicroSludge™-enhanced digester and the control digester is greater than what would be expected by the difference in digester VS destruction. The MicroSludge™ process may help to reduce biosolids cake odors by destroying a larger proportion of the odor precursors during digestion, since the downstream dewatering process was held constant for each side-by-side digestion and dewatering trial.

4.3.3 Effects of Advanced Digestion Processes

The research team found that advanced digestion processes that result in significantly higher volatile-solids destruction may also help to reduce the level of cake odors when compared to the mesophilic-anaerobic digestion process. Based on the first finding of this section, which shows longer digestion SRTs result in better VS destruction, lower RBA, and lower odors as measured by TVOSC, it follows that advanced digestion processes that accelerate and increase the extent of VS destruction may reduce biosolids odors.

The research team found that most advanced digestion processes do appear to result in lower TVOSC emissions from the digested biosolids after dewatering, as compared to mesophilic anaerobic digestion operated at traditional SRTs of 15 to 20 days. Some advanced digestion processes appear to reduce the biosolids cake odors more than others. This appears to offer a strategy to reduce odors through the design and selection of more advanced anaerobic digestion processes and their various configurations. More research and comparisons are needed between different anaerobic digestion processes to help determine their impacts on biosolids cake odors.

4.4 Effects of Biosolids Dewatering

This section summarizes findings of this research with respect to the impacts of biosolids dewatering processes on odors, with an emphasis on the effects of horizontal, solid-bowl centrifuges because of their reported detrimental impacts on biosolids cake odors. The findings summarized below are based on a number of different research trials conducted on different centrifuge effects such as the effects of operational adjustments to high-solids dewatering centrifuges, findings from the results of biosolids samples taken inside an operating centrifuge, and comparisons of centrifuges of different makes and models.

4.4.1 Effects of Dewatering Centrifuges

In full-scale trials conducted at two different WWTPs on multiple high-solids centrifuges, it was found that different centrifuge types can vary significantly with respect to cake odor reduction, and cake odors from a single centrifuge can be impacted by adjusting its operating parameters, such as bowl speed and torque. These full-scale tests, which investigated several centrifuges of different types, as well as variable operating parameters on identical centrifuges, showed the following trends with respect to operational adjustments:

- Reducing centrifuge bowl speed by 10% resulted in lower TVOSC emissions (by an average of 15% reduction) in the dewatered cake in one test, with no significant reduction in cake solids concentration.

- Reducing centrifuge torque by 20% resulted in lower TVOSC emissions (by an average 45% reduction) in another test, but with an average 12% reduction in biosolids cake solids concentration.
- Lower-speed centrifuges operating at 1200 rpm generated about half the level of TVOSC emissions from biosolids cake, when compared to high-speed centrifuges operating at 2200 rpm, but with a 20% reduction in cake dryness compared with the high-speed machines.
- TVOSC emissions increased substantially as cake samples passed the conical section of the high-solids centrifuge scroll. Higher shear at the centrifuge discharge port increased the TVOSC emissions from that sample, even as the cake solids concentration had decreased, as compared to the previous sample taken along the scroll.

In general, it was found that dewatered cakes from higher centrifuge torque settings contain greater concentrations of available substrate for odor production, in the form of bioavailable protein. The highest torque settings had the greatest amount of easily extracted protein, while the lower two torque settings had lower amounts of bioavailable protein. These findings indicate that higher centrifuge torque settings lead to higher biosolids cake odors. This trend is likely a result of greater shear forces on biosolids during centrifugal dewatering. The higher shear appears to inhibit methanogenic activity and increase the amount of TVOSC production in the dewatered cake.

The research team concludes from the series of dewatering centrifuge tests conducted in this study that higher centrifuge bowl speeds and centrifuge torque levels result in higher emissions of TVOSC from dewatered biosolids. This trend is attributed primarily to the higher shear imparted on biosolids during centrifuge dewatering, as a result of higher bowl speeds and/or torque conditions. Higher centrifuge bowl speeds and higher torque levels sometimes, but not always, resulted in incremental increases in cake solids concentration.

In general, the increases in cake solids concentrations from higher bowl speed and torque were not nearly as consistent or dramatic as the increases in TVOSC emissions associated with higher bowl speeds and torque.

4.4.2 Comparison of Centrifuges with Other Dewatering Equipment

Several trials were conducted to examine the impact of different dewatering equipment on TVOSC production during cake storage. These trials included:

- Side-by-side testing of belt filter press and high-solids centrifuge;
- Side-by-side testing of a rotary press and high-solids centrifuge.

Findings of the trials are summarized below.

4.4.2.1 Side-By-Side Comparison of Belt Filter Press and Centrifuge

At one facility where a belt filter press was compared side-by-side with a high-speed centrifuge, the TVOSC from the belt-filter-pressed biosolids were lower (by about 75%) than from the centrifuged biosolids. Headspace TVOSC from the belt filter press was less than 20% of the TVOSC levels measured from the centrifuge. However, dryness of the biosolids from the belt filter press was only 17% solids as compared with 25% solids produced by the centrifuge.

In this trial, cake samples were collected from a belt filter press and high-solids centrifuge that were operating side-by-side, both dewatering the same feed source of anaerobically digested biosolids. The belt filter press produced no measurable concentrations of TVOSC emissions during cake storage, while the centrifuge produced a peak TVOSC concentration slightly greater than 1000 ppmv. The cake solids concentration from the centrifuge was 56% greater than that from the belt filter press, however.

Based on previous research, an important cause for increased TVOSC emissions from dewatered cake is increased inhibition of methanogens, which are known to degrade odorous TVOSC compounds. The methane production was much lower in the high-solids centrifuge compared to the belt filter press. It is not known whether the main cause for methane inhibition was higher shear or higher cake solids concentration produced by the high-solids centrifuge. Whatever its cause, the apparent inhibition of methanogens created by the high-solids centrifuge apparently provided a mechanism for increased TVOSC production.

4.4.2.2 Side-By-Side Comparison of Centrifuge and Rotary Press

Another side-by-side dewatering trial was performed to compare the production of odorous compounds from a rotary press and a high solids centrifuge. The cake from the rotary press produced much lower TVOSC concentrations compared to the high-solids centrifuges, although throughput of solids from the rotary press was also much lower. The cake solids from the rotary press ranged from 16-21% solids depending on operating conditions. This comparison suggests that, since a rotary press imparts lower shear forces during dewatering than centrifuges, the resulting TVOSC emissions from the dewatered cake are significantly reduced.

The rotary press also resulted in much greater methane production compared to the high-solids centrifuges, irrespective of centrifuge manufacturer. Higher methane production could result from lower shear during dewatering, since higher shear forces during dewatering appear to inhibit methanogenic activity. Greater methane production appears to help reduce the production of TVOSCs, since methanogens have been shown to degrade the odor-causing TVOSC emissions and their precursors.

4.5 Effects of Chemical Addition

Several trials were performed in the field to examine the impact of low dose lime addition and/or ferric chloride addition on TVOSC production. The trials were performed at a plant (Site 3) that employs mesophilic anaerobic digestion and dewatering with high solids centrifuges. Ferric chloride ($FeCl_3$) was added prior to the centrifuge and prior to polymer addition, while lime was added to the cake. The lime was mixed with the cake in screw conveyors leading to biosolids storage bins. All samples were taken after the screw conveyors, as shown in Figure 4-4.

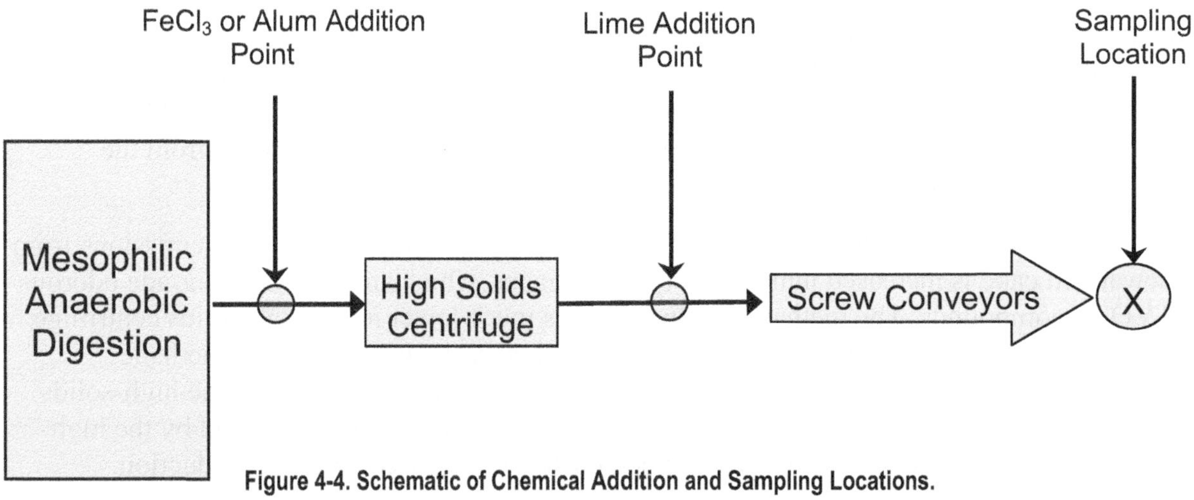

Figure 4-4. Schematic of Chemical Addition and Sampling Locations.

Research has shown the calcium and iron play significant roles in the binding of biopolymeric material. The research team concluded that lime and/or iron addition may reduce odorant production through protein binding, which makes the protein less bioavailable. This decrease in bioavailability may then reduce odorant production, since many of the odor-causing compounds are produced by the biodegradation of protein.

4.5.1 Post-Dewatering Lime Addition

During the same trial, a study was performed to examine the impact of hydrated lime addition on the production of TVOSCs (lime addition at this WWTP was practiced for fecal coliform control). The lime dosage ranged from 0 to 5% by mass of lime to dry mass of biosolids. The hydrated lime was added directly to the cake in slurry form and mixed in a vertical screw conveyor after dewatering, followed by a horizontal screw conveyor. The cake samples were then collected at the end of the horizontal screw conveyor. The ferric chloride dose was held constant for this trial.

In general, as the lime dosage increased, the concentration of TVOSC emissions from the cake also increased. This trend is likely due to the increased pH associated with the lime addition. As the pH increased above pH = 8, the TVOSC headspace concentrations also increased. This appears to be due to an inhibition of the methanogens that degrade the organic sulfur compounds. As the lime dosage increased, methanogenic activity decreased. The microbes that produce TVOSC emissions did not appear to be as inhibited by the higher pH values in the cake, so net TVOSC production was not reduced, and in fact it increased, as a result of lime addition.

The research team concluded from the lime-addition trial that methanogenic activity is inhibited by lime addition (or perhaps more directly by an increase in pH), so odors are not consumed by methanogens during storage, resulting in higher cake odors after low-level lime addition. The addition of low levels of lime (in the form of $Ca(OH)_2$ at a dosage of 5% or less by mass of lime to dry mass of solids) appears to increase cake odors as measured by TVOSC.

4.5.2 Post-Digestion (Pre-Dewatering) Iron Addition

The addition of ferric chloride to digested solids prior to dewatering helps to bind proteins in the cake, thereby reducing cake odors as defined by TVOSC. In the full-scale trial of iron addition, however, the lower doses of ferric chloride resulted in higher odors as measured by TVOSC, than with no iron addition at all. When iron addition exceeded 3% by mass of biosolids, TVOSC started to decrease significantly.

The trend became evident from other experiments that iron addition had to reach at least 3% by mass before any odor reduction in the biosolids cake was achieved, as compared with no iron addition. The research team concluded that the chemical demands of iron in the sludge (in this case, the neutralizing of alkaline conditions) are satisfied after the iron concentration reaches about 3%, above which the protein-binding properties of iron start to make it more effective in the reduction of TVOSC emissions from dewatered cake.

4.5.3 Post-Digestion (Pre-Dewatering) Aluminum Addition

Several laboratory-scale trials were performed with aluminum sulfate (alum) addition to the feed of laboratory centrifuges. The trials used a high-solids centrifuge simulation procedure to produce cake that had a similar TVOSC production as cake samples from full-scale high solids centrifuges. The alum is typically added prior to polymer addition, and the samples are mixed and then dewatered using a laboratory centrifuge.

In the first trial, the polymer dose was kept constant since polymer dose has been shown to impact the production of odor causing compounds. In the second trial, the optimum polymer dose was found for each alum dose. The conclusions from these two trials are provided in the following two sections.

4.5.3.1 Laboratory Trial 1 – Alum Addition with Constant Polymer Dose

It was found that the addition of alum at constant polymer dose substantially reduced the production of TVOSCs compared to the control with no alum. In fact, the lowest dosage examined of 0.5% reduced the peak TVOSC concentration by about 83%. The addition of alum did not appreciably affect the methanogenic activity. Therefore, the addition of alum could have reduced the bioavailability of protein, and thereby decreased the biodegradation of protein that leads to the production of odorous compounds.

4.5.3.2 Laboratory Trial 2 – Alum Addition with Optimum Polymer Dose

For the second trial, the optimum polymer dose was determined for each subsequent alum dose, and the same sample was used as in the previous trial. The increased alum dose resulted in a substantial decrease in the optimum polymer dose, which could represent an important reduction in costs for utilities.

Similar to the previous trial, alum addition resulted in a significant reduction in the peak TVOSC concentration. In these trials, the methane production increased compared to the control for the first two dosages (0.5 and 2%), and decreased slightly at the highest dosage of 4% compared to the control. The results indicate that alum addition in the dosage range used for these trials does not seem to negatively impact beneficial methanogenic activity.

4.5.3.3 Full-Scale Alum Addition Trial

Based on laboratory testing, the addition of alum during dewatering may significantly reduce the production of TVOSCs. As a result, a full-scale investigation was performed to determine if alum addition would reduce the production of odorous compounds under field conditions. The field trial was conducted at a facility which dewaters digested sludge from two different WWTPs.

In the alum trial on biosolids from the first WWTP, lower dosages of alum resulted in a slight reduction in the peak TVOSC, but at the higher dosages of 2% and 4%, the TVOSC increased. The higher dosages were also associated with a substantial decrease in the pH of the feed, which may have contributed to a negative impact on methanogenic activity which has been shown to be beneficial to reducing TVOSCs. Methane production showed that as the Al dosage increased, the methane production decreased.

In the alum trial from the second WWTP, the peak TVOSC decreased as the alum dosage increased. The one exception was the 1% dosage, which was considered anomalous, due to the very low cake solids and poor solids capture. The highest dosage of 4% Al reduced the peak TVOSC by greater than 50% compared to the control sample. The pH also decreased substantially at the 2 and 4% dosage, as was seen in the first trial; however, this did not have a concurrent, negative impact on the TVOSC production. At the highest dosages of 2 and 4%, the methane production was much lower, which would normally be expected to increase the TVOSC production. Instead, the low TVOSCs might indicate some success in removing odor precursor thus not impacted by the reduced methanogenic activity.

In summary, the field trial alum addition improved odor quality, but did not achieve the level of TVOSC reduction as the laboratory trials. The potential reason could be associated with the length of chemical contacting time, in which alum had a longer contacting time with biosolids under the laboratory protocol than the actual field process. Further investigation of an improved field application protocol is therefore recommended to achieve better odor control of biosolids.

4.5.4 Chemical, Enzymatic, and Biological Agents (CEBAs)

CEBAs Nos. 1, 4, 5, and 6 were tested for the recommended addition points in the anaerobic digestion process. Following is a summary of conclusions for each of these CEBAs:

1. CEBA No. 1 resulted in very little TVOSC generated from the biosolids cake after digestion. It appears that the organisms that generate digester gas were severely inhibited by CEBA No. 1. This seems to be true for organisms that generate methane and CO2 and also for organisms that generate TVOSC. Since digestion was inhibited and no VS were destroyed, CEBA No. 1 would not be acceptable for adding to the anaerobic digestion process at the prescribed dosage.

2. CEBA No. 4 appeared to have no effect on digestion or odors, as similar data were found in the CEBA tests and the control.

3. CEBA No. 5 resulted in much improved solids destruction (31% for TS and 35% for VS), but did not reduce organic sulfur emissions from the dewatered sludge cake. It is concluded that the added VS reduction from CEBA No. 5 was not due to protein degradation, but was instead likely to be some other organic component of sludge, possibly polysaccharides.

4. CEBA No. 6 had little impact on the generation of actual odor-causing compounds at the prescribed dose. It was concluded that, for the chemical dose of CEBA No. 6 applied, there was no benefit for reduction of the TVOSC odor potential.

The team concluded that none of the CEBAs added to the anaerobic digestion process showed any significant improvement in odor reduction at the recommended dosage rate of each CEBA. Different dosage rates might lead to different results.

CEBAs Nos. 2, 3, 7, and 8 were tested for the recommended addition points downstream of anaerobic digestion. CEBA Nos. 2 and 3 were added directly to the dewatered cake. CEBA No 7 and 8 were added to the sludge prior to conditioning and dewatering. Following is a summary of conclusions for each CEBA tested in Phase 3:

5. CEBA No. 2 appeared to have no effect on digestion or odors, as similar data were found in the CEBA tests and the control.
6. CEBA No. 3 appeared to have no effect on digestion or odors, as similar data were found in the CEBA tests and the control.
7. CEBA No. 7 increased odors in the dewatered cake and showed a significant inhibition of methanogens.
8. CEBA No. 8 slightly increased odors in the dewatered cake and showed an inhibition of methanogens.

The team concluded that none of the CEBAs added downstream of digestion showed any significant improvement in odor reduction at the recommended dosage rate of each CEBA. Different dosage rates might lead to different results.

4.6 Summary of Findings and Conclusions

The Phase 3 WERF study produced "process" and "mechanistic" conclusions, where process conclusions are those for which the cause and effect relationships are fully known and understood, and mechanistic conclusions are those that have been indicated by WERF Phase 3 research, but for which the cause and effect relationships are unknown or not fully understood. Following is a summary of process findings and conclusions from the Phase 3 WERF study:

1. TVOSC emissions from digested biosolids dewatered by a belt filter press were substantially lower than the TVOSC emissions from the same biosolids dewatered by high-solids centrifuge, but dry solids concentration in biosolids dewatered by the belt filter press were also substantially lower than the centrifuged cake. These results indicate that lower shear in dewatering reduces cake odors as measured by TVOSC.
2. TVOSC emissions from digested biosolids dewatered by a rotary press were substantially lower than the TVOSC emissions from biosolids dewatered by a high-solids centrifuge. Dry solids concentrations of biosolids dewatered by the rotary press were also substantially lower than the dewatered cake from a high-solids centrifuge. These results support our conclusion that lower dewatering shear reduces cake TVOSC.
3. Samples taken from the scroll of a dewatering centrifuge showed that TVOSC emissions increase dramatically when cake solids entered the conical section of the centrifuge, and increase even more in a cake sample taken at the centrifuge discharge port (even though cake solids concentration was lower at the cake discharge port compared to the scroll). Results from this trial showed that higher shear forces along the centrifuge scroll were the likely cause of higher TVOSC emissions from the cake. The researchers concluded that reducing shear during biosolids dewatering results in lower TVOSC emissions from the cake.
4. Based on side-by-side comparisons of different centrifuges dewatering the same feed biosolids, the research team observed the following effects of centrifuge

characteristics and operating parameters on TVOSC emissions from the dewatered biosolids cakes:

- TVOSC emissions from cakes dewatered by medium-solids centrifuges (operating at 1800 rpm) were about half of TVOSC emissions from the cake dewatered by high-solids centrifuges (operating at 2200 rpm). Dry solids concentrations of the biosolids dewatered by the medium-solids centrifuges were slightly lower than the biosolids dewatered by high-solids centrifuges, but the reduction in TVOSC was much greater by proportion to reduction in solids concentration.
- A 10% reduction in centrifuge bowl speed (from 2200 rpm to 2000 rpm) on one high-solids centrifuge resulted in a 20% reduction in TVOSC emissions from the dewatered cake, with no reduction in cake solids concentration.
- A 15% reduction in torque on one high-solids centrifuge resulted in almost a 40% reduction in TVOSC emissions from the dewatered cake, but with a concurrent 15-20% reduction in cake solids concentration.

The research team concluded that operational adjustments to high-solids centrifuges, such as reduction of bowl speed or torque, can reduce TVOSC emissions from dewatered cakes without substantial reductions in cake solids concentrations.

5. Longer solids retention times (SRT) during mesophilic anaerobic digestion result in lower TVOSC emissions from dewatered biosolids cakes. Test data from laboratory-scale digesters indicate that a one-third reduction in peak TVOSC emissions could be achieved by increasing digestion SRT from 15 days to 30 days.

6. Volatile solids (VS) destruction during mesophilic anaerobic digestion increases with longer SRTs, but the increase is not dramatic. The research team found an increase in VS reduction from about 50% at 10 days SRT to 58% at 40 days SRT. The reduction in TVOSC at longer SRTs is more dramatic than VS reduction at the same SRT.

7. Results of trials comparing more advanced anaerobic digestion processes with conventional anaerobic digestion indicate that multi-phase digestion, egg-shaped digesters, and thermophilic digestion can all help to reduce biosolids TVOSC emissions when compared with the conventional, mesophilic anaerobic digestion process. The mechanism for TVOSC reduction once again appears to be better destruction of proteins as odor precursors.

8. The MicroSludge™ Process can help to reduce the peak TVOSC emissions of digested and dewatered biosolids by 50% or more, compared with mesophilic anaerobic digestion. It appears that the mechanism for TVOSC reduction in this case is a higher destruction rate of odor precursors during digestion, which is enhanced by the MicroSludge™ treatment of WAS prior to digestion.

9. Low-level lime addition (up to 5% by weight of lime to dry weight of solids) to anaerobically digested and dewatered biosolids substantially increased the level of TVOSC emissions from the cake sample headspace.

10. Addition of ferric chloride to anaerobically digested biosolids prior to dewatering did not reduce TVOSC emissions from the cake until the ferric chloride dose was

at least 8% on a dry mass-to-mass basis, a substantial dosage rate of iron that was not cost effective, and also required lime addition to maintain a near-neutral pH. Another full-scale trial at a different WWTP indicated that iron addition may be effective at lower dosage rates on some biosolids. The effectiveness of iron addition in reducing biosolids odors depends on unique biosolids characteristics and other factors.

11. Addition of aluminum sulfate (alum) at a laboratory scale to samples of anaerobically-digested biosolids from two WWTPs prior to dewatering reduced TVOSC emissions from the cake substantially by at lower dosage rates than iron, in the range of 0.5% to 1% dry mass of aluminum to mass of dry solids
12. Results of full-scale alum addition to a high-solids dewatering centrifuge indicate that alum addition can reduce TVOSC emissions from dewatered biosolids, although the full-scale trial did not achieve the TVOSC reduction shown in laboratory trials. Shorter contact time and less efficient mixing of alum at full scale compared to laboratory scale. A substantial decrease of pH in the feed sludge at higher alum dosages is another reason for the less successful performance of alum in the full-scale trial.

Following is a summary of mechanistic findings and conclusions from the Phase 3 WERF study:

1. Methanogenic bacteria play an important role in the cycling of organic sulfur compounds, and hence, TVOSC production from dewatered biosolids. Cake samples to which a methanogenic organism inhibitor was added produced greater amounts of TVOSC, which persisted in high concentrations over several weeks of incubation. When the inhibitor was not added to dewatered biosolids cake, TVOSC emissions from most biosolids tended to increase during the first 7-10 days of sample incubation, and decrease to relatively low TVOSC concentration after 21 days of sample incubation. The literature shows TVOSCs are degraded into hydrogen sulfide by methanogens. This sulfide can be bound by iron if present in the cake and precipitate as FeS. If the iron concentration is low, it may be released as H_2S instead.
2. The relationships between iron, aluminum, and the odor potential of anaerobically-digested biosolids are complex. Both iron and aluminum will form bonds with proteins (the primary odor precursors) in biosolids, but iron-bound protein is released more easily than aluminum-bound protein during centrifugation of digested biosolids.
3. TVOSC emissions from anaerobically digested biosolids increase with increasing iron content of the biosolids cake. This trend is consistent with prior research indicating that biopolymer associated with iron is not completely degraded during anaerobic digestion. Shear during centrifugation subsequently releases the iron-bound biopolymer, which leads to increased odors in the biosolids cake.
4. TVOSC in dewatered biosolids headspace samples decrease as aluminum concentration in the biosolids increases. It appears that aluminum-bound biopolymer is not as readily released during centrifugation as iron-bound

biopolymer, leading to lower odors in biosolids cakes with higher aluminum content.

5. None of the CEBAs added to the anaerobic digestion process showed any significant improvement in TVOSC reduction from dewatered cake, at the recommended dosage rate of each CEBA. Different dosage rates of CEBAs might lead to different results.
6. None of the CEBAs added downstream of digestion showed any significant improvement in TVOSC reduction from the dewatered cake, at the recommended dosage rate of each CEBA. Different dosage rates of CEBAs might lead to different results.

CHAPTER 5.0

RECOMMENDATIONS

The purpose of this section is to present the recommendations resulting from the research that has been presented in the preceding chapters. The recommendations of this chapter are in grouped into two major focus areas:

1. The research team's recommendations for putting the findings and conclusions of the research into practice in the design and operation of both existing and new WWTPs
2. The research team's recommendations for future research and development on the reduction of odors from biosolids

5.1 Recommendations for the Design and Operation of WWTPs

The findings of this body of research have focused on the causes of biosolids odors and how those odors may be reduced by process modifications at the WWTP, so that biosolids products are less likely to generate odors during storage, transit, and ultimate use or disposal. The focus of this research has been 1) to identify odor-generating mechanisms in anaerobically-digested and dewatered biosolids and 2) to use that knowledge to reduce the odors from those types of biosolids. Anaerobically digested and dewatered biosolids represent the majority of biosolids generated and processed in the U.S. and Canada, since they are the processes employed at most large, municipal WWTPs.

These research findings apply primarily to Class B dewatered biosolids that are stored onsite at WWTPs and transported in bulk to agricultural land application sites, because those are the biosolids characteristics and handling practices that are known to have generated the most odor complaints. There is increasing evidence, however, that reduction of odors from dewatered biosolids cakes also tends to reduce odors from downstream processing of cakes into other dewatered-biosolids products such as compost or heat-dried pellets.

Chapter 4.0 described the research findings as grouped into the following three basic categories, each of which represents a different set of odor-reduction measures that may be applicable to WWTPs in operation or under design:

- The effects of anaerobic digestion and various enhancements to the anaerobic digestion process on biosolids cake odors
- The effects of dewatering practices on biosolids cake odors, with an emphasis on high-solids centrifuges, because of their wide usage in North American WWTPs and the frequency of their association with biosolids cake odors
- The effects of chemical addition on biosolids cake odors, with an emphasis on aluminum and iron addition after digestion

Figure 5-1 illustrates the potential reduction measures for biosolids odor control that have been explored in the project. As shown in Figure 5-1, the potential measures for reducing biosolids odors are categorized in two basic subject areas:

- Changes that could be practiced during the design of new WWTPs, new unit processes for existing WWTPs, or engineered modifications to existing processes
- Changes that would involve modifying process operations at an existing WWTP, either by adjusting operational parameters of processes or equipment, or via chemical addition.

In virtually all cases, "trial-and-error" laboratory or pilot approaches will likely be needed when using Figure 5-1 to optimize reduction of biosolids cake odors. Also, the overall amount of odor reduction needed should be determined ahead of applying these schemes and then incrementally measured after each one is implemented. Odor measurements before and after process modifications provide for more accurate evaluations of a scheme's true odor reduction potential.

The research team has conducted experiments that show similar or better cake odor reduction than reported herein, but the researchers caution against assuming that similar results will be obtained elsewhere due to inherent differences in biosolids characteristics, treatment processes, and operating conditions. Upon first glance, Figure 5-1 may appear overwhelming based on all the suggested recommendations and options shown, but one or two carefully chosen options might be all that is needed for sufficient odor reduction in a given case. Thus, a careful, planned approach is suggested that accounts for the extent of cake odor reduction needed to satisfy its end use. The resulting options for odor reduction will therefore be focused on approaches that have the greatest chance for success in each case.

Recommendations for biosolids cake odor reduction related to each of these subject areas are summarized in the following sections.

5.1.1 Recommendations for Design of WWTPs and Biosolids Facilities

Potential design changes for odor reduction are represented by options shown on the right-hand side of Figure 5-1, under the process-design areas of anaerobic digestion, followed by dewatering and cake storage of biosolids. It must be noted that these recommendations are not ranked or prioritized according to their estimated effectiveness or potential costs. For example, this research has shown that longer digestion SRT in the range of 30 days will reduce odors compared to more conventional digestion SRTs such as 15 days. But a 30-day SRT requires twice as much digestion volume as a 15-day SRT, which might represent a substantial increase in capital cost. So each recommendation needs to be evaluated and compared to other recommendations in terms of their relative costs, which will vary at each WWTP.

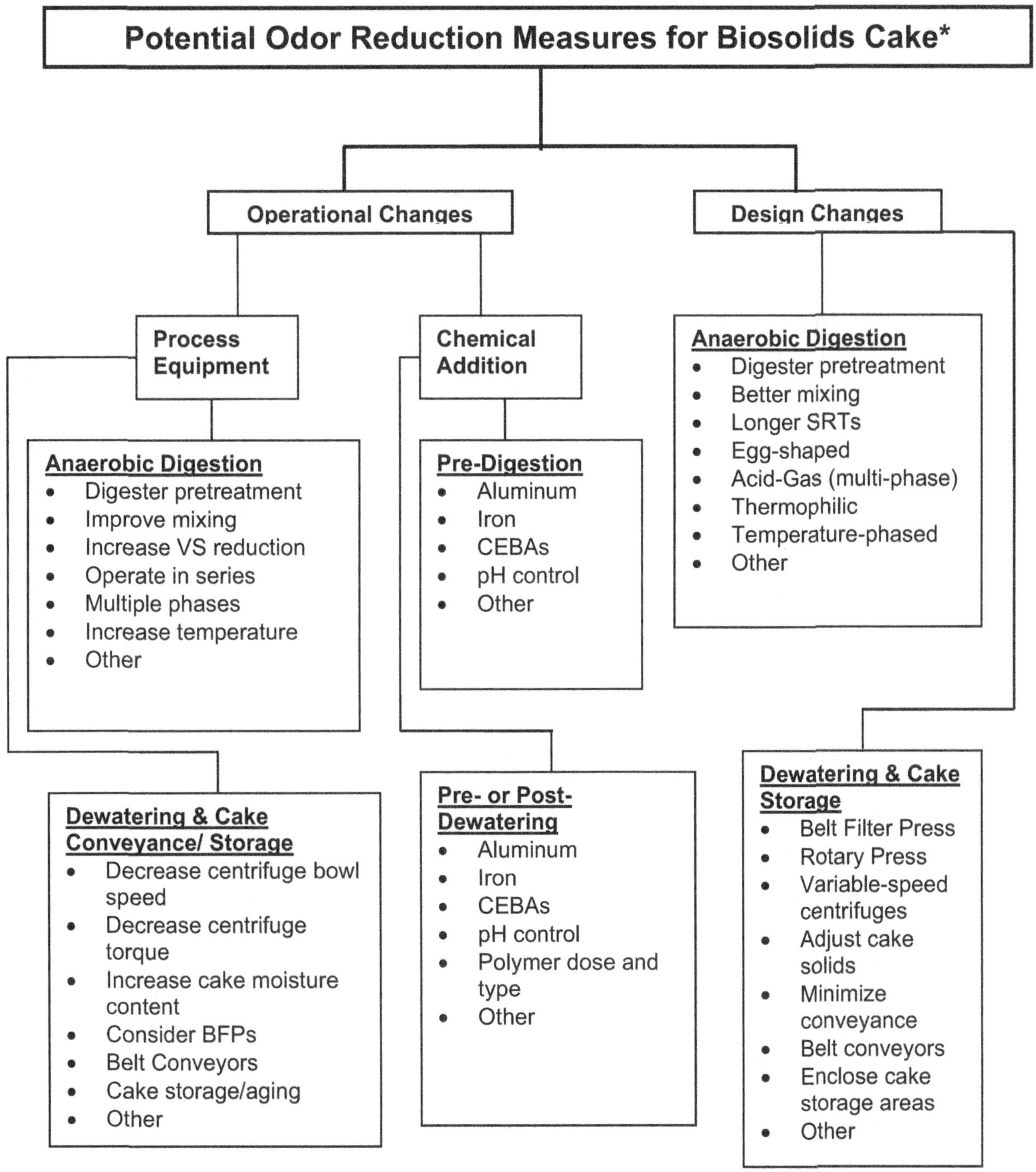

***Note**: None of these options should be considered independently of the others, and odor reductions in one area may impact treatment processes and odors in other areas. Therefore, a consolidated and customized approach is required for each WWTP. Refer to the text of this chapter for additional explanations.

Figure 5-1. Options Summary Chart for Biosolids Odor Reduction.

5.1.1.1 Anaerobic Digestion Processes

Design recommendations for anaerobic digestion processes are based on three significant findings related to digestion, as described below:

1. **Finding:** Longer SRT during mesophilic anaerobic digestion result in lower TVOSC emissions from dewatered biosolids cakes. Test data from laboratory-scale digesters indicate that a one-third reduction in peak TVOSC emissions could be achieved by increasing digestion SRT from 15 days to 30 days.

   Design Recommendation: Design conventional mesophilic anaerobic digestion processes for a minimum SRT of 20 days during peak loading periods and 30 days SRT for average design loading rates. Alternatively, evaluate whether advanced digestion or digestion-enhancement processes may be more cost effective at shorter SRTs, per Item No. 3 below.

2. **Finding:** The MicroSludge™ Process reduced the peak TVOSC emissions of digested and dewatered biosolids cake by 50% at one WWTP site, as compared with conventional, mesophilic anaerobic digestion. It appears that destruction of odor precursors such as bioavailable protein during digestion is enhanced by the MicroSludge™ treatment of WAS prior to digestion. It follows that various other processes that promote more effective destruction of proteins during digestion are also likelier to help reduce odors in digested and dewatered biosolids.

   Design Recommendation: Consider piloting the MicroSludge™ Process or similar type of digestion pre-treatment step in lieu of conventional, mesophilic digestion at longer SRTs (per Finding #1 above). By breaking down the digester feed sludge and making it more amenable to anaerobic digestion, a pre-treatment enhancement such as MicroSludge™ may well be more cost-effective than increasing SRT by adding more digestion volume (Option 1 above), depending on site-specific conditions. There are a number of proprietary sludge-enhancement processes prior to digestion that warrant investigation for their potential benefits to biosolids odor reduction in addition to the MicroSludge™ Process that was investigated in this study.

3. **Finding:** Multi-phased digestion, egg-shaped digesters, and thermophilic digestion all appear to be effective to some degree in reducing biosolids TVOSC emissions when compared with the conventional, mesophilic anaerobic digestion process. The mechanism for TVOSC reduction once again appears to be more complete destruction of proteins as odor precursors. It is evident through results of a number of trials on these processes that a number of factors during digestion, in addition to longer SRTs as described in Item 1 above, could contribute to better protein destruction. Some of those factors are:

   a. More efficient and vigorous mixing, such as found in egg-shaped digesters.
   b. Higher digestion temperatures, as provided by thermophilic and temperature-phased digestion, which speed up anaerobic biological activity.
   c. Phased digestion in its various configurations, such as two-phase (acid-gas) or temperature-phased (thermophilic followed by mesophilic) digestion.
   d. It is also apparent that series operation of digesters is more effective at protein destruction than single-stage digestion in general, due to better mixing and avoidance of short-circuiting.

   Design Recommendation: Consider the potential advantages of implementing a more advanced digestion process that incorporates one or more of the odor-reduction measures discussed above, especially if a new anaerobic digestion system is called for, rather than modifications to an existing digestion system. In general, this research of advanced digestion processes indicates that it is possible to achieve greater than 50% VS destruction with most advanced digestion processes, and even 55% to 60% VS destruction has been observed in several cases. The research indicates that a proportionately higher percentage of odor precursors are destroyed when VS destruction in digestion exceeds 50%, so processes should be sought that can achieve VS destruction at this level or higher on a consistent basis. Site-specific bench-scale testing and evaluations of advanced digestion processes, followed by onsite piloting of the selected process where possible are strongly recommended prior to design of full-scale digestion facilities.

5.1.1.2 Dewatering Processes

Design recommendations for dewatering processes are based on two significant findings related to biosolids dewatering, as described below:

1. **Finding:** Based on side-by-side comparisons of different centrifuges dewatering the same feed biosolids, the research team concluded that reduction of bowl speed or torque can reduce TVOSC emissions from dewatered cakes without substantial reductions in cake solids concentrations.

   Design Recommendation: When centrifuges are selected for biosolids dewatering, perform onsite pilot testing with a centrifuge that is capable of varying bowl speeds and torque levels in order to reduce shear during dewatering. Follow the cake sampling and analytical protocols of this project, and characterize the dewatered biosolids in terms of odor characteristics as well as cake dryness. Decisions regarding the selection and specifying of centrifuges can then be made based on other aspects of cake quality besides only dry solids concentration, such as odor production potential. The research indicates that at least 20% TVOSC reduction can be achieved be reducing the main bowl speed only 10% with no reduction in cake dryness. TVOSC reduction in the range of 40% has been observed by making a 20% reduction in centrifuge torque, but cake dryness has also decreased in this case.

2. **Finding:** TVOSC emissions from digested biosolids dewatered by belt filter presses and rotary presses were substantially lower than the TVOSC emissions from the same biosolids dewatered by high-solids centrifuges, but dry solids concentration in biosolids dewatered by the belt filter press and rotary press were also substantially lower than the centrifuged cake.

   Design Recommendation: When there is a need for new biosolids dewatering equipment, conduct a life-cycle cost evaluation of belt filter presses and/or rotary presses in addition to centrifuges, and consider side-by-side pilot-testing of different dewatering equipment to predict the levels of cake solids concentrations that can be achieved by each of these dewatering devices. Similar to the recommendations above for centrifuges, WWTP operators will need to follow the cake sampling and analytical protocols of this project to evaluate dewatered biosolids from a standpoint of odor potential as well as cake solids content, and incorporate those findings into the selection process for a biosolids dewatering device.

5.1.1.3 Biosolids Cake Conveyance and Storage

Recommendations from the findings of this research related to cake conveyance and storage are as follows:

1. **Design Recommendation - Cake Conveyance:** This research has found that most of the shear on biosolids after digestion occurs during centrifugal dewatering, but one experiment found that screw conveyance of dewatered cake increased bioavailable protein and headspace TVOSC emissions. Based on the finding that increasing shear on dewatered cake may increase its odor potential, consider conveyance devices that are lower in shear. For instance, belt conveyors are generally preferred over screw conveyors or cake pumps. When pumping biosolids cakes, consider incorporating lubrication rings in discharge piping to reduce pipeline pressure and hence, reduce shear during conveyance.

2. **Design Recommendation - Cake Storage:** The experiments in this study and in previous research have shown that odors from dewatered biosolids can be reduced during product storage due to the growth of odor-consuming methanogens in the stored biosolids cake. This trend has been evident from observing the rise and fall in TVOSC concentration in sample bottle headspaces when they are analyzed over a bottle incubation period of 2-4 weeks.

   In most experiments involving mesophilic anaerobic digestion, peak TVOSC emission rates occur after a storage period of 5-9 days and then decrease to levels of only 20% of peak after 15-20 days. Therefore, onsite provisions are recommended for storing and aging dewatered biosolids for a 2-4 week period prior to transport and land application. It is also recommended to conduct bottle storage and headspace analyses of dewatered biosolids daily over a 3-4 week period in order to determine the rise-and-fall characteristics of TVOSC for each type of biosolids from each WWTP. Experiments conducted in Phases 2 and 3 of this project have shown variability in the storage and odor characteristics of biosolids, so these types of experiments are necessary to assess the potential for cake storage to reduce biosolids odors, in each specific case.

5.1.2 Recommendations for WWTP Biosolids Handling Operations

Potential operational changes for odor reduction are represented by options shown in the left-hand and middle sections of Figure 5-1, under the process-design areas of anaerobic digestion, followed by dewatering, and finally cake storage of biosolids. In Figure 5-1, the operational changes are grouped in the two basic categories of process equipment or chemical addition. Process equipment changes are further separated into modifications involving anaerobic digestion (either prior to or during digestion), and modifications to the dewatering process or beyond, such as cake conveyance and storage. Modifications related to chemical additions are further separated into pre-digestion or post-digestion points of chemical addition in the treatment process.

An example decision-process flow chart of potential operational changes is presented in Figure 5-2. This flow chart follows a typical process flow of wastewater and biosolids through a WWTP, with unit processes represented by rectangular boxes and decision steps represented by diamonds.

Potential Operation Changes at WWTPs to Enhance Biosolids Odor Reduction

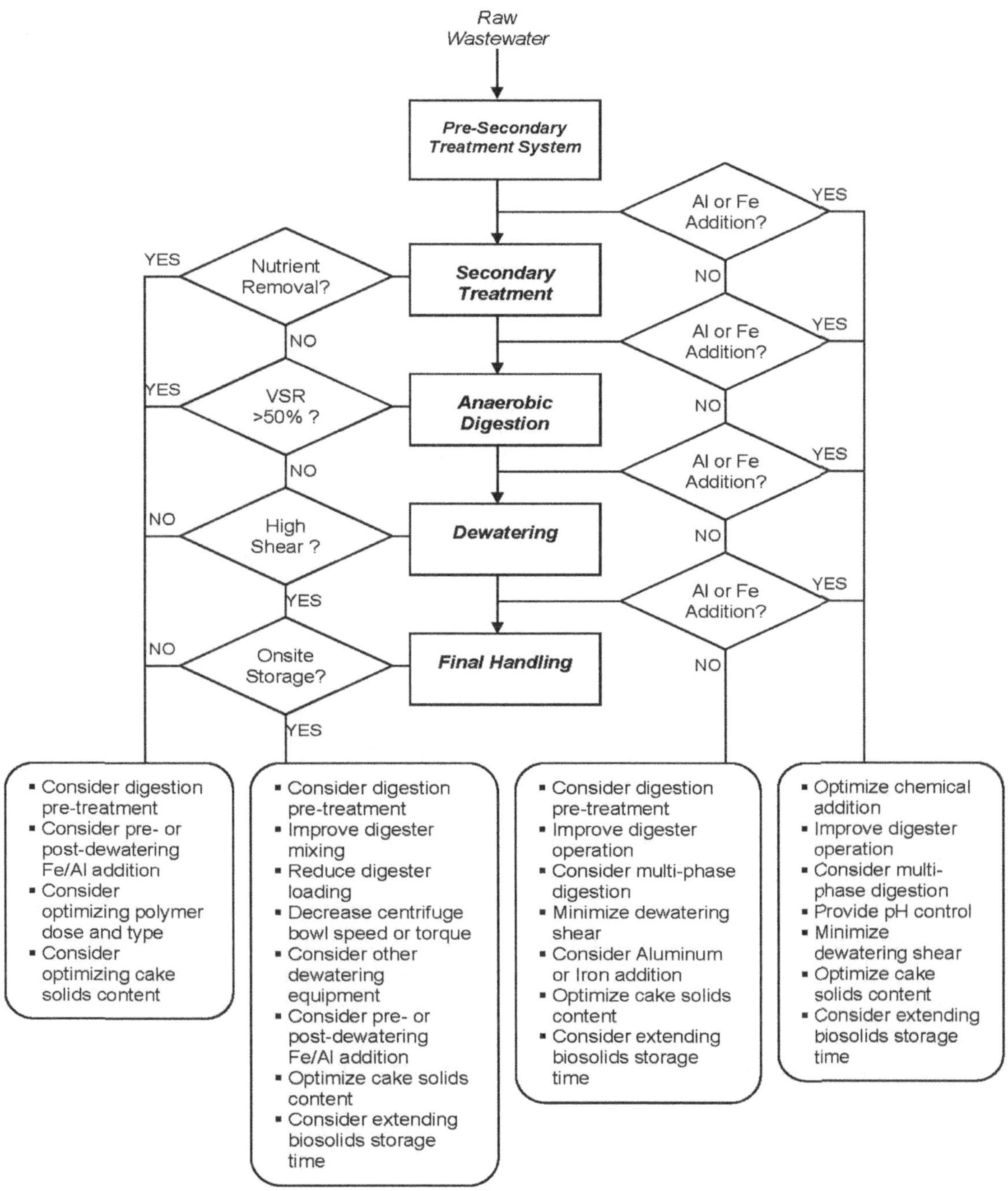

Figure 5-2. Decision Flow Chart of Potential Biosolids Odor-Reduction Options.

There are potential chemical addition points between each process step, and also an operational decision related to each process that may impact biosolids odor characteristics. Using this sample decision flow chart and customizing it for the processes at each WWTP under consideration for biosolids odor reduction, one can define the decision path that best describes the operations at a particular POTW. The intent for this type of decision flow chart is to assist the WWTP manager in evaluating the most promising solutions for a particular facility. The solution boxes were compiled to fit each decision path they are connected to. Since this research has not yet definitively identified the multi-effect correlations between the various listed operations conditions and the biosolids odor potential, the reader is cautioned to use this flow chart for one condition at a time.

5.1.2.1 Processes Upstream of Anaerobic Digestion

There are not many opportunities to offset the odor impacts of dewatered biosolids in conventional treatment processes upstream of the anaerobic digestion process. The primary option researched in this project was chemical addition to the primary or secondary treatment processes. Generally, chemical addition at these points in the wastewater treatment process is practiced for nutrient removal or odor control of wastewater treatment processes. Findings of this research indicate that, depending on the chemical used and the process point applied, biosolids odor characteristics may be enhanced or worsened by chemical addition.

The findings of this research relate to the effects of iron and aluminum as active ingredients in conventional wastewater treatment chemicals. As discussed in Chapter 4.0, it appears that the addition of iron upstream of the digestion process does not improve odor characteristics of dewatered biosolids after digestion. In fact, iron addition upstream of digestion may even cause the biosolids odor characteristics to worsen. Its effects on biosolids odors appear to influence protein destruction during digestion, followed by the release of bioavailable protein after digestion that contributes to odor emissions. This finding appears contrary to the widely known benefits of iron addition to digesters, which can significantly reduce the concentration of hydrogen sulfide in digester gas. WWTP operators must weigh the benefits of hydrogen sulfide reduction with the potential downside of higher cake odors, before deciding on the use of iron for digester gas hydrogen sulfide control.

It appears based on this research that iron binds bioavailable protein and keeps more of thc protein from decomposing during digestion. Depending on the pH, valence state, and tightness of the iron-to-protein bond, the protein that remains after digestion may become bioavailable later due to shear in the dewatering process. The bioavailable protein in dewatered cake is a precursor to biosolids odors. Iron addition prior to digestion may increase the amount of protein in digested solids, which later becomes bioavailable to serve as food for odor-forming bacteria during or after dewatering. For these reasons, iron addition to sludges prior to digestion may increase biosolids cake odors rather than reduce them.

On the other hand, aluminum addition prior to digestion tends to form tighter bonds with protein that appear to persist through dewatering and cake storage, keeping proteins from becoming odor precursors in the process. Possibly a scheme to add aluminum prior to digestion for cake odor reduction, and then add iron after aluminum for digester gas hydrogen sulfide reduction could provide the desired, combined benefits in reducing hydrogen sulfide in digester gas as well as reducing biosolids cake odors. More research on the effects of both aluminum and iron is recommended as stated in the next section under Recommendations for Further Research.

5.1.2.2 The Anaerobic Digestion Process

Most of the recommendations related to anaerobic digestion pertain to design considerations as discussed previously in this chapter, but there are also some operational measures to consider, as shown in Figure 5-2.

It was found that for a single sludge, increasing the VS destruction by increasing the digester SRT also resulted in decreased TVOSC. This suggests that by increasing the VS destruction, odors will be reduced. Methods to increase VS destruction such as predigestion sludge thickening, better mixing or sludge pretreatment need to be investigated to determine the correlation between VS destruction and odor reduction. The data from the digestion SRT study suggests that increased VS destruction can lead to reduced dewatered cake odors.

Other recommendations such as adding a digestion pre-treatment step or considering more advanced digestion processes are more thoroughly discussed in design considerations above. It must be cautioned that some sludges have a greater propensity toward odors than others, and in some cases this propensity will result in high odor levels from dewatered biosolids cake, despite improvements in the digestion process. In these cases, it will likely be necessary to implement post-digestion measures such as chemical addition or dewatering process changes, in order to make a noticeable improvement in odor emissions from the final biosolids product of the WWTP.

5.1.2.3 Operational Considerations After Digestion

The operational considerations after digestion are based on several significant findings of the project, as discussed below:

1. **Finding:** TVOSC emissions from cakes dewatered by medium-solids centrifuges (1800 rpm bowl speed) were about half of TVOSC emissions from the cake dewatered by high-solids centrifuges (2200 rpm bowl speed). Dry solids concentrations of biosolids dewatered by the medium-solids centrifuges were slightly lower than the biosolids dewatered by high-solids centrifuges, but the reduction in TVOSC was much greater in proportion to the reduction in solids concentration. A 10% reduction in centrifuge bowl speed (from 2200 rpm to 2000 rpm) on one high-solids centrifuge resulted in a 20% reduction in TVOSC emissions from the dewatered cake, with no observed reduction in cake solids concentration. A 15% reduction in torque on one high-solids centrifuge resulted in almost a 40% reduction in TVOSC emissions from the dewatered cake, with a concurrent 15-20% reduction in cake solids concentration.

   Recommendation: If high-solids centrifuges are used for dewatering, investigate the feasibility of adjusting the centrifuge main bowl speed and torque levels. This is best done by setting up a field-testing protocol similar to the centrifuge testing and sampling protocols that were performed at the Philadelphia Biosolids Recycling Center, in which various combinations of bowl speed and torque were evaluated and samples analyzed for dry solids content and odor characteristics. Based on the results, more permanent adjustments to the centrifuges to reduce odor (and potentially save energy as well) may be warranted.

2. **Finding:** Addition of ferric chloride ($FeCl_3$) to anaerobically digested biosolids prior to dewatering did not reduce TVOSC emissions from the cake until the ferric chloride dose was at least 8% on a dry mass-to-mass basis, a substantial dosage rate of iron that was clearly not cost effective, and also required supplemental lime addition to maintain a near-neutral pH. However, another full-scale trial at a different WWTP indicated that iron addition may indeed be effective at lower dosage rates on some biosolids. In conclusion, the effectiveness of iron addition in reducing biosolids odors seems to depend on unique biosolids characteristics and other factors.

Recommendation: If ferric chloride is an option for chemical feed to the dewatering centrifuges, conduct laboratory scale tests of ferric chloride addition to find out, through centrifuge-simulation, cake sampling, and TVOSC-headspace analyses, if there might be any advantage to ferric chloride addition at the WWTP. If the laboratory results look promising, set up a full-scale field test using the example test protocols in Appendix A.

3. **Finding:** Results of full-scale alum ($Al_2(SO_4)_3$) addition to a high-solids dewatering centrifuge indicate that alum addition can reduce TVOSC emissions from dewatered biosolids, although the full-scale alum trial did not achieve the dramatic TVOSC reduction shown in laboratory trials. Shorter contact time and less efficient mixing of alum at full scale compared to laboratory scale were the primary factors attributed to the differing results between the two trials. Adjustments to polymer dose are also required since alum addition will usually decrease the amount of polymer needed for optimum dewatering. A substantial decrease of pH in the feed sludge at higher alum dosages is another reason for the less successful performance of alum in the full-scale trial. In that case, addition of a pH buffering agent may be necessary.

   Recommendation: If alum (or another aluminum-based chemical such as poly-aluminum chloride) is an option for chemical feed to the dewatering centrifuges, conduct laboratory scale tests of the aluminum addition to evaluate if there might be an odor-reduction advantage to aluminum addition at the WWTP. For each alum dose tested, determine the optimum polymer dose required. Also vary the mixing time to determine the minimum required contact time prior to dewatering. Evaluate the impacts of aluminum addition on the pH chemistry of the dewatering process, and add a buffering agent if the pH does not remain within normally acceptable ranges of 6-8. If the laboratory results look promising, set up a full-scale field test using applicable test protocols from Appendix A, with pH adjustment if necessary.

5.2 Recommendations for Future Research

A WERF Program Area meeting for Wastewater and Biosolids Research was held on December 7, 2006, during which the results and findings of this project were presented and discussed among WERF subscribers and biosolids odor researchers. As a result of that discussion, needs for future research related to biosolids cake odor reduction were identified. Those needs are summarized as follows:

1. Need to investigate the presence of other odorous compounds besides TVOSC, to confirm whether TVOSC alone is a reliable indicator and surrogate for odors from biosolids that have been anaerobicly digested and dewatered.

2. Need to investigate further how the presence or absence of ferric or ferrous iron in biosolids at different points in wastewater treatment and biosolids handling process train may affect biosolids cake odors.

3. Need to explore how some of the laboratory findings of Phase 3 related to aluminum addition, digestion, and dewatering can be optimized in the field to reduce odors.

4. Need to explore odor reduction benefits of chemicals that are being added to WWTPs for phosphorus removal (ferric and aluminum compounds), and how they can be optimized for both phosphorus removal and odor reduction goals.

5. Need to take a step back from all of the data and try to identify the real causative mechanisms for odor production, and how those mechanisms can be neutralized via process changes or chemical addition. There appears to be a knowledge gap in the research with regard to the biological and microbiological aspects of the mechanisms at work. An environmental engineer with a molecular biology background might be a good candidate to help fill the gap.

6. Need to identify a WWTP where lessons learned in this research can be applied full-scale to reduce odors and find out if the 50-75% odor reduction observed in some of the laboratory studies and demonstrations can be realized full scale at operating WWTPs. Would combinations of two or more odor-reduction techniques be required to solve a biosolids-odor problem at a given WWTP?

The findings and recommendations of the current biosolids-odor research do not take into account the relative costs and benefits of the odor control measures discussed herein. Costs and benefits of odor-reduction measures are likely to vary substantially at each WWTP due to a number of factors, including sludge and biosolids characteristics, digestion process parameters, chemical-addition practices, dewatering processes, and cake storage procedures. Hence, the costs of options evaluated must be considered on a site-specific basis.

Most experts in the field agree that the anaerobic digestion process will continue to be the biosolids stabilization method for many WWTPs, due to the capital investments that many POTWs have already made and the resource potential that anaerobic digestion provides through recovery of digester gas and beneficial use of stabilized biosolids. The WERF Program Area meeting participants suggested a general three-pronged approach to subsequent research, either specific to this project or combined with other WERF research efforts:

1. Demonstrations of biosolids odor reduction in full-scale settings at several WWTPs using the Phase 3 approaches in order to apply findings and broaden the body of knowledge on this subject.

2. Production of a practice manual with a step-by-step approach for odor reduction in digested and dewatered biosolids, providing results and summarizing recommendations described in this report.

3. Teaming with WEF or the National Biosolids Partnership to make biosolids cake-odor reduction techniques more widely known and applied in the industry.

The WERF Program Area meeting participants also discussed the manner in which subsequent biosolids odor-reduction work could be conducted. There are two other ongoing WERF research studies that have some practical connections to this research:

1. A WERF-sponsored waste-sludge reduction study. Two areas are being considered in the research: liquid treatment and digestion pre-treatment. This study's researchers are evaluating the impacts of higher VS destruction on minimizing biosolids volume. Meeting that objective also tends to improve biosolids odors.

2. A pathogen regrowth study in biosolids, being conducted on biosolids that have been dewatered by high-solids centrifuges, in which some of the results have shown striking similarities with the results of this project in terms of the correlation between odor levels and fecal coliform concentrations.

One possible direction for future research on this topic would be to phase it into directed research combined with the above-described WAS reduction and pathogen regrowth studies. This would allow for potential expansion of the project team and sharing of the findings from each project to be incorporated into each of the other individual studies.

REFERENCES

Adams, G.M., Witherspoon, J., Card, T., Erdal, Z., Forbes, R.H., Geselbracht, J., Glindemann, D., Hargreaves, R. Hentz, L., Higgins, M., McEwen, D., and Murthy, S. (2003) "Identifying and Controlling the Municipal Wastewater Odor Environment Phase 2: Impacts of In-Plant Operational Parameters on Biosolids Odor Quality." *Water Environment Research Foundation, Report No.00HHE5T,* Alexandria, VA.

American Public Health Association (1998) *Standard Methods for Examination of Water and Wastewater; 20th Edition.* Eds. L.S. Clesceri, A.E. Greenberg, and A.D. Eaton. American Public Health Association, Washington, D.C.

American Public Health Association; American Water Works Association; and Water Environment Federation (1995) *Standard Methods for the Examination of Water and Wastewater. 19th Ed.*, Washington, D.C.

Castillo-Gonzalez, H.A. and Bruns, M.A. (2005) Dissimilatory iron reduction and odor indicator abatement by biofilm communities in swine manure microcosms. *Applied and Environmental Microbiology* 71, 9, 4972-4978.

Dubois, M., Gilles, K.A., Hamilton, J.K., Rebers, P.A., and Smith, F. (1956) Colorimetric Methods for Determination of Sugars and Related Substances. *Analytical Chem.*, 28, 350.

Erdal, Z.K., Mendenhall, T.C., Neely, S.K., Wagoner, D.L., and Quigley, C. (2003) Implementing Improvements in A North Carolina Residuals Management Program. *Proceedings of WEF/AWWA Joint Residuals and Biosolids Management Conference.* Baltimore, MD.

Erdal, Z.K., Wagoner, D.L., Quigley, C., Mendenhall, T.C., and Neely, S.K. (2004) Maintaining Class B Biosolids Post-Dewatering Through Low-Level Lime Dosing. *Proceedings of Water Env. Fed. 77th Annual Conf. WEFTEC*, New Orleans.

Frølund, B., Palmgren, R., Keiding, K., and Nielsen, P.H. (1996) Extraction of Extracellular Polymers from Activated Sludge Using a Cation Exchange Resin. *Water Res.*, 30, 1749.

Glindemann, D., Murthy, S.N., Higgins, M.J., Chen, Y.C., and Novak, J.T. (2006) Biosolids incubation method for odorous gas measurement from dewatered sludge cakes. *Journal of Residuals Science & Technology* 3(3): 153-160.

Hartree, E.F. (1972) Determination of Protein: A Modification of the Lowry Method That Gives a Linear Photometric Response. *Anal. Biochem.*, 48, 422.

Higgins, M.J. and Novak, J.T. (1997a) The Effect of Cations on the Settling and Dewatering of Activated Sludges. *Water Environ. Res.*, 69, 215.

Higgins, M.J. and Novak, J.T. (1997b) Characterization of Exocellular Protein and Its Role in Bioflocculation. *J. Environ. Engr*. ASCE, 123, 479-485.

Higgins, M.J., Murthy, S.N., Yarosz, D.P., Novak, J.T., Glindemann, D., Toffey, W.E. and Abu-Orf, M. (2002) 'Effect of Chemical Addition on Production of Volatile Sulfur Compounds and Odor from Anaerobically Digested Biosolids', WEFTEC 2002 Annual Conference, Chicago, IL.

Higgins, M.J., Chen, Y.C., Maas, N.A., Troxell, J., Murthy, S.N. (2004) WERF Phase 1: Factors affecting conditioning during dewatering: Sludge and Polymer Characteristics. *Proceedings of Water Env. Fed. 77th Annual Conf*., New Orleans

Higgins, M.J., Hamel, K., Chen, Y.C., Murthy, S.N., Barben, E.J., Livadaros, A., Travis, M., Maas, N.A. (2005) Part II of Field Research: Impact of Centrifuge Torque and Polymer Dose on Odor Production from Anaerobically Digested Biosolids. *Proceedings of WEF/AWWA Joint Residuals and Biosolids Management Conference*. Nashville, Tennessee.

Higgins, M.J., Yarosz, D.P., Chen, Y.C., Murthy, S.N., Maas, N., Cooney, J., Glindemann, D., Novak, J.T. (2006) Cycling of volatile organic sulfur compounds in anaerobically digested biosolids and its implications for odors. *Water Env. Research*.78, 243-252.

Lomans, B.P., Luderer, R., Steenbakkers, P., Pol, A., Van Der Drift, C., Vogels, G.D., and Op Den Camp, J.J.M.. (2001) Microbial Populations Involved in Cycling of Dimethyl Sulfide and Methanethiol in Freshwater Sediments. *Applied and Env. Microbiol*., 67, 1044-1051.

Lovley, D.R. and Phillips, E.J.P (1987) Competitive Mechanisms for Inhibition of Sulfate Reduction and Methane Production in the Zone of Ferric Iron Reduction in Sediments. *Applied and Environmental Microbiology*, 1987, p. 2636-2641.

Lowry, O.H., Rosebrough, N.J., Farr, A.L., and Randall, R.J. (1951) Protein Measurement with the Folin Phenol Reagent. *J. Biol. Chem*., 193, 265.

Metcalf and Eddy Inc. (2003) *Wastewater Engineering: Treatment and Reuse*. 4th Edition. McGraw-Hill, New York, NY.

Mikkelsen L.H. and Keiding, K. (2002) Physico-chemical characteristics of full scale sewage sludges with implications to dewatering. *Wat. Research* 36, 2451–2462

Muller C.D., Verma, N., Higgins, M.J., and Novak, J.T. (2004) The role of shear in the generation of nuisance odors from dewatered biosolids. *WEFTEC 2004 Conference and Exhibition Proceedings*, October, 2004, New Orleans, LA.

Muller, C.D. (2006) *Shear Forces, Floc Structure and their Impact on Anaerobic Digestion and Biosolids Stability*. PhD dissertation, Civil Eng. Department, Virginia Polytechnic Institute and State University, Blacksburg, VA.

Murthy, S.N., Forbes, B., Burrowes, P., Esqueda, T., Glindemann, D., Novak, J.T., Higgins, M.J., Mendenhall, T. and Toffey, W. (2002) Impact of High Shear Solids Processing on Production of Volatile Sulfur Compounds from Anaerobically Digested Biosolids, *Proceedings of WEFTEC 2002 Annual Conference*, Chicago, IL.

Murthy, S.N., Higgins, M.J., Chen, Y., Toffey, W., and Golembeski, J. (2003) Influence of Solids Characteristics and Dewatering Process on Volatile Sulfur Compound Production from Anaerobically Digested Biosolids. *Proceedings of WEF/EPA Residuals and Biosolids Conference.*

Nielsen, P.H., Frølund, B., and Keiding, K. (1996) Changes in the Composition of Extracellular Polymeric Substances in Activated sludge during Anaerobic Storage. *Appl. Microbiol. Biotechnol.*, 44, 823-830.

Novak, J., Glindemann, D., Murthy, S.N., Gerwin, S.C., and Peot, C. (2002) Mechanisms for Generation and Control of Trimethyl Amine and Dimethyl Disulfide from Lime Stablilized Biosolids. *Proceedings of WEF Odors and Toxic Air Emissions 2002 Conferenc*e. Albuquerque, NM: WEF.

Novak, J.T., Sadler, M.E., and Murthy, S.N. (2003) Mechanisms of Floc Destruction During Anaerobic and Aerobic Digestion and The Effect on Conditioning and Dewatering of Biosolids. *Water Environ. Res.*, 37, 3136 - 3144.

Novak, J.T., Adams, G., Chen, Y.-C., Erdal, Z., Forbes, Jr., R.H., Glindemann, D., Hargreaves, J.R., Hentz, L., Higgins, M.J., Murthy, S.N., Witherspoon, J., and Card, T. (2004) Odor Generation Patterns Form Anaerobically Digested Biosolids. *Proceedings WEF/A&WMA Specialty Conf.* April 18-21, Seattle, WA.

Novak, J.T., Adams, G., Chen, Y-C., Erdal, Z., Forbes, R.H. Jr., Glindemann, D., Hargreaves, J.R., Hentz, L., Higgins, M.J., Murthy, S.N., and Witherspoon, J. (2006) "Odorous Gas Generation Patterns From Anaerobically Digested Biosolids" *Wat Envr Res*. 78, 8, 821-827.

Park, C. Abu-Orf, M.M., and Novak, J.T. (2004) Analysis of Floc Structure and Predicting Sludge Digestibility Using Different Cation-Associated EPS Extraction Methods. *WEFTEC 2004 Conference and Exhibition Proceedings*, October, 2004, New Orleans, LA

Park, C., Abu-Orf, M.M., and Novak, J.T. (2006) Predicting the Digestability of Waste Activated Sludges. *Wat Envr Res*. 78, 1, 59-68.

Rosenfeld P.E., Henry C.L., Dills R.L., and Harrison R.B. Comparison of Odor Emissions from Three (2001). Different Biosolids Applied to Forest Soil. *Water, Air, and Soil Pollution* 127: 173-191, 2001

U.S. Environmental Protection Agency (1993) *Standards for the disposal of sewage sludge. Fed. Regist.*, 58, 32, 40 CFR Part 503.

U.S. Environmental Protection Agency (1996) Method 3050 Acid Digestion of Soils, Sediments, and Sludges. U.S. EPA, 2000.

U.S. Environmental Protection Agency (1999) *Environmental regulations and technology: Control of pathogens and vector attraction in sewage sludge.* Office of Research and Development, Washington, D.C.

U.S. Environmental Protection Agency (2003) *Control of Pathogens and Vector Attraction in Sewage Sludges* - Under 40 CFR Part 503. EPA/625/R-92/013, Revised July, 2003. Washington, D.C.

Wilén, B.-M., Jin, B., and Lant, P. (2003) The Influence of Key Chemical Constituents in Activated Sludge on Surface and Flocculating Properties. *Water Res.*, 37, 2127-2139.

Webster, N.A., Worley, S., Robinson, N., and Govind, R. (2006) A Neutral pH Biotrickling Scrubber Replaces Thermal Oxidizer, Saving $600,000 Annually. *Proceedings of WEF/AWWA Odors and Air Emissions Conference.*

WERF SUBSCRIBERS

WASTEWATER UTILITY

Alabama
Montgomery Water Works & Sanitary Sewer Board

Alaska
Anchorage Water & Wastewater Utility

Arizona
Glendale, City of, Utilities Department
Mesa, City of
Peoria, City of
Phoenix Water Services Dept.
Pima County Wastewater Management
Safford, City of

Arkansas
Little Rock Wastewater Utility

California
Central Contra Costa Sanitary District
Corona, City of
Crestline Sanitation District
Delta Diablo Sanitation District
Dublin San Ramon Services District
East Bay Dischargers Authority
East Bay Municipal Utility District
Eastern Municipal Water District
El Dorado Irrigation District
Fairfield-Suisun Sewer District
Fresno Department of Public Utilities
Inland Empire Utilities Agency
Irvine Ranch Water District
Las Virgenes Municipal Water District
Livermore, City of
Los Angeles, City of
Los Angeles County, Sanitation Districts of
Napa Sanitation District
Orange County Sanitation District
Palo Alto, City of
Riverside, City of
Sacramento Regional County Sanitation District
San Diego Metropolitan Wastewater Department, City of
San Francisco, City & County of
San Jose, City of
Santa Barbara, City of
Santa Cruz, City of
Santa Rosa, City of
South Bayside System Authority
South Coast Water District
South Orange County Wastewater Authority
Stege Sanitary District
Sunnyvale, City of
Union Sanitary District
West Valley Sanitation District

Colorado
Aurora, City of
Boulder, City of
Greeley, City of
Littleton/Englewood Water Pollution Control Plant
Metro Wastewater Reclamation District, Denver

Connecticut
Greater New Haven WPCA
Stamford, City of

District of Columbia
District of Columbia Water & Sewer Authority

Florida
Broward, County of
Fort Lauderdale, City of
Miami-Dade Water & Sewer Authority
Orange County Utilities Department
Reedy Creek Improvement District
Seminole County Environmental Services
St. Petersburg, City of
Tallahassee, City of
Tampa, City of
Toho Water Authority
West Palm Beach, City of

Georgia
Atlanta Department of Watershed Management
Augusta, City of
Clayton County Water Authority
Cobb County Water System
Columbus Water Works
Fulton County
Gwinnett County Department of Public Utilities
Savannah, City of

Hawaii
Honolulu, City & County of

Idaho
Boise, City of

Illinois
American Bottoms Wastewater Treatment Plant
Greater Peoria Sanitary District
Kankakee River Metropolitan Agency
Metropolitan Water Reclamation District of Greater Chicago
Wheaton Sanitary District

Iowa
Ames, City of
Cedar Rapids Wastewater Facility
Des Moines, City of
Iowa City

Kansas
Johnson County Unified Wastewater Districts
Unified Government of Wyandotte County/ Kansas City, City of

Kentucky
Louisville & Jefferson County Metropolitan Sewer District
Sanitation District No. 1

Louisiana
Sewerage & Water Board of New Orleans

Maine
Bangor, City of
Portland Water District

Maryland
Anne Arundel County Bureau of Utility Operations
Howard County Department of Public Works
Washington Suburban Sanitary Commission

Massachusetts
Boston Water & Sewer Commission
Massachusetts Water Resources Authority (MWRA)
Upper Blackstone Water Pollution Abatement District

Michigan
Ann Arbor, City of
Detroit, City of
Holland Board of Public Works
Saginaw, City of
Wayne County Department of Environment
Wyoming, City of

Minnesota
Rochester, City of
Western Lake Superior Sanitary District

Missouri
Independence, City of
Kansas City Missouri Water Services Department
Little Blue Valley Sewer District
Metropolitan St. Louis Sewer District

Nebraska
Lincoln Wastewater System

Nevada
Henderson, City of
Las Vegas, City of
Reno, City of

New Jersey
Bergen County Utilities Authority
Ocean County Utilities Authority
Passaic Valley Sewerage Commissioners

New York
New York City Department of Environmental Protection

North Carolina
Charlotte/Mecklenburg Utilities
Durham, City of
Metropolitan Sewerage District of Buncombe County
Orange Water & Sewer Authority

Ohio
Akron, City of
Butler County Department of Environmental Services
Columbus, City of
Metropolitan Sewer District of Greater Cincinnati
Northeast Ohio Regional Sewer District
Summit, County of

Oklahoma
Oklahoma City Water & Wastewater Utility Department
Tulsa, City of

Oregon
Albany, City of
Clean Water Services
Eugene, City of
Gresham, City of
Portland, City of Bureau of Environmental Services
Water Environment Services

Pennsylvania
Hemlock Municipal Sewer Cooperative (HMSC)
Philadelphia, City of
University Area Joint Authority

South Carolina
Charleston Water System
Mount Pleasant Waterworks & Sewer Commission
Spartanburg Water

Tennessee
Cleveland, City of
Knoxville Utilities Board
Murfreesboro Water & Sewer Department
Nashville Metro Water Services

Texas
Austin, City of
Dallas Water Utilities
Denton, City of
El Paso Water Utilities
Fort Worth, City of
Houston, City of
San Antonio Water System
Trinity River Authority

Utah
Salt Lake City Corporation

Virginia
Alexandria Sanitation Authority

Arlington, County of
Fairfax County
Hampton Roads Sanitation District
Hanover, County of
Henrico, County of
Hopewell Regional Wastewater Treatment Facility
Loudoun County Sanitation Authority
Lynchburg Regional WWTP
Prince William County Service Authority
Richmond, City of
Rivanna Water & Sewer Authority

Washington

Everett, City of
King County Department of Natural Resources
Seattle Public Utilities
Sunnyside, Port of
Yakima, City of

Wisconsin

Green Bay Metro Sewerage District
Kenosha Water Utility
Madison Metropolitan Sewerage District
Milwaukee Metropolitan Sewerage District
Racine, City of
Sheboygan Regional Wastewater Treatment
Wausau Water Works

Australia

ACTEW (Ecowise)
South Australian Water Corporation
Sydney Water Corporation
Water Corporation of Western Australia

Canada

Lethbridge, City of
Regina, City of, Saskatchewan
Toronto, City of, Ontario
Winnipeg, City of, Manitoba

New Zealand

Watercare Services Limited

United Kingdom

United Utilities North West (UUNW)

STORMWATER UTILITY

California

Fresno Metropolitan Flood Control District
Los Angeles, City of, Department of Public Works
Monterey, City of
Sacramento, County of
San Francisco, City & County of
Santa Rosa, City of
Sunnyvale, City of

Colorado

Aurora, City of
Boulder, City of

Georgia

Griffin, City of

Iowa

Cedar Rapids Wastewater Facility
Des Moines, City of

Kansas

Overland Park, City of

Kentucky

Louisville & Jefferson County Metropolitan Sewer District

Maine

Portland Water District

North Carolina

Charlotte, City of, Stormwater Services

Pennsylvania

Philadelphia, City of

Tennessee

Chattanooga Stormwater Management

Texas

Harris County Flood Control District, Texas

Washington

Bellevue Utilities Department
Seattle Public Utilities

STATE

Arkansas Department of Environmental Quality
Connecticut Department of Environmental Protection
Kansas Department of Health & Environment
Kentucky Department of Environmental Protection
New England Interstate Water Pollution Control Commission (NEIWPCC)
Ohio River Valley Sanitation Commission
Urban Drainage & Flood Control District, CO

CORPORATE

ADS Environmental Services
Advanced Data Mining International
Alan Plummer & Associates
Alpine Technology Inc.
Aqua-Aerobic Systems Inc.
Aquateam–Norwegian Water Technology Centre A/S
ARCADIS
Associated Engineering
Black & Veatch
Blue Water Technologies, Inc.
Boyle Engineering Corporation
Brown & Caldwell
Burgess & Niple, Ltd.
Burns & McDonnell
CABE Associates Inc.
The Cadmus Group
Camp Dresser & McKee Inc.
Carollo Engineers Inc.
Carpenter Environmental Associates Inc.
CDS Technologies Inc.
CET Engineering Services
Chemtrac Systems Inc.
CH2M HILL
CRA Infrastructure & Engineering
CONTECH Stormwater Solutions
D&B/Guarino Engineers, LLC
Damon S. Williams Associates, LLC
Earth Tech Inc.
Ecovation
EMA Inc.
Environmental Operating Solutions, Inc.
Environ/The ADVENT Group, Inc.
Fay, Spofford, & Thorndike Inc.
Freese & Nichols Inc.
ftn Associates Inc.
Gannett Fleming Inc.
Garden & Associates, Ltd.
Geosyntec Consultants
GHD
Golder Associates Ltd.
Greeley and Hansen LLC
Hazen & Sawyer, P.C.
HDR Engineering Inc.
HNTB Corporation
Hydromantis Inc.
HydroQual Inc.
Infilco Degremont Inc.
Jacobson Satchell Consultants, Inc.
Jacques Whitford NAWE, Inc.
Jason Consultants LLC Inc.
Jordan, Jones, & Goulding Inc.
KCI Technologies Inc.
Kelly & Weaver, P.C.
Kennedy/Jenks Consultants
KMK Consultants
Komline Sanderson Engineering Corporation
Larry Walker Associates
Limno-Tech Inc.
The Low Impact Development Center Inc.
Malcolm Pirnie Inc.
Material Matters
McKim & Creed
Metcalf & Eddy Inc.
Monteco Corporation
MPR Engineering Corporation, Inc.
MWH
NewFields Water Resources, LLC
O'Brien & Gere Engineers Inc.
Odor & Corrosion Technology Consultants Inc.
Original Engineering Consultants, Ltd.
Oscar Larson & Associates
Parametrix Inc.
Parsons
Post, Buckley, Schuh & Jernigan
Praxair, Inc.
Ring Industrial Group
RMC Water & Environment
Ross & Associates Ltd.
Rothberg, Tamburini & Windsor, Inc.
SAIC
Savin Engineers
Siemens Water Technologies
Stantec Consulting Inc.
Stearns & Wheler, LLC
Stone Environmental Inc.
Stratus Consulting Inc.
Synagro Technologies Inc.
Tetra Tech Inc.
Trojan Technologies Inc.
Trussell Technologies, Inc.
URS Corporation
Wade-Trim Inc.
Westin Engineering Inc.
Weston Solutions Inc.
Woodard & Curran
Zenon Environmental Inc./GE Water Process Technology
Zoeller Pump Company

INDUSTRY

American Electric Power
American Water
ChevronTexaco Energy Research & Technology Company
The Coca-Cola Company
Dow Chemical Company
DuPont Company
Eastman Chemical Company
Eli Lilly & Company
Merck & Company Inc.
Premier Chemicals LLC
Procter & Gamble Company
Thames Water Utilities
Severn Trent Services Inc.
Suez Environment
United Water Services LLC

Note: List as of 10/31/07

WERF Board of Directors

WERF Research Council

IWA Publishing's authorised EU representative for General Product Safety Regulations is Diane D'Arras, 15 rue Duret, 75116 Paris, France, e-mail: safety@iwap.co.uk.

Printed and bound by CPI Group (UK) Ltd, Croydon, CR0 4YY
12/05/2026
02108892-0004